1

Biology

Galapagos Island marine iguanas which survive today as living proof of the diversity of nature when man and man-made selection pressures are curbed (see chapter 6 Evolution).

DAVID ABBOTT

HEADMASTER
NORTHAMPTON SCHOOL FOR BOYS

formerly Deputy Headmaster
Garth Hill School Berkshire

Biology

UNIVERSITY TUTORIAL PRESS

Published by University Tutorial Press Ltd
9/10 Great Sutton Street, London EC1V 0DA

Published 1976

ISBN 0 7231 0629 0 Full bound

ISBN 0 7231 0726 2 Limp bound

Printed in Great Britain by
Burlington Press
Foxton, Royston, Herts.

Acknowledgements

Thanks are due to the following:
For photographs:
Dr H. B. D. Kettlewell, University of Oxford, *cover photograph* and *Fig 6.1;* Eric Hosking, *frontispiece*; Cambridge Scientific Instruments and Long Ashton Research Station, *Fig 1.43*; Professor M. H. F. Wilkins, Dept. of Biophysics, Kings College London, *Fig 1.53*; Brian Chapman, University of Cambridge, *Fig 2.2*; Philip Harris Biological Limited, *Figs 2.7, 2.9, 3.3, 3.6, 3.8, 3.9, 3.11, 3.25, 9.25*; East Malling Research Station, *Figs 3.14, 3.19, 3.46e, 4.1b, 5.16*; Dr J. David George, *Fig 3.15*; Trustees of the British Museum (Natural History), *Fig 3.21*; Oxford Scientific Films Ltd, *Figs 3.22, 6.12;* Zoological Society of London, *Figs 3.26a, 3.26b, 6.8, 6.9, 6.10*; Lowson's Textbook of Botany, W. O. Howarth and L. G. G. Warne, *Fig 3.29*; Rothamsted Experimental Station and the Annals of Applied Biology *Fig 3.46a*; Glasshouse Crops Research Institute and the Annals of Applied Biology, *Figs 3.46b, 3.46d*; National Vegetable Research Station and the Annals of Applied Biology, *Fig 3.46c*; J. L. Mason and Ardea, *Fig 4.1a*; Biophoto Associates, *Fig 4.1c*; Rothamsted Experimental Station, *Fig 4.1d*; Professor J. Heslop-Harrison and Dr Y. Heslop-Harrison, Royal Botanic Gardens Kew, *Fig 4.6*; A. L. Primavesi BSc and Philip Harris Biological Limited, *Fig 4.7*; Central Press Photos Ltd, *Figs 4.14, 10.1*; John Clegg and Ardea, *Figs 4.16a, 4.16b, 4.16c, 4.16d, 6.12*; Royal Free Hospital School of Medicine, *Figs 4.22, 6.12*; W. J. Garnett MSc, *Fig 4.23*; Horse and Pony, *Fig 6.2*; Provincial Press Agency, *Fig 6.3*; Director, Institute of Geological Sciences, London, NERC Copyright, *Fig 6.5*; Royal Society for the Protection of Birds, *Fig 7.14*; Dr B. E. F. Gunning, Queens University Belfast, *Fig 9.19*; M. I. Walker BSc and Philip Harris Biological Ltd, *Fig 9.29*; Queen Mary College, London, *Fig 10.13*; G. F. Sheard, *Fig 10.21*; Dr E. H. Mercer, *Fig 10.23;* Medical Research Council, *Figs 10.26, 11.1a*; Cambridge Scientific Instruments Ltd, *Figs 11.1b, 11.8*; Dr S. M. Lewis, Royal Postgraduate School Medical School, *Fig 11.1c*.

Thomas Nelson & Sons Ltd for allowing us to base *Figs 2.4, 10.9, 10.12* on drawings from 'Biology: A Functional Approach' by M. B. V. Roberts.

Peter Fitzjohn for drawing the diagrams.

The following examining bodies for allowing us to use examination questions:
Oxford and Cambridge Schools Examination Board (O and C); Cambridge University Local Examinations Syndicate (C); Northern Ireland Schools Examinations Council (NI); The Associated Examining Board (AEB); University of London School Examinations Council (L); Birmingham Joint Matriculation Board (JMB) and the Oxford Delegacy of Local Examinations (O).

The author's personal thanks are due to his colleagues Mr D. I. Deadman and Mr R. Fay of Northampton School for Boys and Dr J. K. Scott for their encouragement.

Contents

1 **Biochemistry** p. 7
2 **Cytology and Histology** p. 27
3 **Life Forms** p. 41
4 **Reproduction and Growth** p. 71
5 **Genetics** p. 89
6 **Evolution** p. 95
7 **Ecology** p. 107
8 **Respiration** p. 123
9 **Nutrition** p. 131
10 **Response and Locomotion** p. 151
11 **Body Hydraulics** p. 173
Questions *p. 182*
Bibliography *p. 185*
Index *p. 186*

Foreword

The writer of this Foreword belongs to a generation of biologists who suffer from a well-known condition. This is due to the yearly reception of degree students wishing to read biological sciences, the vast majority of whom have been trained on purely descriptive biology. These students have to make the adjustment, often very difficult for them, that degree level biology is a quantitative subject and that qualitative statements are no longer sufficient in many cases. They have to adjust to a subject that involves mathematics and a considerable amount of statistics, as well as containing a lot of chemistry and physics. For some this is too much, and they wish that their earlier training had been along more rigorous quantitative lines.
I most heartily welcome David Abbott's concise and methodical treatment of the subject at A-level and wish it all the success that it deserves. It should prove to be a most useful course for use at this level.

JOHN SCOTT, BSc, PhD

Head of Biological Sciences
Cambridgeshire College of
Arts and Technology

Preface

This book aims to cover everything *central* in an 'A' level Biology course. The changed approach to sixth form biology over the last few years, emphasising inquiry and experimentation, means that it is now impossible, in a single book, to cover every topic in enough depth to satisfy *all* students. The book thus makes no claim to be comprehensive, and the teacher and student are left to follow up individual topics in depth. This is consistent with the idea that sixth form courses should be flexible and, wherever possible, satisfy individual interests.

The book is based on topics rather than organisms. Thus, after dealing with the fundamentals of biochemistry and the principles of biology (such as classification), there follow chapters on reproduction, genetics, evolution, ecology, respiration, and nutrition.

The suggestions for practical work given are brief and are only intended as a guide. Obviously, teachers will devise other practical work, and if the students are working in the spirit of the Nuffield course, they themselves will be suggesting other experiments.

A knowledge of 'O' level biology is assumed.

DAVID ABBOTT

1 Biochemistry

1.1 Chemical basis of life

The major features of living organisms are as follows: they feed, grow, respire, reproduce themselves, move excrete and respond to stimuli (*e.g.* to noise, to touch, to light). In order to understand these biological processes the biologist has to be a chemist too, and so the science of biochemistry has developed.

Spallanzani (1783) first showed that the process of digestion of meat by hawks was essentially a liquefaction, caused by the interaction of certain substances in the gastric juices with the meat. These substances were later called *enzymes*. Enzymes act rather like *catalysts* in that they accelerate processes in organisms but themselves remain unchanged. Much of biochemistry is concerned with *reaction kinetics*, in which energy against reaction path diagrams are drawn for major pathways by which chemical reactions occur in the living cells (the fundamental constituents of organisms). The mechanisms of these reactions are then interpreted. For any such reaction to proceed, the reactants must overcome an energy barrier tending to oppose the process trying to take place, the *energy of activation*. What a catalyst or enzyme does is to lower the energy of activation of the process it is catalysing.

Enzymes are specific; most reactions in a living organism need their own specific enzymes to catalyse them. Consider the sequence of events taking place from the time starch in the diet of a mammal is first broken down during digestion into simple sugar molecules; these are then absorbed into cells and polymerised to glycogen or degraded to carbon(IV) oxide and water. A whole range of enzymes would be required here, acting one after the other. Sequences of changes like this are referred to as *metabolic pathways*. Processes that result in an energy release are *catabolic* (ΔG negative to the chemist), *e.g.* respiration of sugars to carbon(IV) oxide and water, while those that need energy to take place are *anabolic* (ΔG positive), *e.g.* synthesis of proteins. In *cybernetics*, the complex mechanisms in cells are studied by computers. *Cytology* is the science in which the biology and chemistry of cells is studied.

The main elements in the human body are carbon, hydrogen and oxygen (together over 90% by mass), but there are smaller amounts of nitrogen, calcium, phosphorus, sulphur, potassium, sodium, chlorine, magnesium, iron, manganese, copper, iodine, zinc and cobalt; the last five are present only as minute traces. Much of the carbon, hydrogen and oxygen is present as water (H and O only), in carbohydrates and in lipids. Nitrogen, carbon, hydrogen, oxygen and sulphur are present as protein. In addition, the body contains a small percentage of less complex organic and inorganic molecules (other than H_2O).

NOMENCLATURE The problem of chemical nomenclature is a difficult one. There is a general move in chemistry towards following the recommendations of the Commissions of the International Union of Pure and Applied Chemistry whose reports outline a system of nomenclature, symbols and terminology for both organic and inorganic compounds. This system is described in an excellent booklet *Chemical Nomenclature, Symbols and Terminology* (Association for Science Education, 1972). However, the approach adopted in this book is entirely pragmatic. As far as possible IUPAC nomenclature is used, but with trivial names given in brackets. But where the IUPAC name is likely to cause confusion to the reader the trivial name only is used.

1.2 Carbohydrates

Carbohydrates contain carbon, hydrogen and oxygen, the ratio of hydrogen to oxygen generally being the same as in water, *i.e.* $C_x(H_2O)_y$. Some carbohydrates however, contain N and S, and some do not have the H and O in the above ratio, *e.g.* deoxyribose, which is present in the nucleic acid DNA (deoxyribonucleic acid). In living organisms, carbohydrates have roles as structural materials (*e.g.* cellulose) and as energy-givers (*e.g.* oxidation of glucose is the main source of energy of the cell). Carbohydrates can be subdivided into monosaccharides, oligosaccharides and polysaccharides. Monosaccharides and oligosaccharides are classed as *sugars*, having a crystalline nature, being sweet to taste and being soluble in water.

MONOSACCHARIDES These are simple sugars, having between 3 and 6 carbon atoms. Those with 6 carbon atoms are the *hexoses* ($C_6H_{12}O_6$) and those with 5 are the *pentoses* ($C_5H_{10}O_5$), etc. Examples of hexoses are *glucose* and *fructose*, while *ribose* and *deoxyribose* are examples of pentoses. Glucose is an *aldehyde* and fructose is a *ketone*, and they provide an example of *isomerism of structure*; they both have the formula $C_6H_{12}O_6$. It is customary to write their structural formulae in a simplified manner as shown in Fig 1.1.

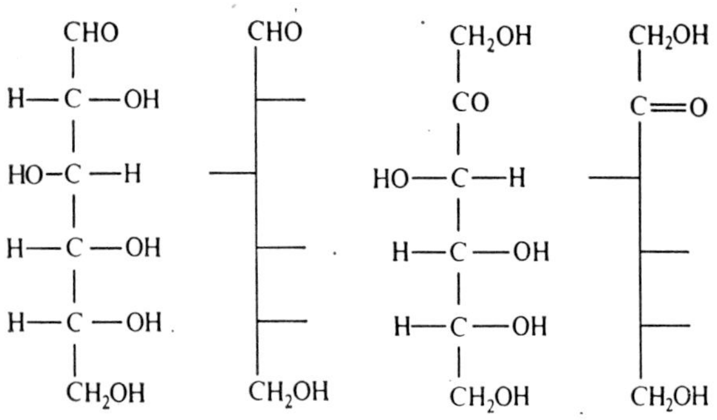

Fig 1.1 **Left, glucose; right, fructose**

Some sugars rotate the plane of polarisation of polarised light to the right and are said to be dextro-rotatory; others rotate the plane to the left and are said to be laevo-rotatory. We would write this as *d*— (or +ve) and *l*— (or —ve), respectively; the type of rotation would have to be found by using a *polarimeter* (an instrument used for studying optical rotations). If *d*— and *l*— forms of a sugar exist the two molecules are said to exhibit *stereo-isomerism* and one molecule is the mirror image of the second one.

When talking about the above molecule of glucose we would refer to it as D-configuratory, meaning that it is related to the standard structure shown in Fig 1.2. If it happened that this glucose also rotated the plane of polarisation of polarised light to the right, we would write it as D(*d*—or+ve) glucose. If in Fig 1.2 the hydroxyl had projected in the opposite direction, we would refer to the molecule as L-configuratory, and so on.

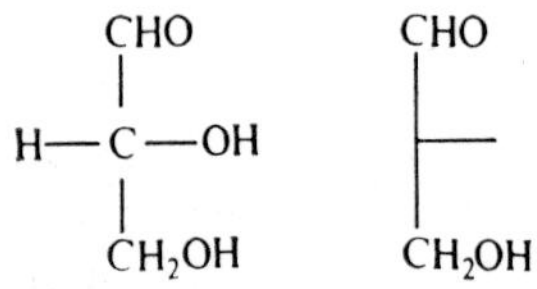

Fig 1.2 **D-configuratory structure**

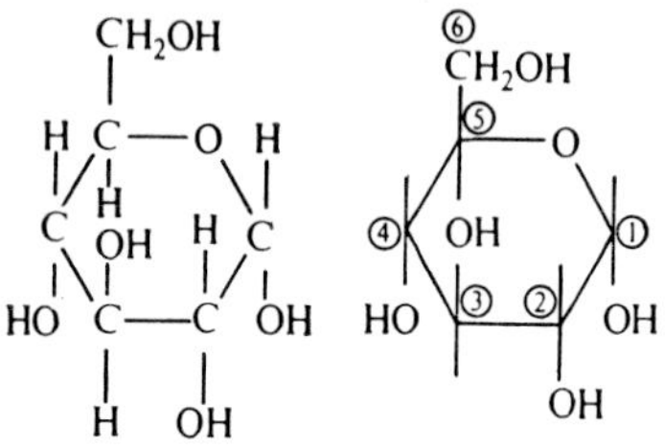

Fig 1.3 **D-glucopyranose**

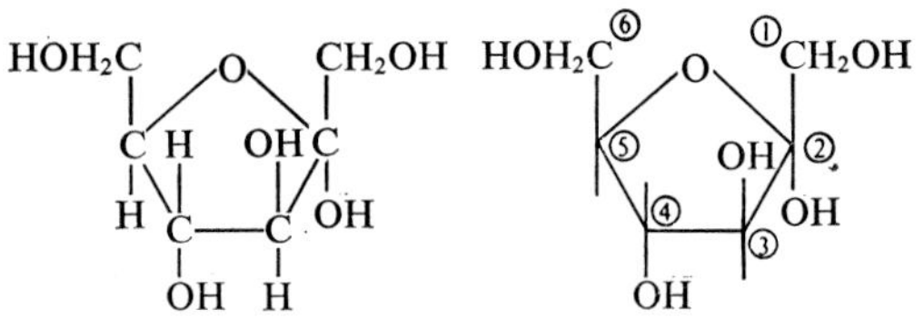

Fig 1.4 **D-fructofuranose**

The forms of hexose sugars, glucose and fructose, so far described, are referred to as 'open-chain' and are present as these structures in solution. But the normal forms of glucose and fructose involve rings of atoms, glucose with a *pyranose* and fructose with a *furanose* ring (Figs 1.3 and 1.4). A pyranose ring contains five $-CH_2$ groups and an oxygen atom. In sugars some of the H's are replaced by —OH or $-CH_2OH$ groups. A furanose ring contains four $-CH_2$ groups and an oxygen atom. Again, in sugars, some of the H's are replaced.

These ring forms are designated α– or β– according to the positioning of the OH group on the first carbon atom (for glucose) or second carbon atom (for fructose), as shown in Figs 1.5 and 1.6.

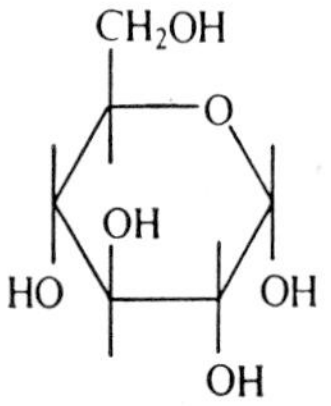

Fig 1.5 α—**D-glucopyranose**

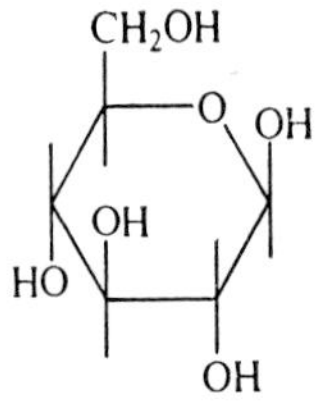

Fig 1.6 β—**D-glucopyranose**

Two other important monosaccharides are *ribose* and *deoxyribose*, which have furanose ring structures (Figs 1.7 and 1.8).

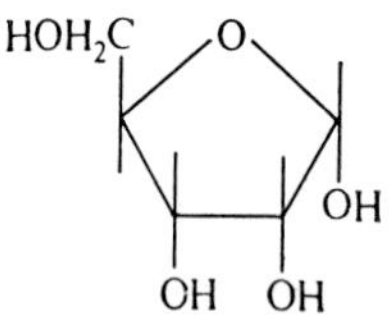

Fig 1.7 **D-ribofuranose**

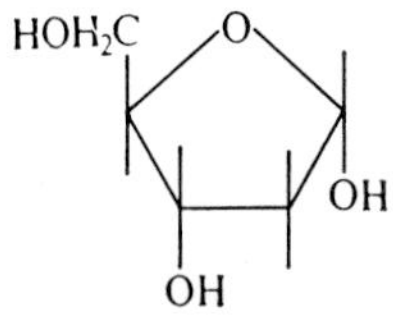

Fig 1.8 **D-deoxyribofuranose**

When the OH group on the first carbon atom of a pyranose, or the second carbon atom of a furanose, condenses with an OH group in another compound (with the splitting off of H_2O), a *glycoside* is formed. The molecule with which the first sugar condensed is called the *aglycone*. If the glycoside contains glucose as its sugar it is called a *glucoside*. An example is glucose 1-phosphate(V) (Fig 1.9), the formation of which is the start of most processes in the cell in which glucose takes part.

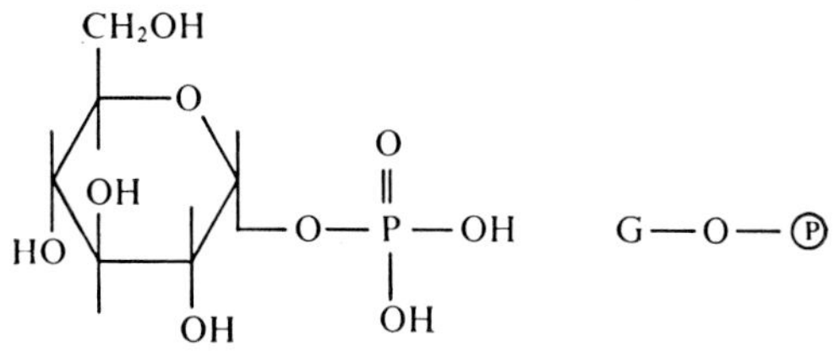

G—O—Ⓟ

Fig 1.9 **Glucose 1-phosphate(V)**

The monosaccharides are the building blocks used to construct more complex carbohydrates. The relationship between the monosaccharides and polysaccharides is shown schematically in Fig 1.10.

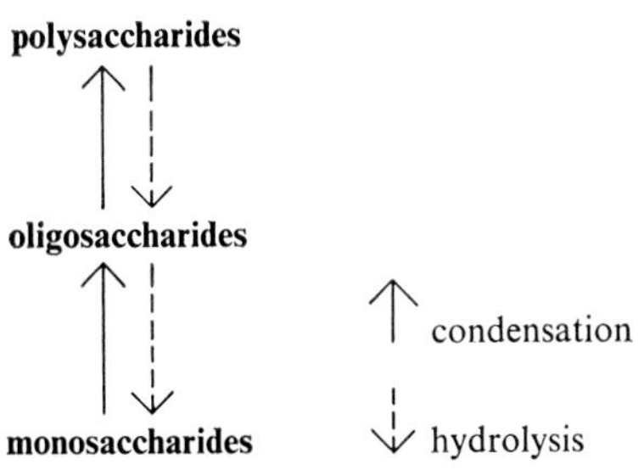

Fig 1.10 **Relationship between monosaccharides and polysaccharides**

OLIGOSACCHARIDES When a glycoside contains between 2 and 6 monosaccharide molecules condensed together it is called an *oligosaccharide*. Some common examples of disaccharides (compounds containing two sugar molecules) are as follows:

Maltose (4-α-D-glucopyranosylglucopyranose) is obtained from starch (Fig 1.11). It is hydrolysed to glucose by the enzyme maltase.

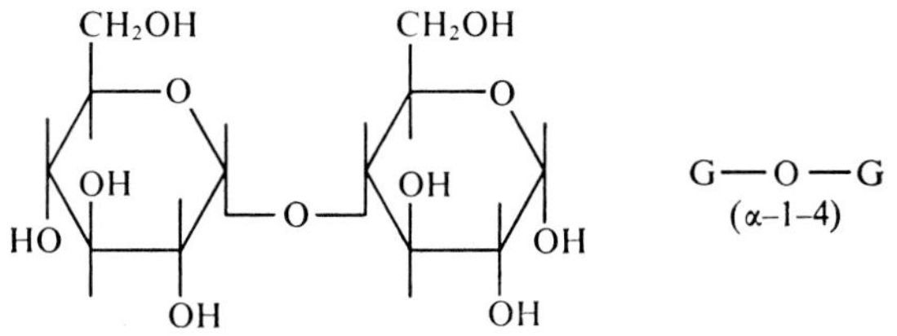

G—O—G (α-1-4)

Fig 1.11 **Maltose**

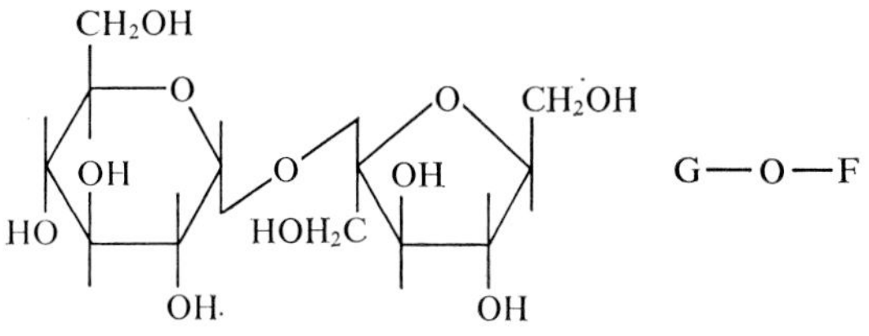

G—O—F

Fig 1.12 **Sucrose**

Sucrose (Fig 1.12) is a glycoside in which the glucose unit has an α-configuration and the fructose unit has a β-configuration, and the linkage is through the 2-position of the fructose unit. It is an important foodstuff (table sugar), and is obtained from sugar cane or beet. Sugars in plants are transported as sucrose. It is hydrolysed to glucose and fructose by the enzyme sucrase.

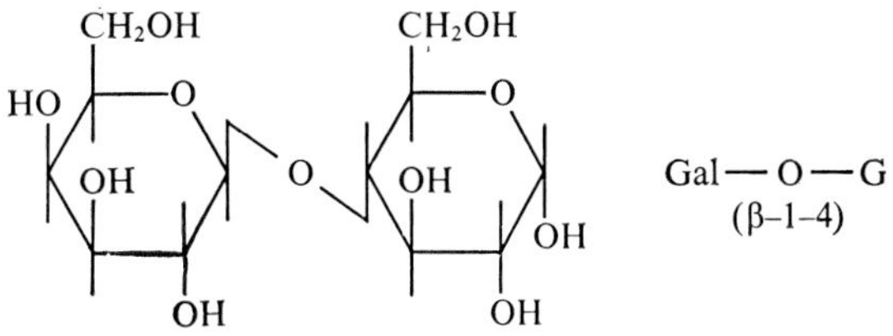

Gal—O—G (β-1-4)

Fig 1.13 **Lactose**

Lactose (Fig 1.13) (4-β-D-galactopyranosylglucopyranose) occurs in the milk of mammals. It is hydrolysed to one molecule of glucose and one of galactose by the enzyme lactase.

Cellobiose (Fig 1.14) (4-β-D-glucopyranosylglucopyranose) is obtained by careful hydrolysis of cellulose and hydrolysed by emulsin, an enzyme which only hydrolyses β-glucosides.

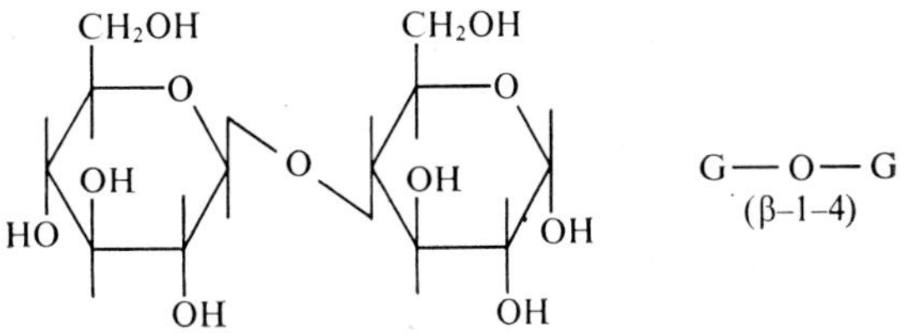

G—O—G (β-1-4)

Fig 1.14 **Cellobiose**

POLYSACCHARIDES These have structures in which many simple sugar molecules are condensed together. Unlike the mono- and oligosaccharides from which they are constructed they are insoluble, which means that they make excellent storage carbohydrates.

Starch is a common storage polysaccharide in plants (Fig 1.15). It is hydrolysed by the enzyme α–amylase to give maltose. It can be separated by treatment with hot water into two fractions, a soluble component called amylose, and a sparingly soluble residue called amylopectin. Amylose is the component which gives a blue colour (due to complex formation) with iodine. The relative molecular masses range from 10 000 to 50 000 (amylose) and 50 000 to 100 000 (amylopectin). Amylose is a straight-chain polymer (Fig 1.16).

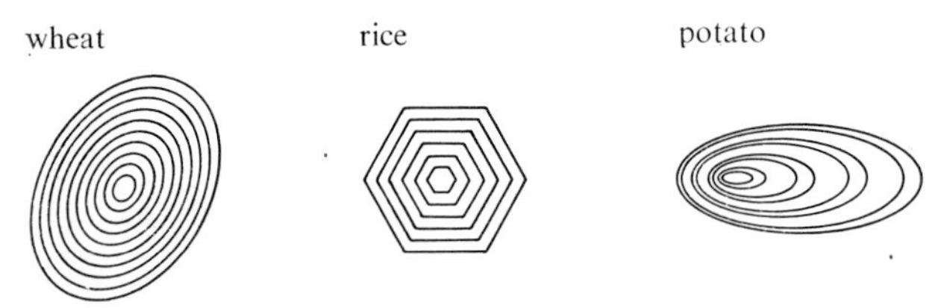

Fig 1.15 **Some characteristic starch grains**

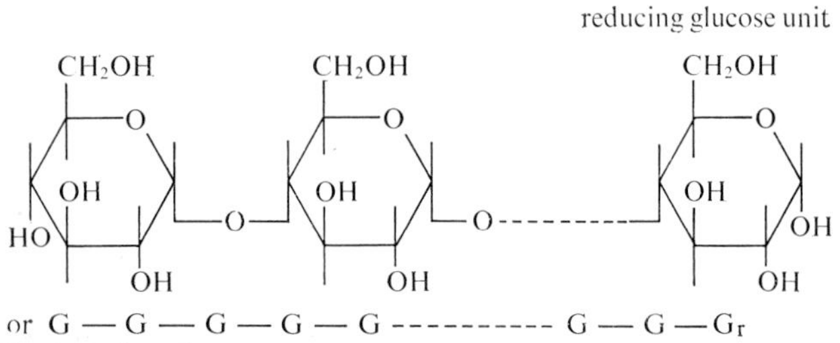

reducing glucose unit

or G — G — G — G — G ---------- G — G — G_r

Fig 1.16 **Amylose**

Amylopectin is a branched polymer, branching taking place through the 6-positions. Its structure is tree-like (Fig 1.17).

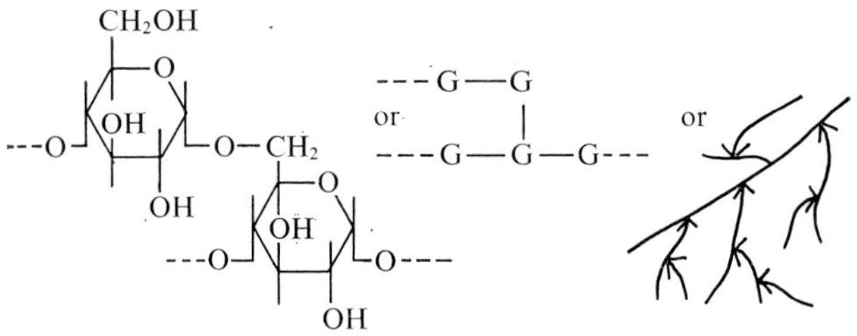

Fig 1.17 **Amylopectin**

Glycogen is a branched polymer like amylopectin, but has a greater degree of branching. It is the main storage compound of animals and fungi.

Cellulose (Fig 1.18) is the most widely distributed skeletal polysaccharide. It makes up about half of the cell-wall material of wood and other plant products. Cotton is pure cellulose. Wood cellulose is found in association with *lignin*, a polymer containing benzene rings and branched, rather like amylopectin, although it is not a polysaccharide. Lignin is a support compound. Cellulose and lignin can be separated by converting the lignin to a sulphonated compound using calcium hydrogen sulphate(IV). The relative molecular mass of cellulose is 20 000 to 40 000.

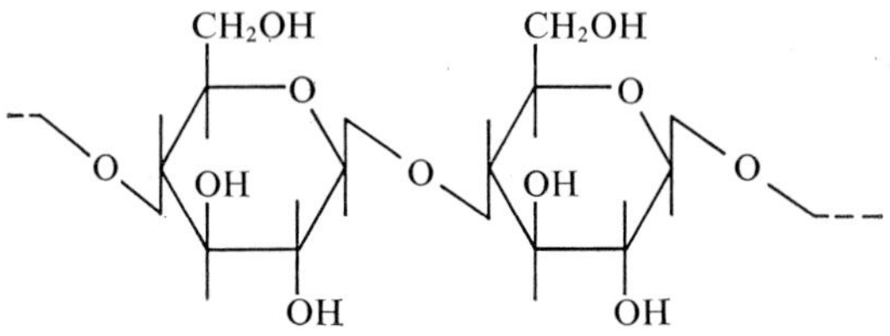

Fig 1.18 **Cellulose**

Cellulose is always associated with *hemicelluloses*, which are structurally like cellulose but have structural sugar units other than glucose.

Chitin (Fig 1.19) is a polysaccharide which forms, for example, the hard exo-skeleton of crustaceans and insects, and is in the cell walls of fungi. Glucosamine, the 2-amino derivative of glucose, in which the NH_2 group is ethanoylated [acetylated] to $NHCOCH_3$ (compare phenylamine [aniline] to N-phenylethanamide [acetanilide]), is present. Chitin is hydrolised to N-ethanoylglucosamine with an enzyme in the intestine of the snail. Chitin is structurally like cellulose, but it probably has a lot of cross-links between adjacent chains and is very tough.

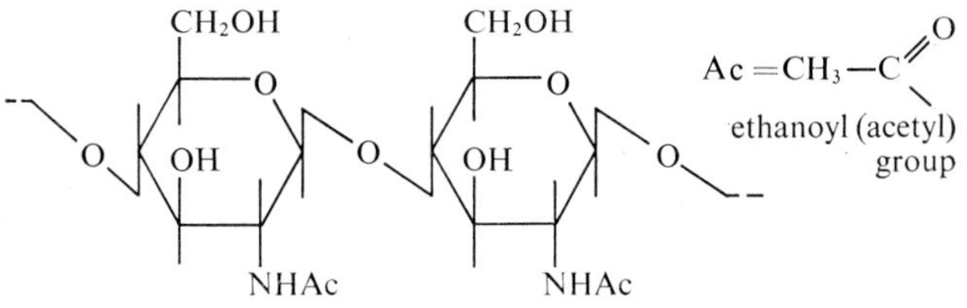

Ac = CH_3—C(=O)— ethanoyl (acetyl) group

Fig 1.19 **Chitin**

Inulin (Fig 1.20) is a storage polysaccharide of many plants (*e.g.* Compositae, daisy family, including the dahlia, dandelion, and also the Jerusalem artichoke). Its structure consists mainly of fructose units in the furanose form,

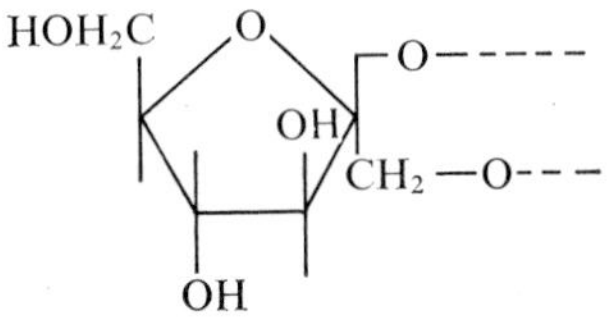

Fig 1.20 **Inulin**

and with a few glucose units in the chain and at the terminal positions.

Ascorbic acid (vitamin C) (Fig 1.21) is a carbohydrate, unsaturated with a double covalent bond. It is in fresh fruits, *e.g.* lemons, and prevents scurvy – a disease characterised by a tendency to haemorrhage and structural changes in bones, teeth and cartilage.

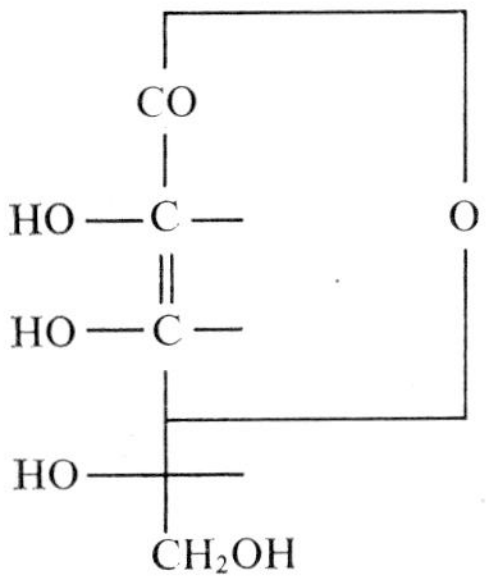

CO
HO—C—
HO—C—
HO—
CH_2OH
O

Fig 1.21 **Ring form of L-ascorbic acid**

Heteropolysaccharides are those which contain more than one type of monosaccharide unit. In general, the linkage is from the 1–C of one unit to the 2, 3 or 6-C of the other unit. Their structures are not known as well as those of *homopolysaccharides* like glycogen and starch, but some well-known ones are the *pectic acids*, polymers of uronic acids combined with other substances (a uronic acid of, say, glucose would be the one in which the CH_2OH group is oxidised to COOH). Pectic acids are responsible for the solidity of many fruits and they are used in making jams, jellies, etc. The *mucopolysaccharides* are found associated with proteins. On hydrolysis, these yield mainly uronic acids and amino-sugars. Mucopolysaccharides are found in *heparin*, an anti-coagulant of blood, in cartilage, and the *hyaluronic acid* which acts as a 'cement' between cells. The structures of polysaccharides can be investigated using techniques of acid hydrolysis, ethanolysis [acetolysis] (ethanoylation followed by hydrolysis), methylation, enzymic hydrolysis, infrared and NMR spectroscopic analysis, optical rotation studies, oxidation studies, and so on. The techniques of *chromatography* and *electrophoresis* are most important in such studies.

1.3 Synthesis and degradation

Glucose is formed during the photosynthesis of green plants, and is used partly as an immediate source of energy, both in sunlight and in the dark (when no more photosynthesis occurs). In animals, glucose is formed during digestion. All cells need glucose for the generation of ATP (*adenosine triphosphate*). Breakdown of ATP provides most of the energy required for metabolic reactions in living cells. In plants those cells that have chloroplasts synthesise glucose from carbon(IV) oxide. If it is not needed at once, the glucose is stored, after conversion to starch (in plants) and glycogen (in animals). The plant reserves of starch provide the source of glucose for animals which eventually eat them – animals have enzymes capable of degrading the starch to glucose again. In animals, glycogen is stored in the liver and muscles where there are enzymes capable of polymerising glucose called *polymerases*. These link together glucose molecules with the elimination of water (condensation) to form the large molecule or *polymer*.

When there is a great demand for glucose in the metabolism, the starch (plant) or glycogen (animal) are once again degraded to glucose which enters the correct sequence of chemical changes leading to its oxidation (see §8.2 on p 123). The breakdown requires the presence of a phosphorylase enzyme that hydrolyses the polymer back to the monomer units. In fact, the monomer units are released as *glucose phosphate(V)* and this means that phosphate(V) ions are needed in the degradation (Fig 1.22). If, for example, the brain needed glucose, it would have to withdraw it from the bloodstream. There

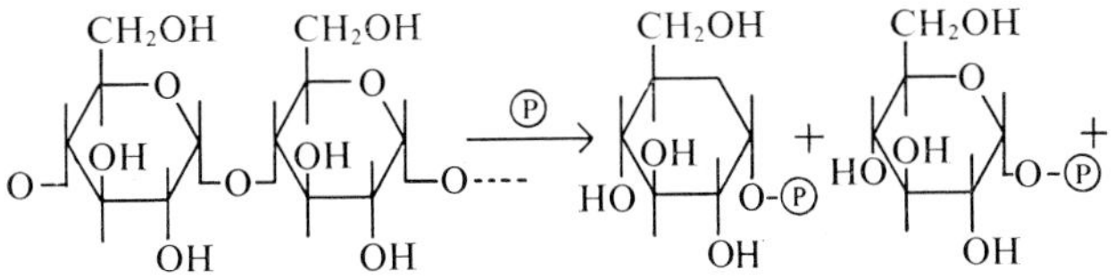

CH_2OH CH_2OH CH_2OH CH_2OH
O OH O OH O---- ⓟ → HO OH O-ⓟ + HO OH O-ⓟ +
OH OH OH OH

Fig 1.22 **Degradation of a polymer**

is an *equilibrium* of glucose, and if it is removed from the bloodstream by the brain or another organ, some of the glycogen in the liver would be degraded so as to raise the glucose level in the blood again. If the amount of glycogen in the liver is at its maximum level, glucose is converted to fats which are stored – for example, beneath the skin. If the intake of carbohydrates in the diet is low, then fats from the storage places will be oxidised to form ATP for cellular activity and 'slimming' will result.

1.4 Nitrogen compounds

AMINO-ACIDS These are the units of which proteins are polymers; they contain both acidic (–COOH) and basic (—NH_2) groupings. Some contain just one of each type, some contain two acidic groups to one basic group, and others contain the reverse. The basic structure of an amino-acid is shown in Fig 1.23. The structure of the R group varies from amino-acid to amino-acid. Some amino-acids contain benzene rings or other types of ring, and some contain sulphur atoms. The structures of the 20 known amino-acids are shown in Fig 1.24. Amino-acids are the building blocks from which proteins are constructed.

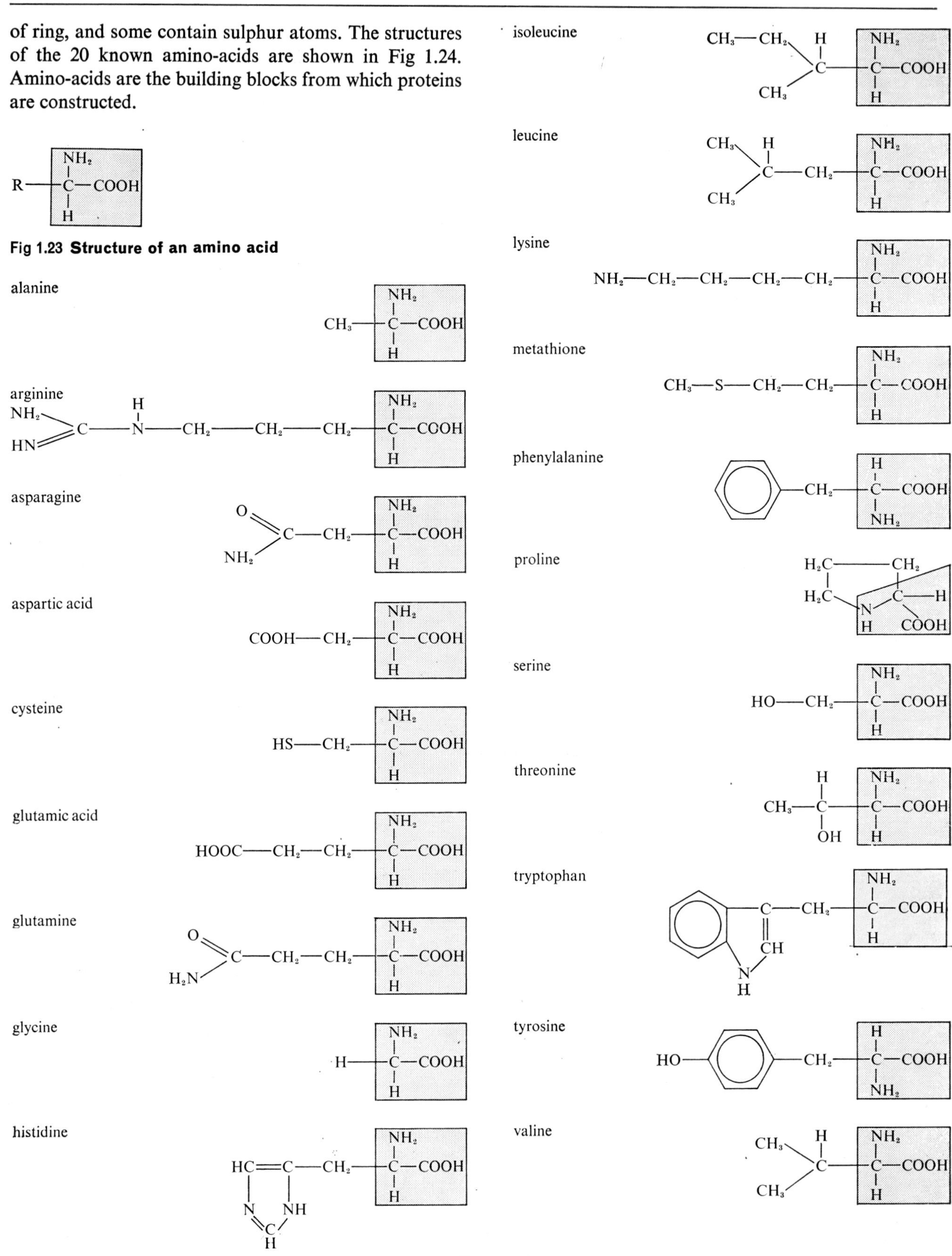

Fig 1.23 Structure of an amino acid

Fig 1.24 **Structure of 20 common amino acids (non-IUPAC names)**

$$\mathrm{H{-}N(H){-}C(R)(H){-}CO{\cdot}OH} + \mathrm{H{-}N(H){-}C(R_1)(H){-}CO{\cdot}OH} \longrightarrow \mathrm{{-}N(H){-}C(R)(H){-}\underbrace{C(=O){-}N(H)}_{\text{peptide link}}{-}C(R_1)(H){-}C(=O){-}}$$

Fig 1.25 **Formation of protein**

PROTEINS In proteins, a COOH group of one amino-acid unit condenses (with the elimination of H_2O) with the NH_2 group of another unit, forming a *peptide linkage* (Fig 1.25). In Fig 1.25, R, R_1, etc, are definite groupings, as in Fig 1.24.

Haemoglobin, a blood protein, contains about three hundred amino-acids. Changing round the sequence of amino-acids in just one case in this molecule is sufficient to make the haemoglobin perform incorrectly, giving the disease *sickle-cell anaemia.* The enzymes which mediate complex metabolic changes are proteins. Sometimes proteins act as storage compounds and they are also structural compounds. Proteins can be *fibrous* (*e.g.* keratin in hair) or *globular* (*e.g.* albumen in egg white). In some cases a globular protein can change to a fibrous one, *e.g.* in clotting of blood, fibrinogen to fibrin. Fibrous proteins have long chain molecules, globular proteins have folded molecules. The former are insoluble in aqueous solutions and the latter are soluble. Proteins can be hydrolysed to amino-acids by protease enzymes (proteolytic enzymes) and the acids can then be separated by chromatography or electrophoresis.

In the living cell, the *ribosome* is the organelle concerned with protein synthesis by the cell cytoplasm, aided by *nucleic acids.*

1.5 Lipids

Lipids contain only carbon, oxygen and hydrogen and include the fats, oils and waxes. Unlike carbohydrates, the O and H are not present in the proportions in which they are present in water.

FATS These are *esters* of the trihydric alcohol propan-1,2,3-triol [*glycerol*] with alkanoic [*carboxylic*] *acids* (Fig 1.26). If only one OH group of glycerol is esterified, a *monoglyceride* is formed, if two OH's are esterified a *diglyceride* is formed and if all three OH's are esterified a *triglyceride* is formed. Mutton fat is glyceryl tristearate, in which the acid responsible for esterification is octadecanoic [stearic] acid ($C_{17}H_{35}COOH$).

OILS These are like fats, except that the particular esterifying acid gives a liquid product instead of a solid. Thus, palm oil contains octadecenoic [oleic] acid,

Table 1.1 **Some typical proteins**

protein	*approximate molecular mass*	*found in*	*function*
insulin	6 000	pancreas	regulates blood sugar level
lysozyme	15 000	egg white	enzyme: aids breakdown of bacterial cell walls
α-chymotrypsin	25 000	pancreas	enzyme: aids breakdown of esters and peptide bonds
haemoglobin	67 000	blood (red corpuscles)	respiration
catalase	250 000	liver, kidney	enzyme: aids breakdown of hydrogen peroxide to water
fibrinogen	320 000	blood (plasma)	blood clotting
collagen	350 000	connective tissue of animals	structure
keratin	very large	hair/skin of animals	structure

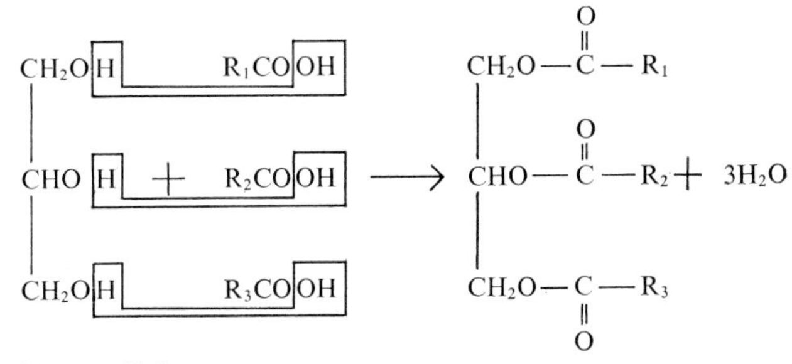

Fig 1.26 **Fats**

$C_{17}H_{33}COOH$. This, like the acids in many oils, is *unsaturated.* In the production of margarine, an oil containing an unsaturated acid is *hydrogenated* (reacted with hydrogen in the presence of nickel as catalyst) to form a solid edible fat (Fig 1.27).

$$\mathrm{{-}C(H){=}C(H){-}} + H_2 \longrightarrow \mathrm{{-}C(H)(H){-}C(H)(H){-}}$$

Fig 1.27 **Hydrogenation of oil to a fat**

WAXES These are esters of high relative molecular mass monohydric alcohols with high relative molecular mass alkanoic [carboxylic] acids. Waxes are found in animals (*e.g.* in the ear, as beeswax, etc.) and in plants (*e.g.* on leaves). The East African 'honey guide' bird can digest beeswax.

PHOSPHOLIPIDS These are the essential structural materials of many cell membranes, and have structures like that shown in Fig 1.28.

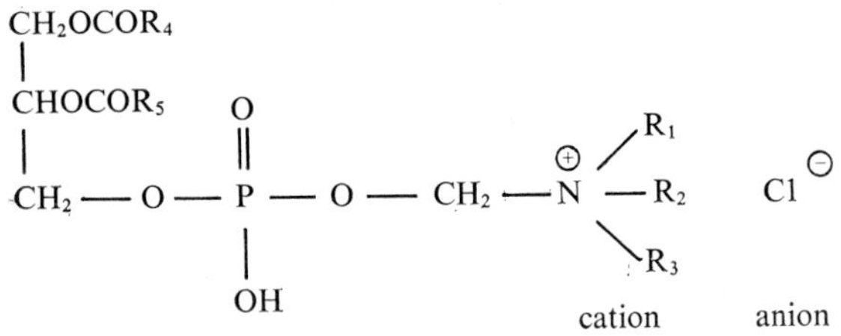

Fig 1.28 **Phospholipid**

As well as their function in respiration, lipids function as storage materials capable of yielding high energy, *e.g.* in hibernating animals, in seeds of plants such as castor oil.

1.6 Steroids

The steroid compounds are extremely important in both animals and in plants, and have structures based on the parent hydrocarbon *cyclopentanoperhydrophenanthrene* (Fig 1.29).

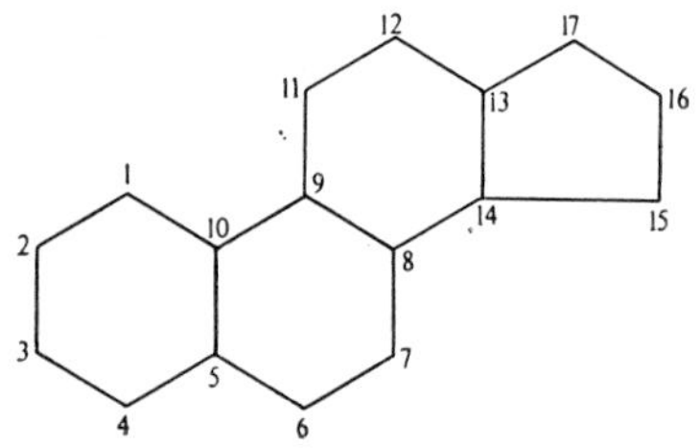

Fig 1.29 **Cyclopentanoperhydrophenanthrene**

STEROLS Those found in animals are called zoosterols and those in plants phytosterols. The best known animal sterol is *cholesterol*, $C_{27}H_{45}OH$, which occurs, sometimes as an ester, in most organs of the body, particularly in the brain and nerves. Gallstones are mainly cholesterol. The plant sterols are also widely distributed in nature. *Sitosterol* from the wheat embryo has the formula $C_{29}H_{50}O$. *Ergosterol* is in yeast. *Stigmasterol* is also found in many plants.

BILE ACIDS These are found in the bile of man and many animals, and occur along with cholesterol. Human bile contains:

cholic acid $C_{23}H_{36}(OH)_3COOH$;
deoxycholic acid $C_{23}H_{37}(OH)_2COOH$;
anthropodeoxycholic acid $C_{23}H_{37}(OH)_2COOH$;
lithocholic acid $C_{23}H_{38}(OH)COOH$;

Cholanic acid (not an alcohol) may also be present. The bile acids conjugate with amino-acids to form peptide-like compounds. They facilitate hydrolysis of fats. Some emulsify fats (bringing them into the most favourable condition for enzymic attack) and others combine with fats to form colloidally soluble molecules.

SAPONINS These occur widely in plants and produce foams in water. They are glycosides. They are powerful poisons.

CARDIAC POISONS These occur in the 'digitalis' plants, *e.g.* foxglove (Apocyanaceae) and lily of the valley (Liliaceae). They are used in the treatment of heart disease, slowing down the beat.

TOAD VENOMS These are also steroids but are not glycosides. Examples are bufotoxin and bufitalin.

SEX HORMONES These will be discussed on p 77.

GENERAL POINTS ON STEROID COMPOUNDS If alkyl groups R_1, R_2 and R_3 are at positions 10, 13 and 17, respectively (Fig 1.29), then we can say that

a R_3 can vary in size quite considerably
b R_2 is usually methyl, CH_3, and exceptionally an aldehyde, CHO
c R_1 is usually methyl and exceptionally an aldehyde
d There is variability in unsaturation (*e.g.* in the sex hormones two rings assume aromatic character)
e The favoured position for the OH group is at 3.

1.7 Carotenoid pigments

Lycopene and *carotene*, two yellow pigments, belong to a group of unsaturated hydrocarbons of terpene-like character. Lycopene is chiefly responsible for the red colour of the tomato, the hips of the wild rose and other fruits, and has the formula $C_{40}H_{56}$. Carotene is isomeric with lycopene. It can be isolated from the carrot and occurs very widely in association with chlorophylls and xanthophylls in green leaves and in countless fruits and flowers. It also occurs in animal organisms, *e.g.* in milk and in blood serum. There are three isomers of carotene, differing in the site of the alkene bond. Various oxygen-containing plant pigments are related to carotene, their oxygen atoms being present mainly as OH groups.

They are known as 'phytoxanthins' and to this class belong the following:

XANTHOPHYLL, $C_{40}H_{56}O_2$, a yellow chloroplast pigment found in green leaves; there are two OH groups in the rings, occupying the same positions as in the isomer *zeaxanthin*, the pigment in maize.

FUCOXANTHIN, $C_{40}H_{58}O_6$, the brown/yellow pigment of the brown algae (p 56).

CAPSANTHIN, $C_{40}H_{58}O_3$, the pigment in paprika, which has the same carbon skeleton as the carotenes and other phytoxanthins and is a ketone.

ASTACIN, $C_{40}H_{48}O_4$, the red pigment of lobster and crab shells which is abundant in other crustaceans, fish, coelenterates and occurs mainly as an ester. Astacin is derived from the pigment in lobster eggs – *astaxanthin.*

1.8 Terpenes

The carotenes are related structurally to compounds called *terpenes*, of which probably the commonest is turpentine. The terpenes have a distinctive fragrance and can be isolated from natural sources by steam distillation. They are responsible for the characteristic odours of pine trees, citrus fruits and plants like geraniums and celery. Chemically, the terpenes are derived from two isoprene (2-methylbuta-1,2:3,4-diene) units. They have 10 carbon atoms. Other compounds which are classed as terpenes contain 15, 20 and 30 carbon atoms and are known, respectively, as sesquiterpenes, diterpenes and triterpenes. The odour of terpenes attracts insects and so aids fertilisation of the plants. Some examples are limonene in citrus oils, *e.g.* lemon; pinene in pine oil; camphor in *Cinnamonum camphora*; geraniol in rose oil; ionone in violet oil; menthol in mint oil; farnesol in lily of the valley; selinene in celery oil; squalene in shark liver oil.

1.9 Anthoxanthin pigments

These are responsible for many of the yellow to brown

HO ⑦ ⑧ ① O ② ③ ⑥ ⑤ ④ OH O

Fig 1.30 **Chrysin**

O O flavone ring

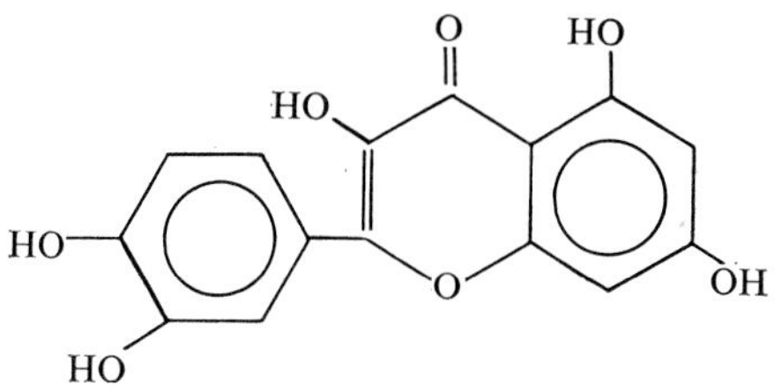

HO O HO HO O OH HO

Fig 1.31 **Quercetin**

colours in nature. They usually have one or more OH groups in the aromatic ring of a *flavone* system. *Chrysin* (Fig 1.30) is found in poplar buds (it is the 5, 7 compound). *Quercetin* (Fig 1.31) is in oak bark.

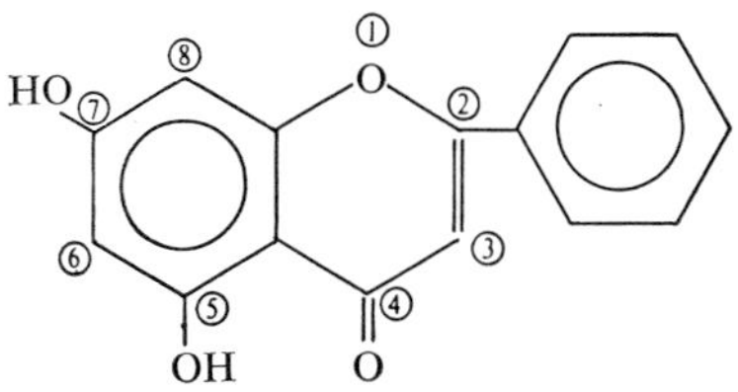

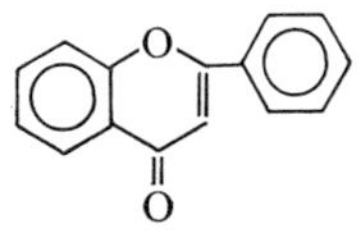

1.10 Anthocyanin pigments

These are responsible for many of the pinks and reds, down to the blues, in nature. They are mainly glycosides, and when the anthocyanin is hydrolysed so as to break the glycosidic bond an *anthocyanidin* is formed. The colour is affected by the pH (acidity or alkalinity) of the cell sap. The pigments are easily extracted. The dried plant material is ground up with ethanol to extract the anthocyanin, dry hydrogen chloride is added to form a salt which crystallises out when ethoxyethane [ether] is added. Examples are *peonidin* which gives the red colour to peonies (Fig 1.32) and *delphinidin* which gives the bright blue colour to delphiniums (Fig 1.33).

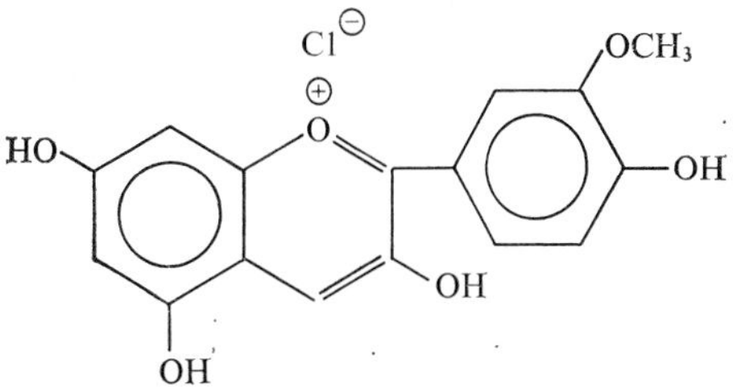

$Cl^{\ominus}$ $\oplus$ O OCH_3 HO OH OH OH

Fig 1.32 **Peonidin**

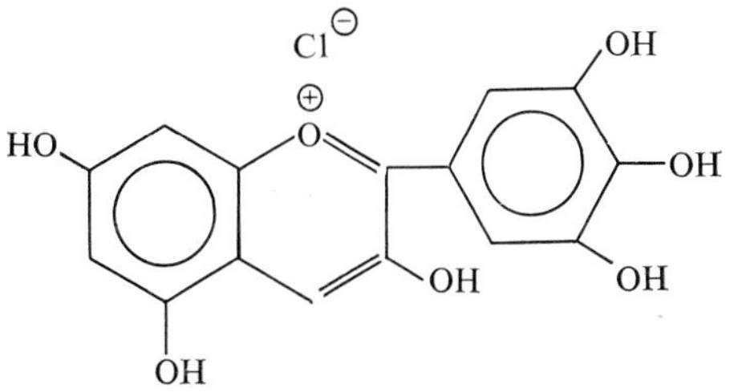

$Cl^{\ominus}$ $\oplus$ O OH HO OH OH OH OH

Fig 1.33 **Delphinidin**

The sugar molecule of the glycoside is always attached to the 3-position, but if this is blocked the 5-position is

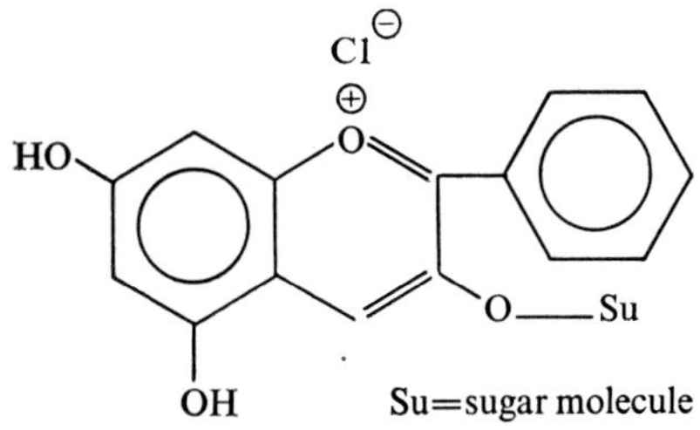

Cl⊖ ⊕ HO O O—Su OH Su=sugar molecule

Fig 1.34 **Glycoside**

used (Fig 1.34). The change of colour with change of pH is due to a change to a quinonoid ring system (Fig 1.35). The sugar residue is either glucose, galactose or 6-deoxy glucose [rhamnose].

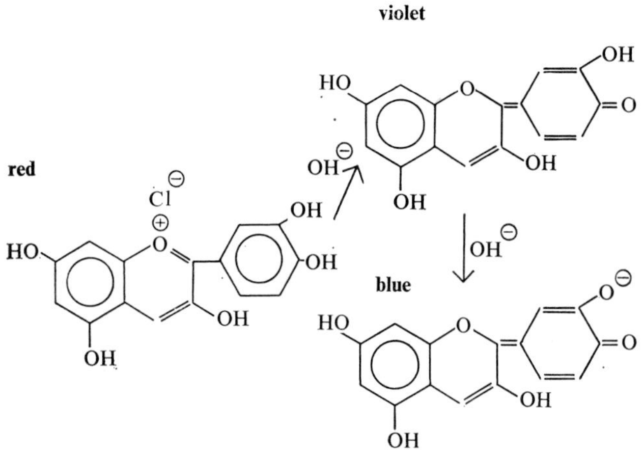

violet red blue

Fig 1.35 **Colour change in anthocyanin**

1.11 Alkaloids

Alkaloids (=alkali-like) are nitrogen-containing bases that occur in plants, particularly those in the families
Papaveraceae (*e.g.* poppies)
Papilionaceae (*e.g.* lupins)
Ranunculaceae (*e.g.* aconite)
Solanaceae (*e.g.* potato, tobacco)
They often have curative or toxic physiological actions on animals. They are normally found localised in the seeds, leaves, bark or roots. *Nicotine* (Fig 1.36), for example, is in the tobacco leaf.

The alkaloid bases occur generally as salts of plant acids, *e.g.* ethanoic [acetic], ethanedioic [oxalic], tartaric, citric, malic. When alkaloids are to be isolated from plants and the material is rich in fats (as in seeds) it is first extracted with petroleum-ether (using a Soxhlet extraction apparatus) before attempting to obtain the alkaloids. The alkaloids are then extracted with methanol, acidified and steam distilled and they are then individually separated by chromatography or fractional crystallisation. Some examples are morphine, atropine from deadly nightshade (*Atropa belladonna*), lysergic acid (from which the hallucinogenic drug LSD is made) and strychnine.

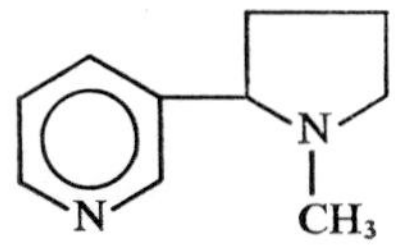

N N CH_3

Fig 1.36 **Nicotine**

1.12 Nucleic acids

Nucleic acids occur in all living things and are responsible for protein synthesis as well as forming the basis of heredity. Their relative molecular masses are several million and they can be 'seen' by using an *electron microscope*, whose principle is similar to the light microscope but which uses an electron beam instead of a light beam (§1.14).

Nucleic acids are of two types: *ribonucleic* acids (RNA) and *deoxyribonucleic* acids (DNA). Nucleic acids are long-chain polymers formed by condensation of smaller units called *nucleotides*. These consist of a 5-carbon (pentose) sugar condensed with a phosphate(V) group and a complex organic base. In RNA the sugar is ribose (Fig 1.7) and in DNA it is deoxyribose (Fig 1.8). In RNA the bases are cytosine, uracil, adenine and guanine (C, U, A, G) whereas in DNA uracil is replaced by thymine (C, T, A, G).

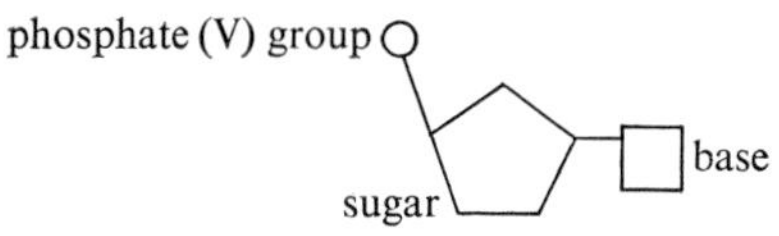

phosphate (V) group sugar base

Fig 1.37 **Nucleotide**

We can represent such a nucleotide diagrammatically (Fig 1.37). The rectangle is the base, the pentagon the sugar (furanose) and the circle the phosphate(V) group. The structure of the phosphate(V) group and the bases is shown in Figs 1.38 and 1.39.

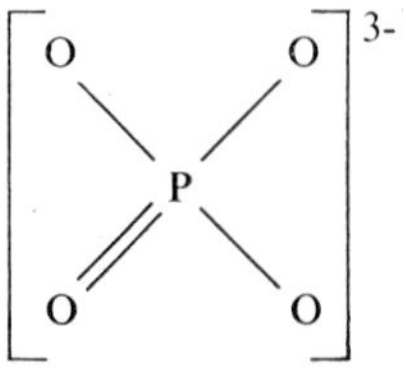

$[PO_4]^{3-}$

Fig 1.38 **Phosphate(V) group**

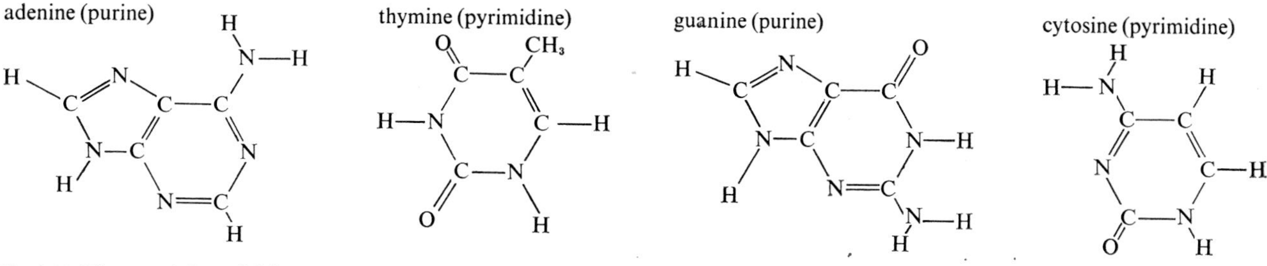

Fig 1.39 **The nucleic acid bases**

Watson and Crick (1953) postulated that the structure of a nucleic acid consists of a double helix of the polynucleotide chains linked across neighbouring sugars by pairs of bases, via *hydrogen bonds* (H—O . . . H—). Wilkins, who shared a Nobel prize with them for this work, used X-ray analysis to study the actual structure.

In DNA each unit is joined to the one above it and below it via phosphate(V) groups, which esterify with sugar molecules. These ester bonds are strong covalent ones and provide a firm backbone to the molecule, and the backbones then twist round each other. In DNA (Fig 1.40) the base pairs are always (for steric reasons) adenine and thymine (A, T) or guanine and cystosine (G, C). The structure of RNA is similar but has only one strand.

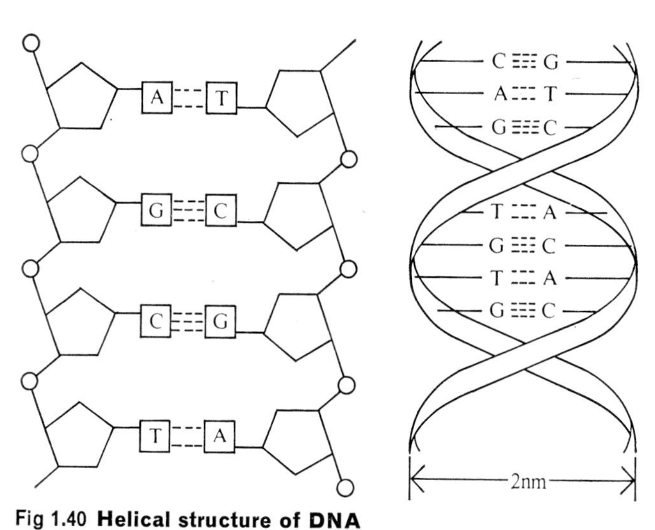

Fig 1.40 **Helical structure of DNA**

To summarise, DNA is found only in the cell nucleus and is concerned with the hereditary coding whereas RNA is both in the nucleus and in the cytoplasm and performs the protein synthesis. Living things have both types of nucleic acid, except for some *viruses* which have only RNA. If the DNA strips in a human cell were pulled out to their full length, they would make a thread about 1 metre long. In general, the length varies according to the complexity of the organism, *e.g.* starfish 0·3 m, bird 0·7 m, but note that the value for a frog is nearly 3 m.

1.13 Summary

We have now considered the main groups of organic chemicals present in living things. These are tabulated in Table 1.2. We now turn to some of the techniques and apparatus used to study them, and the cells in which they occur.

Table 1.2 **Summary of biologically important organic chemicals**

carbohydrates	
monosaccharides	glucose, fructose
oligosaccharides	sucrose, maltose
polysaccharides	starch, cellulose, chitin, glycogen
nitrogen compounds	
amino acids	
globular proteins	albumen
fibrous proteins	keratin
lipids	
fats*	tristearin
waxes	beeswax
oils	palm oil
steroids	
sterols*	cholesterol, sitosterol
bile acids	cholic acid
saponins	
cardiac poisons	digitalis
toad venoms	bufotoxin, bufitalin
sex hormones	oestrone, testosterone, progesterone
pigments	
carotenoids	lycopene, carotene
anthoxanthins	chrysin, quercetin
anthocyanins	peonidin, delphinidin
porphyrins	haem, chlorophyll
terpenes	limonene, pinene
alkaloids	nicotine, strychnine, morphine atropine
nucleic acids	
deoxyribonucleic acid (DNA)	
ribonucleic acid (RNA)	

* many sterols are fatty in nature.

1.14 Microscopes

The development of microscopes has been of crucial importance to the development of biology. When during the late 17th century van Leeuwenhoek discovered the early form of light microscope he opened up a whole new world to scientists. Similarly the development of the electron microscope made possible theoretical advances which would have been impossible if biologists had had to rely on the light microscope.

The *light microscope* (Fig 1.41) uses lenses to magnify objects so that we can see them larger; but in practice it cannot magnify more than about 1500 times without losing detail. It has limited resolution, this limit being imposed by the wavelength of light.

The *electron microscope* (Fig 1.42) overcomes this limitation, and its resolving power is over 500 times as great as that of the light microscope. The electron microscope can magnify an object over 5×10^5 times without losing detail. The electron beam is focused by powerful electromagnets. The image is viewed on a fluorescent screen, or photographed. When electrons in the beam collide with electrons in the specimen they scatter. Where the density of specimen electrons is greatest, scattering is greatest, and the image appears correspondingly darker.

Other important microscopes are the *phase contrast microscope* which enables transparent materials to be observed, and the *polarising microscope* which enables different sorts of material embedded in another substance to be differentiated. The *scanning electron microscope* (Fig 1.43) produces a '3-dimensional' image.

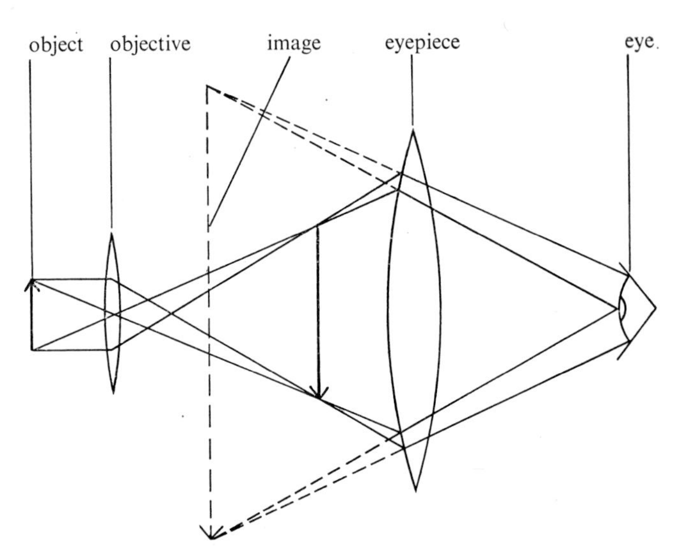

Fig 1.41 **Light microscope**

Fig 1.43 **Scanning electron micrograph showing style from an aubretia flower**

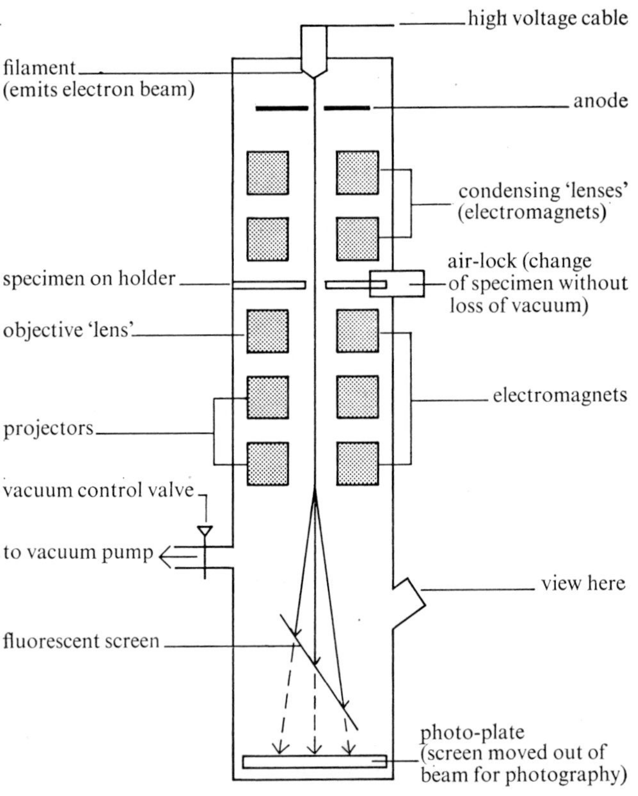

Fig 1.42 **Electron microscope**

1.15 Chromatography and electrophoresis

CHROMATOGRAPHY This is an analytical technique in which separation of solutes is effected by their differential movements, in a flow of a suitable solvent, through some porous medium. In biology it is used to separate, isolate, purify and identify the components of complex mixtures, *e.g.* a mixture of sugars from hydrolysis of a polysaccharide, a mixture of amino-acids from hydrolysis of a protein or the mixture of pigments obtained by grinding up green leaves with a suitable solvent such as propanone (acetone).

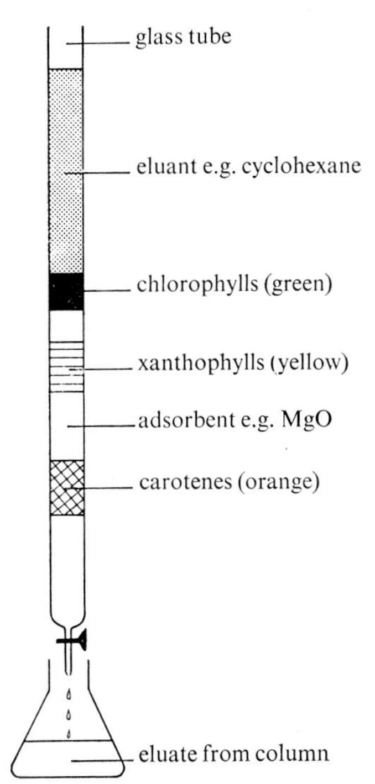

Fig 1.44 Tswett's separation of green plant pigments

The Russian biologist Tswett was probably the first person to appreciate the possibilities of the method, in 1906. He noticed that when a solution of green plant pigments was applied to the top of a column of powdered calcium carbonate, and the column was washed with light petroleum, a series of horizontal coloured bands (green, yellow, orange) began to separate down the length of the column (Fig. 1.44). When the base of the tube containing the column was opened and further solvent was added to the top the different coloured components separated further. The word chromatography means literally 'colour writing' since the first substances to be separated were all coloured. The method is now applied to the separation of colourless materials, but the name remains. The separation of Tswett is an example of *adsorption* chromatography (adsorption means the taking of a substance on to the surface of another, the *adsorbent*). Martin and Synge introduced *partition columns*, packed with cellulose powder. The seemingly dry powder contains some moisture attached to the individual particles and separation is effected by the continuous distribution of the components in the mixture between the water held and the solvent as it sinks down the column. Column chromatography is still used for separation of large amounts of material, since the separated components can be eluted separately from the end of the column. For example, a mixture of glucose and fructose obtained by hydrolysis of sucrose might be separated on a column packed with special charcoal (mixed with a celite packing material) by eluting with aqueous ethanol. The separation could be followed *polarimetrically*, *i.e.* by measurement of the optical rotation of polarised light in successive volumes of eluate collected—the two peaks in the optical rotation/volume eluate graph would correspond to the glucose and fructose.

PAPER CHROMATOGRAPHY With the column method it is difficult to obtain reliable identification of small quantities of materials and so, in 1944, Consden, Gordon and Martin introduced *paper chromatography*, still working on the same principle as the partition columns; filter paper is made of cellulose fibres. Samples to be separated are applied in solution to a spot on a base line drawn on the sheet of filter paper; application is normally made with a micro-pipette so as to keep the diameter of the spot as small as possible. After drying the spot, and respotting if necessary, the chromatogram is developed in a suitable solvent by the ascending or descending technique (Fig 1.45). When the solvent front has almost reached the end of the paper, the position is marked and the paper is allowed to dry in air. If the components are colourless, they then have to be located with a selective spray reagent which converts them to coloured spots. The chlorophyll/xanthophyll/carotene pigments extracted from green leaves (*e.g.* spinach) using propanone, can be separated into green, orange and yellow spots using petroleum/ether as the solvent for chromatography.

A mixture of glucose and fructose (hydrolysis of sucrose with a trace of hot dilute mineral acid) can be separated using as solvent ethyl ethanoate [acetate]/glacial ethanoic [acetic] acid/water (9/2/2 by volume). Location can be effected by dipping the dried paper in a saturated solution of silver nitrate (V) in propanone, allowing to dry and then dipping again in ethanolic sodium hydroxide or by spraying with it (Fig. 1.46); the sugars appear as dark brown spots. The background colour can be removed by dipping in sodium thiosulphate (VI) solution; this then gives a permanent record.

A mixture of the amino-acids glycine, aspartic acid, glutamic acid and tyrosine can be separated using

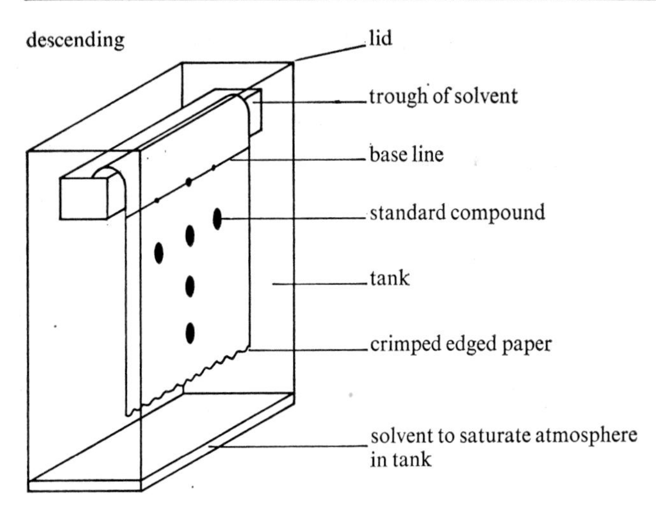

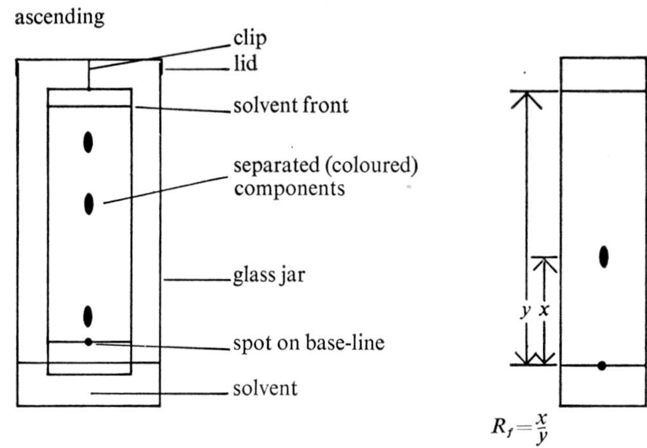

Fig 1.45 **Paper chromatography**

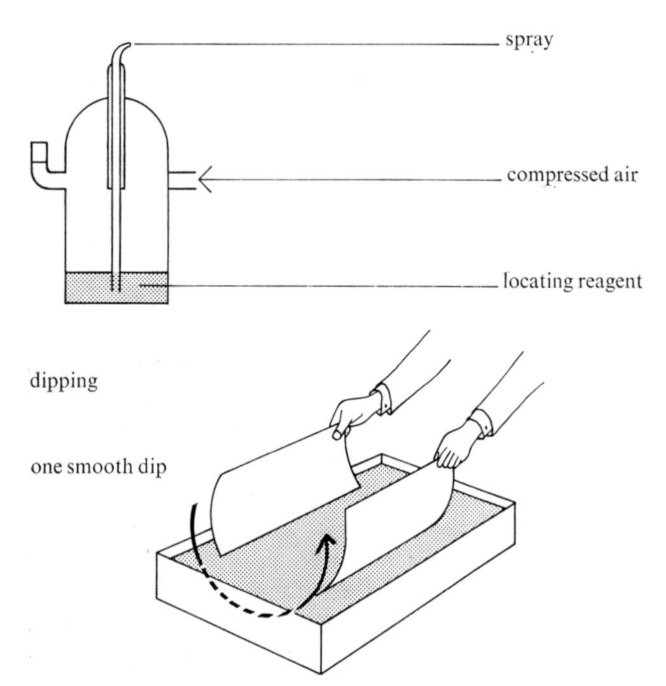

Fig 1.46 **Locating components on chromatograms**

butanol/ethanoic (acetic) acid/water (6/1/2 by volume). Location is effected by spraying with ninhydrin (triketo-hydrindene hydrate), which can be purchased as an aerosol pack, and then heating carefully in an oven at 378 K (or warming over a red-hot gauze) – the acids are converted to purple, yellow or orange derivatives.

Note Ninhydrin is carcinogenic, so be careful when using it.

It is possible to save time in chromatography if you think beforehand, as illustrated by the following. Suppose you extract the materials from rat gut and wish to chromatograph the sugars and amino-acids. The solution of the extract is streaked (rather than spotted) on to the base line and the chromatogram developed in a carefully chosen solvent – one that will give good resolution of both amino-acids and sugars. The paper is dried and then cut lengthwise so as to form two chromatograms. Each half is subjected to its own specific locating reagents, so that when fitted together the sugars and amino-acids can both be seen.

In the paper chromatography technique, there are many variations which are used. In two-way chromatography, wider separation is achieved by first developing a square chromatogram in one solvent and drying, and then turning the paper through a right angle and developing in a second solvent (Fig 1.47). Separations can be made quantitative by cutting sections out of chromatograms and taking the isolated component into solution again, or by eluting components from the chromatograms, and carrying out some form of quantitative analysis, for instance spectrophotometry. Sometimes, a photoelectric absorptiometer might be used on the actual chromatogram to determine the concentration of substance in a particular spot.

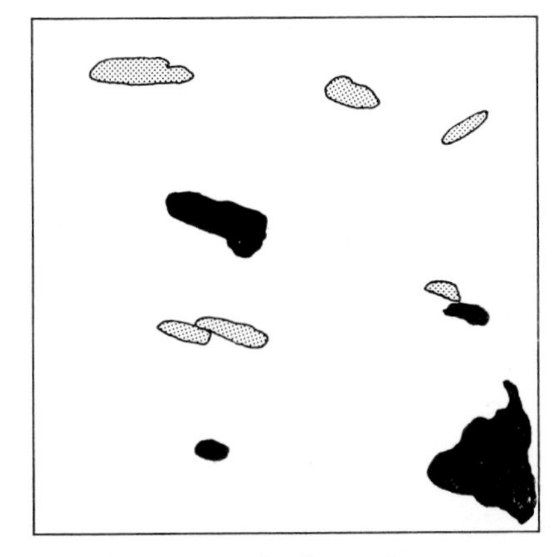

Fig 1.47 **A two-way chromatogram**

Migration of a component on a paper chromatogram is normally measured and quoted as the R_f value, which is the distance migrated relative to the solvent front (Fig 1.46). Sometimes the solvent has to be run off the end of the paper, using descending chromatography, to achieve a good separation and then the migration is quoted as an R_X value, where X is a standard compound, *e.g.* glucose in separation of sugars.

Of special interest is the paper chromatogram where radioactive materials are used, as in photosynthesis. The labelled compounds can be detected after separation on chromatograms using special Geiger-counters or the chromatogram can be exposed to a specially sensitised plate on which dark spots appear where exposure to radioactivity has taken place. The plate is called an *autoradiograph*.

THIN-LAYER CHROMATOGRAPHY The use of this type of chromatography can be extended to adsorption separations by use of the *thin-layer chromatoplate* technique. A uniform thin layer of the adsorbent is spread thinly on a small plate or plastic strip (many types are commercially available) and the mixture to be separated is applied on a base line and 'run' in the solvent as in paper chromatography. Location of colourless substances is achieved by spraying, rather than dipping (which might damage the thin layer) and the plate can be preserved permanently by overspraying with some commercial fixing agent. Separation of the chlorophyll and chloroplast pigments can be done nicely by this method (Fig 1.48).

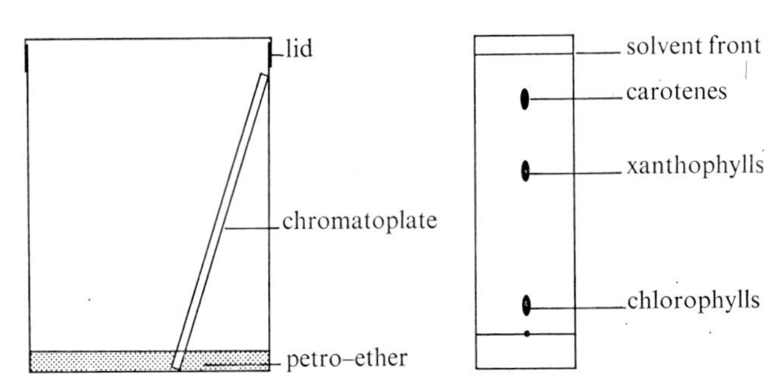

Fig 1.48 **Thin-layer chromatography**

GAS-LIQUID CHROMATOGRAPHY is a logical develop ment from column chromatography and can be used both for preparative work and for identification. A mixture of gases or vapours is carried by an inert 'carrier gas' (*e.g.* nitrogen or argon) through a column of an adsorbent, which can be either a solid (*e.g.* activated charcoal) or a liquid (*e.g.* a liquid silicone supported on a porous material such as kieselguhr). The separation can be followed by measuring the thermal conductivity of the eluant gas. Generally a Wheatstone bridge is first balanced with just the carrier gas flowing through the column. Then when the eluant also contains one of the components from the mixture, the thermal conductivity is altered, the bridge is unbalanced, and the voltage produced is made to drive a self-recording pen across a moving roll of paper (Fig 1.49).

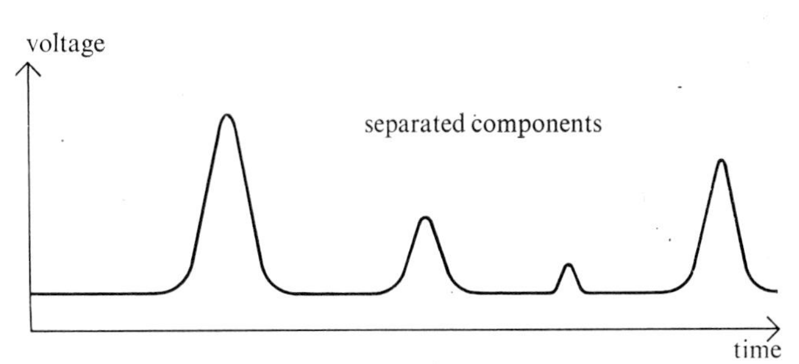

Fig 1.49 **Gas-liquid chromatography**

ION-EXCHANGE columns can be used for the separation of ionic materials.

GEL-FILTRATION chromatography is a particular type of column chromatography in which separation of molecules is effected by their different sizes. The medium used is produced from a polysaccharide called *dextran* (like amylose, but with the glucose linked to glucose via α—1→6 backbone structure).

ZONE ELECTROPHORESIS This is the separation of charged particles which have been adsorbed on a suitable supporting medium (*e.g.* silica gel or filter paper) by a potential difference applied across the ends of the moistened supporting medium. The ends of the medium dip into buffer solutions of a suitable pH to effect separation and the whole medium is damped with the same buffer solution (Fig 1.50). It is really an incomplete form of electrolysis in which the components are stopped on their path towards the electrodes, and never allowed to reach them. The substances do not have to be ionic since it is possible to separate molecules which form complexes with ions in the buffer solution. Thus, sugar molecules can form complexes with borate(III) ions in a borate

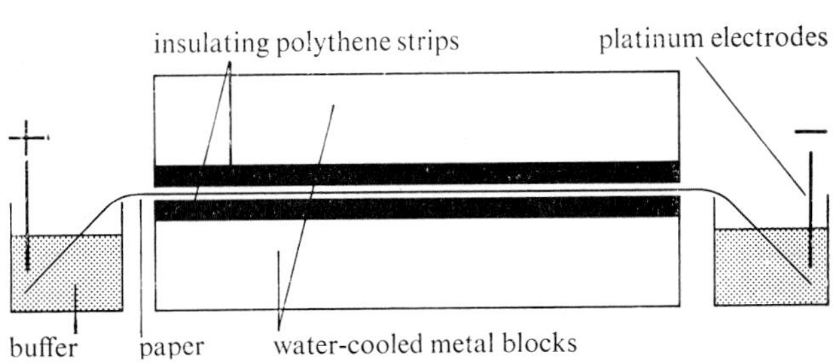

Fig 1.50 **Zone electrophoresis**

buffer and sugar alcohols can be separated by complexing with molybdate or wolframate ions. At the end of the 'run' the *electrophoretogram* (Fig 1.51) is developed in a manner similar to that used in paper chromatography. The direction of migration depends on the sign of the charge on the ion. The rate of migration depends on the applied potential difference, the complexing ions, the structure of the complex (*i.e.* its stereochemistry), and many other factors.

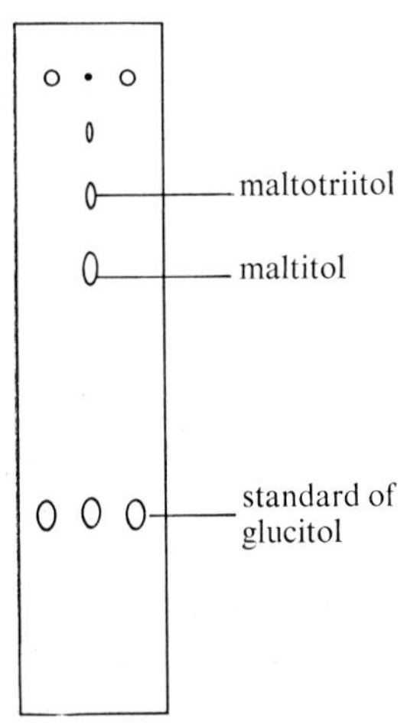

Fig 1.51 **Electrophoretogram**

1.16 Spectroscopic analysis

In recent years, spectroscopic methods have become increasingly important in biology. Thus, in autecology experiments the structure of some chemical compound extracted from a plant might be unknown and this is where a preliminary investigation by means of an *infrared spectrum* proves useful.

Vibrations in the bonds in molecules are affected by absorption of energy from electromagnetic radiation. Since absorption of energy takes place in definite quanta (for details of the 'quantisation of energy' see a textbook of Physical Chemistry), characteristic vibrations are produced only in response to certain frequencies of radiation. If energy is absorbed from the infrared, the molecule vibrates more vigorously (and does not break up, as is the usual occurrence with absorption of ultra-violet). Different quanta of energy are absorbed depending on the atoms forming the bond of the molecule, the multiplicity of the bonding (single, double, triple) and the mode of vibration which ensues (rocking or stretching of the molecule). The infrared spectrophotometer uses a thin film of the substance in solution or in the liquid state, although occasionally the solid is used. When the infra-red radiation passes through it, absorption occurs and if the percentage of radiation transmitted is traced out by a pen on a moving roll of paper there are 'troughs' at the frequency or wavelength where strong absorption takes place (Fig 1.52). Only molecules having atoms bearing partial electrical charges can form infrared spectra; thus O_2 would not absorb, nor would Cl_2, but the groupings in many organic compounds absorb strongly. The presence of groupings such as C=C, C=O, COOH, C—C, C≡N can be recognised in a compound.

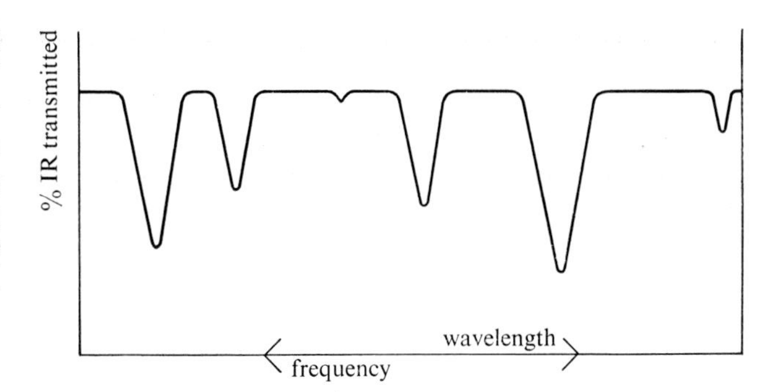

Fig 1.52 **Spectroscopic analysis**

SPECTROPHOTOMETRY This is probably the most accurate way of finding out very small (*e.g.* microgram) concentrations of substances (*e.g.* sugars, ATP, inorganic ions) in solution. Usually a suitable photoelectric cell is used to obtain a direct measure of light intensity, and hence of absorption. The instrument consists of a radiation source, a filter to obtain monochromatic radiation, a glass cell for the solution, the photo-cell and a device to measure the response of the photo-cell. The instrument is first calibrated by plotting a curve showing concentration/absorption for a series of solutions of known concentration. The concentration of the unknown solution is then found by noting the response of the cell and referring to the calibration curve. A *portable absorptiometer* is very useful for biological work. Absorptions in the ultra-violet region of the spectrum are very useful when determining concentrations of sugars, and a hydrogen-discharge lamp is used as source. In the standard method of demonstration of the structure of a polysaccharide, the molecule is oxidised with sodium iodate(VII) and the reduced iodate(VII) is measured spectrophotometrically. This can be related to structural aspects of the molecule.

NUCLEAR MAGNETIC RESONANCE (NMR) spectra are also widely used. The principle is essentially that hydrogen nuclei in atoms can be regarded as tiny bar magnets which take up definite orientations when placed in magnetic flux lines. A rotating magnetic field is then applied at right angles to the main field in such a way that the nuclear magnet resonates with the rotating field, absorbing energy from it. The field strength required for resonance depends on the molecular geometry of the hydrogen atom and so the spectra can be used to elucidate molecular structures.

1.17 X-Ray diffraction

X-ray techniques are often used in the investigation of biological structures, *e.g.* proteins. The 'scattering' (really diffraction) of the X-rays by the atoms and ions in the structure is used to produce an electron map of the molecule. The structure of DNA was studied by this method. Fig 1.53 shows an X-ray diffraction photograph of DNA.

1.18 Centrifugation

This is a technique for the large-scale isolation of sub-cellular materials. The tissue is placed in a suitable medium and ground up so that the cells are not destroyed, forming a *homogenate*. If you had a beaker containing solid particles in water, shook it up and allowed it to settle then the densest and heaviest grains would sediment first to the bottom. The rate of sedimentation would depend on several things, the mass of the particle, the density compared with that of the environment and the gravitational force to which the particles are subjected. One way of increasing the gravitational force would be to whirl the beaker round very fast on the end of a long arm. The centripetal force which must operate to keep circular motion results in an equal and opposite force ('centrifugal') outwards (Newton's third law of motion) and the heaviest particles would tend to be forced to the periphery of the circle. Simple laboratory centrifuges are available for 'spinning down' precipitates. *Ultracentrifuges* operate by applying forces of up to 5×10^5 times gravity on the suspended particles. Their speed of rotation can reach 1 000 Hz and many are refrigerated so that specimens (enzymes, hormones, sub-cellular components) are not destroyed. Very high speed centrifuges can be used in a partial vacuum as well.

DIFFERENTIAL CENTRIFUGATION The homogenate is suspended in a medium of density about the same as water (10^3 kg m^{-3}) and the mixture is centrifuged at particular combinations of speed and time. For the smallest speeds and shortest times the heaviest particles collect at the bottom and as the speed and time increase, the lighter particles come in their turn to the bottom, each being removed at will.

DENSITY GRADIENT CENTRIFUGATION In this method a difference of density, rather than mass, is used. A concentrated (density about $1{\cdot}2 \times 10^3$ kg m^{-3}) sucrose solution is placed at the bottom of the tube and above it are placed a series of sucrose solutions of gradually decreasing density. The suspended material, in a medium of density about 10^3 kg m^{-3} is pipetted on to the top of

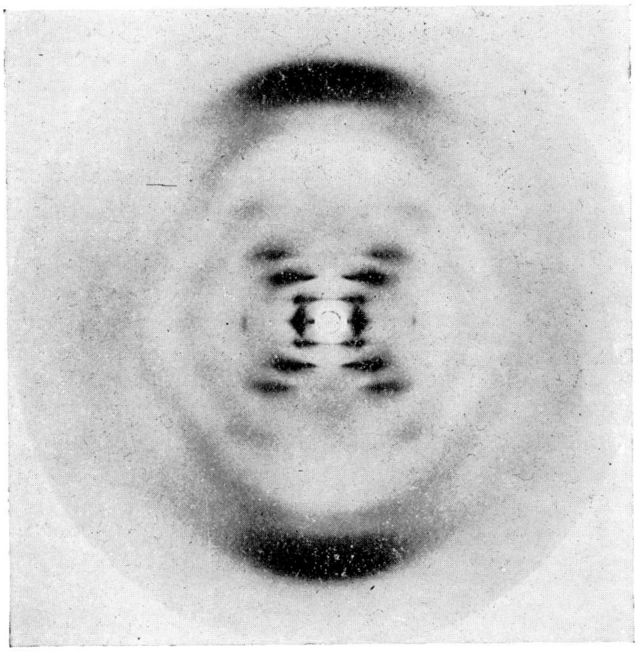

Fig 1.53 **X-ray diffraction pattern of DNA**

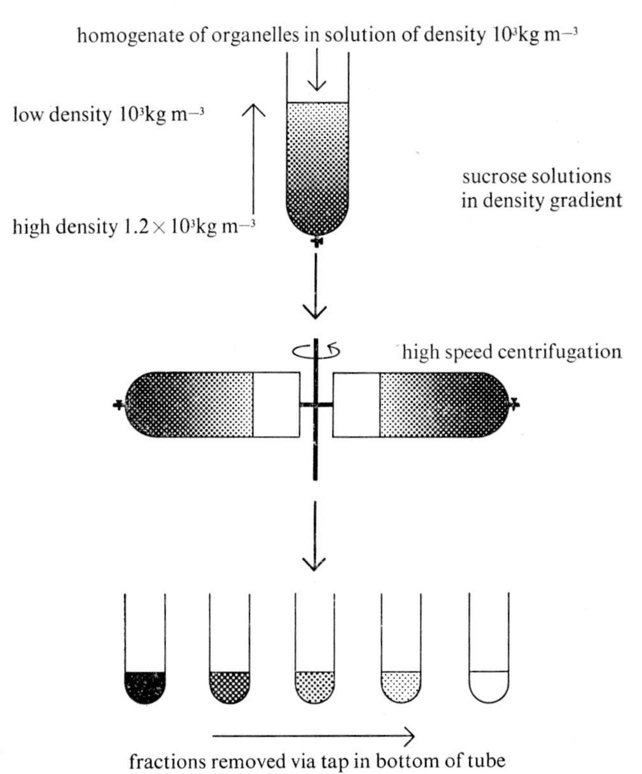

Fig 1.54 **Density gradient centrifugation**

the least dense solution and centrifugation is commenced. The particles settle out in the layer in which their density corresponds to that of the environment (Fig 1.54). Separation of cell organelles by this method is mentioned in Chapter 2.

The solid content of red blood corpuscles of the mammal contains the *chromoprotein* (about 30%) called *haemoglobin*. The red corpuscles are separated from the *plasma* by centrifugation. The plasma is siphoned off, and on addition of ethoxyethane (ether) to the remainder, *haemolysis* occurs with bursting of the red corpuscles. After a further centrifugation a clear red solution is obtained containing the protein. Crystallisation as oxy-haemoglobin can be achieved by addition of ethanol, followed by cooling to 253 K. On careful hydrolysis with dilute hydrochloric acid, two fragments are obtained:

haemin (4%) globin (96%)

Thus, the molecule is a conjugated protein, with a protein linked to a *prosthetic group* containing iron. The product of hydrolysis contains a chloride ion, since hydrochloric acid was used to effect hydrolysis (Fig 1.55). The rings involved are called *pyrrole rings*, and the pattern of these rings involved is referred to as a *porphyrin ring system.*

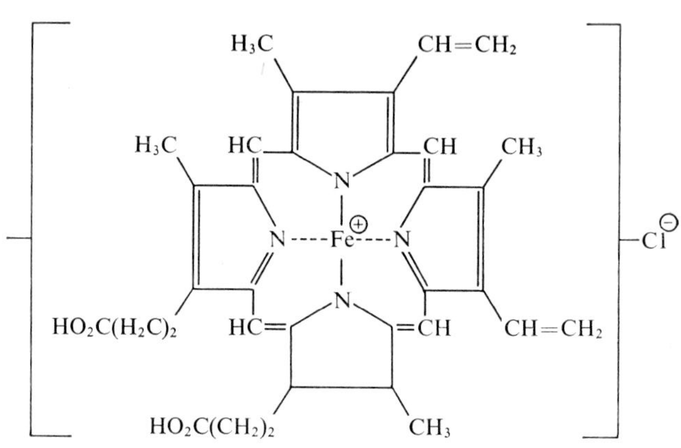

Fig 1.55 **Haemoglobin**

1.19 **Ion-exchange**

Ion-exchange resins consist of small beads made of polystyrene cross-linked in such a way as to produce a sponge-like network, the matrix of which contains chemical groups capable of exchanging cations or anions. *Cation-exchangers* contain ions which are capable of exchanging with cations in the medium in which they are placed. Generally, this ion will be H^+ from a sulphonic acid group, $-SO_2O^-H^+$. *Anion-exchangers* contain ions which are capable of exchanging with anions, *e.g.* OH^- from the grouping $-N^+(CH_3)_3OH^-$. For example if R is the insoluble resin matrix, the reactions would be represented like this:

$$n\,R - H^+ + M^{n+} \rightleftharpoons \underset{\text{cation}}{R_n - M^{n+}} + n\,H^+$$

$$m\,R - OH^- + X^{m-} \rightleftharpoons \underset{\text{anion}}{R_m - X^{m-}} + m\,OH^-$$

Ion-exchange resins are used to free solutions from ions before, for example, running a chromatogram or before freeze drying. *Ion-exchange chromatography*, carried out in columns or on special chromatoplates, can be used to effect many ionic separations. *Ion-exchange gels* and *celluloses* are available for chromatographic separations.

1.20 **Colloids: Dialysis: Freeze-drying**

Colloids are 'solutions' in which the size of the particles is intermediate between those in true solution and in suspension. Approximate size ranges are as follows (particle diameters):

solutions 10 nm (and below)
↑
colloids 10 — 1000 nm
↓
suspensions 1000 nm (and above)

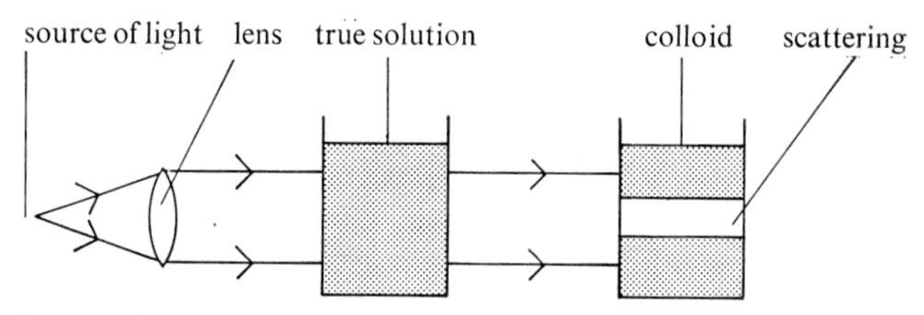

Fig 1.56 **Tyndall cone**

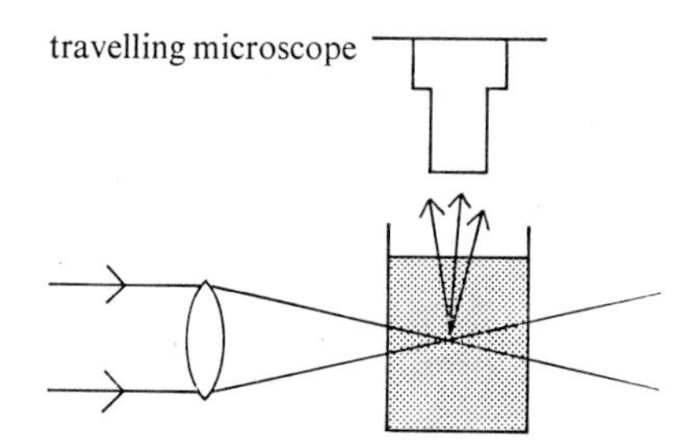

Fig 1.57 **Ultramicroscope**

The colloidal particles cannot be seen with the naked eye, but they are large enough to scatter light incident on them (*Tyndall effect*; Fig 1.56). This fact is made use of in the *ultramicroscope*, in which the scattered light is viewed using a scanning microscope and is seen as bright specks on a dark background (Fig 1.57). These specks are seen to be in continual random motion, due to bombardment of the particles by molecules in the dispersion medium—the *Brownian motion*. Colloids can have a variety of disperse phases and dispersion media, among which are *sols*, in which solid is dispersed in liquid, *emulsions*, when liquid is dispersed in liquid (*e.g.* milk), and *fogs*, when liquid is dispersed in gas. When some sols are concentrated they form *gels*, *e.g.* table jelly, silica gel (used for drying atmospheres in apparatus).

It is frequently necessary to isolate solid biological materials from concentrated aqueous solutions when it is found impossible to crystallise them. An example is in the isolation of certain fungal enzymes. After allowing a mould to grow in a suitable solution to release the enzyme extracellularly, the cell particles are removed by centrifugation and the supernatent colloidal solution is dialysed against a suitable buffer solution in which the enzyme is stable. In *dialysis* the colloidal solution of the enzyme is placed in a dialysis bag (of special mesh polymeric material, like cellophane) and the bag is placed in a beaker of the buffer. The colloidal enzyme molecules cannot pass through the bag but the ions from the buffer enter and the enzyme in the bag eventually becomes suspended in the buffer medium.

The solution is next concentrated by *rotary evaporation* (but not with heating, as this would destroy the enzyme) and the concentrate is frozen by immersion of the flask containing it either in liquid nitrogen or in propanone [acetone]/dry-ice. The flask is then attached to a *freeze-drier* which is an instrument for creating a low pressure inside the flask so that the water may be sublimed off from the frozen suspension medium, leaving the powdered enzyme plus ions from the buffer.

Sols are of two types, *lyophilic* (solvent-loving, *e.g.* starch, gelatin, albumen) and *lyophobic* (solvent-hating, *e.g.* silver sol, gold sol, iron(III) hydroxide sol). The former have a higher viscosity, smaller Tyndall effect and more complex electrical properties. Proteins in colloidal solution, for example, are lyophilic and adsorb ions from their environment. In acid media, they adsorb H^+ ions and in alkaline media they adsorb OH^- ions. Their direction of motion in *electrophoresis* (motion under the action of an electrical field), therefore, is governed by pH, and there is clearly a pH at which the electrophoretic mobility is zero – the *isoelectric point*. The protein is amphoteric, dissociating as both an acid and a base, and it is possible to find the pH at the isoelectric point by finding when its acidic and basic dissociations are equal (Fig 1.58).

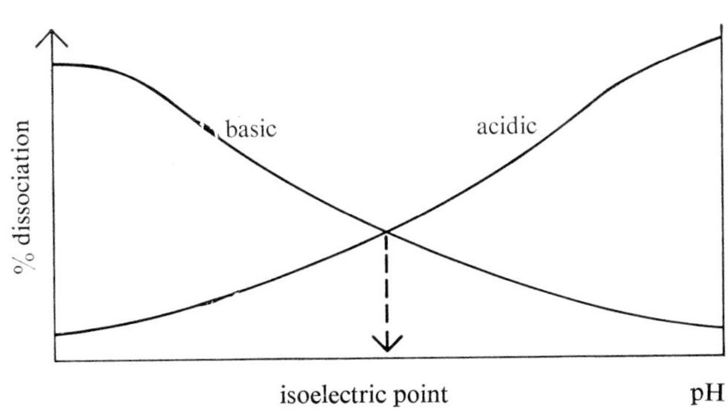

Fig 1.58 **Determining isoelectric point**

Coagulation (precipitation) of a colloidal solution can be effected by ions having an electrical charge opposite to that on the colloidal particles. The coagulating power is proportional to the size of the charge on the ion, this being called the *Hardy-Schulze rule*. For a negative colloid, the order of coagulating power would be

$Al^{3+} > Mg^{2+} > Na^+$

All the particles in a sol have the same sign of charge, due to adsorption of ions from the dispersion medium. The mutual repulsion between the *Helmholtz double layer* of charge stabilises the sol (Fig 1.59).

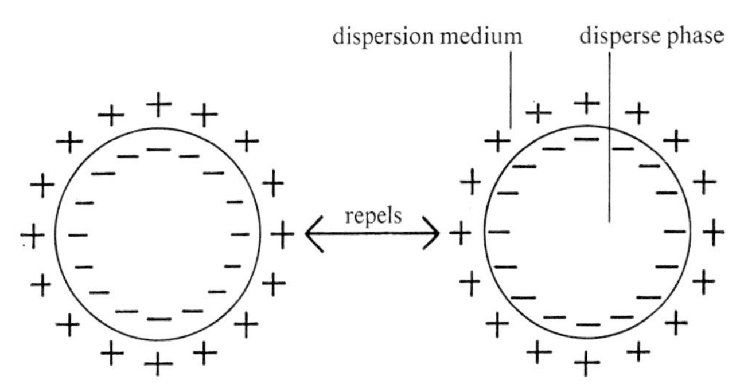

Fig 1.59 **Helmholtz double layer**

It is often found that lyophilic sols can protect lyophobic ones from coagulation. In the river Nile, for example, clay in colloidal solution (lyophobic) is prevented from coagulation by a protective layer of organic matter. Near the sea, the high salt concentration breaks down the protective layer and precipitation of the coagulated clay causes a muddy delta. The *gold number* is a number which measures the effectiveness of protection. A lyophilic sol of high gold number is relatively ineffective and one of low gold number is a good protector.

1.21 Carcinogens

Before leaving the subject of biochemistry mention should be made of one other group of compounds. The discovery that *polynuclear hydrocarbons* (those with several benzene rings) can produce malignant tumours in animal tissues has opened up investigations into cures for various types of cancer. It was Kennaway and Cook (1930) who discovered that the hydrocarbon dibenzanthracene (Fig. 1.60) caused carcinoma. The hydrocarbon benzpyrene (obtained from pitch or coal tar) also causes carcinoma.

It is very interesting that another potent carcinogen, methylcholanthrene, can be obtained from a steroid bile acid.

Tobacco smoke is known to contain several compounds which could be carcinogenic, *e.g.* stigmasterol, phenanthrene, substituted pyrenes, phenolic compounds.

1.22 The computer in biology

Brief mention must be made of the fact that problems in biochemistry in the future will be aided in their solution by the computer. This work would involve both structures and also kinetics of reactions, programming of chromatography 'runs', etc.

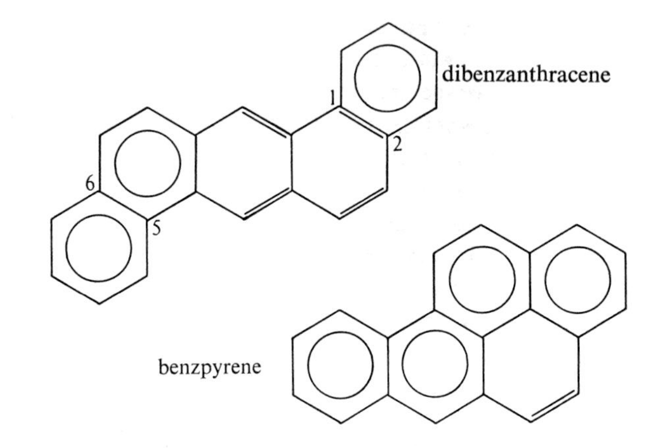

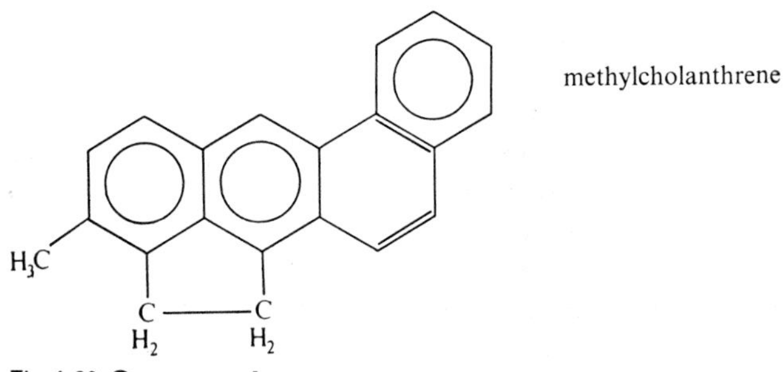

Fig 1.60 **Some carcinogens**

2 Cytology and Histology

2.1 Form and function

By the beginning of the 19th century, Schleiden and Schwann had deduced that the living organism is an aggregate of apparently self-contained units, called cells. In general, plant cells have a thick wall surrounding them whereas animal cells do not; this underlies the basic differences between animals and plants. The inside of the cell was first called *protoplasm* and was seen to contain various granules and bodies. One such large body is the *nucleus* (Brown, 1833) which is present in all cells. The protoplasm was seen to be surrounded by a *cell membrane*, with (for plants cells only) a wall of cellulose around the whole. More recently the material between the nucleus and the membrane has been called the *cytoplasm*.

Cytology (study of cells) and *histology* (study of tissue structure) may be carried out using an optical microscope (Fig 1.41). Cells may be as small as 200 nm in size. Animal cells (*e.g.* cells from inside the cheek) and plant cells (*e.g.* onion cells) can be so studied. When cells are examined using staining techniques, the material must be dead, but with phase contrast and interference microscopes, which make use of the difference in refractive index of materials of the same opacity, living specimens may be examined. The production of the *electron microscope* (Fig 1.42) (Knoll and Ruska, 1930) has made possible the examination of the *ultrastructure* of cells, and their inclusions or *organelles*. Ultra-thin specimens are prepared and are coated or 'stained' with materials that interrupt electron beams (*e.g.* uranium, osmium).

Cells are not always self-contained units; if cells from lung or muscle are removed they usually die quickly. Some cells, *e.g.* cancer cells, can continue to reproduce and grow outside the body. Cells are specialised; for instance, long muscle cells shorten when suitably stimulated. Nerve cells transmit information from, say, the brain to muscles (*e.g.* arm). Nerve cells, too, are long.

The subcellular organelles referred to include the *mitochondria*, *ribosomes*, *nucleus* and *endoplasmic reticulum*. Cells for photosynthesis in plants, which need chlorophyll, have *chloroplasts* as well. In the cytoplasm, around the organelles, there are various fibres and vacuoles (holes) filled with fluid.

When the movement of living cells is studied using the optical microscope, many cells (ranging from amoebal cells to mammalian ones) are seen to take droplets of fluid from outside the cell into the cytoplasm, this being termed *pinocytosis*. Pinocytic vesicles appear to be pinched in the membrane and can be seen in an electron micrograph.

Typical cells (Fig 2.1) are roughly spherical and range in diameter from 200 to 10^5 nm. The membrane is seen under the electron microscope to be double-layered. The most prominent part is the nucleus and this can occupy up to one quarter of the volume of the cell. The cytoplasm contains vacuoles, dark particles (ribosomes) and a number of oval bodies (mitochondria), composed internally within the lumen of membrane-like objects called *cristae*.

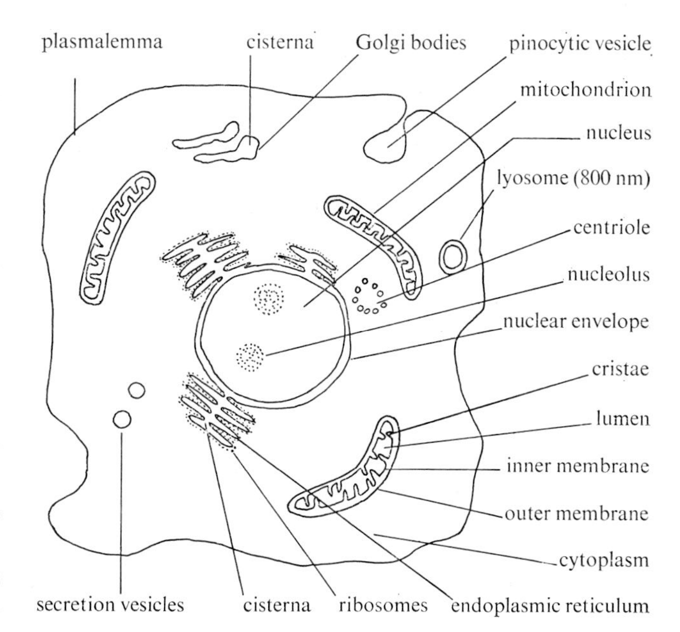

Fig 2.1 **Generalised animal cell**

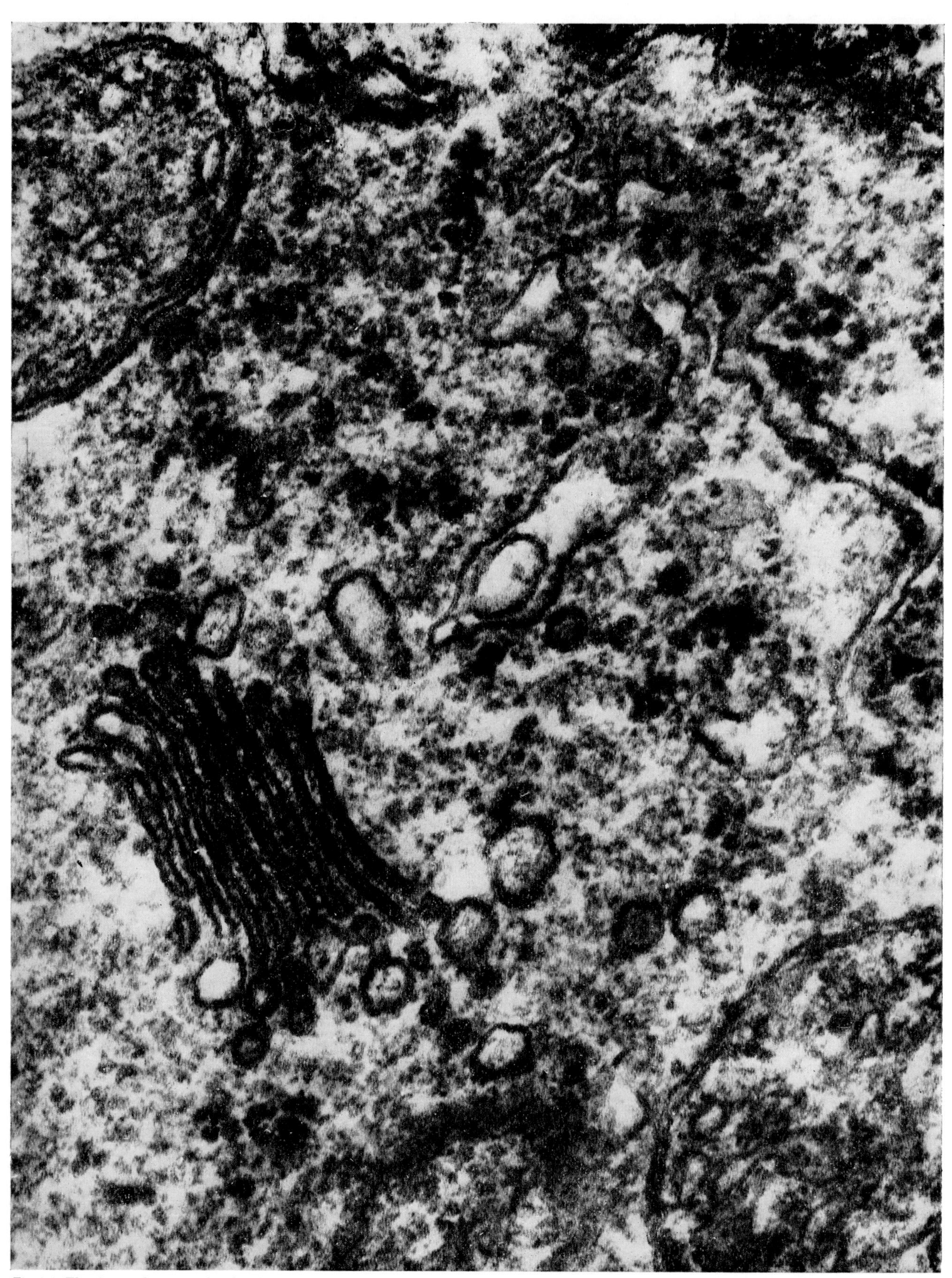

Fig 2.2 Electronmicrograph of cell

It is possible, by use of a microscope, to pick out individual portions of large cells using a fine steel wire. A more modern technique involves *ultracentrifugation* §1.18. The *homogenate* is prepared by gently grinding up the tissue in a suitable medium so that the organelles are released into the supporting fluid. The homogenate is then spun round very fast. Separation depends on the force applied and the density of the particle compared with the surrounding fluid. Using differential centrifugation, the order of ease of separation of organelles is as follows 1. nuclei 2. mitochondria 3. lyosomes 4. membranes 5. ribosomes.

The *plasmalemma* or *plasma membrane* surrounds the cell and is thrown into folds which increase its surface area. The surface properties of the membrane (*e.g.* calcium ion content and peptide bonding) hold individual cells together. Since substances passing in and out of the cell encounter this membrane, it controls exchange of materials. Within the cell there is a network or reticulum called the endoplasmic reticulum (ER) consisting of a system of membranes with spaces in between called *cisternae.* As well as providing a large surface area for the chemical reactions in the cell to take place, the ER may also keep substances separated so that they do not react, and allow areas with different pH value to remain separated. Substances may also move about the cell via the ER. The ribosomes, which contain RNA and are concerned with protein synthesis in the cell, are clustered on the ER. Substances synthesised in the ER are stored in the cisternae of the *Golgi bodies.* The *lyosomes* consist of a vacuole filled with solutions of enzymes capable of hydrolysing proteins, nucleic acids and polysaccharides, all of which are in the cytoplasm, so that the membrane bounding it prevents the enzymes from lysing (digesting) these substances. When cells are shattered or damaged by disease or mechanical injury the liberation of the enzymes into the cells causes *autolysis* (self-digestion).

The nucleus is frequently ovoid in shape and its size is governed by the amount of cytoplasm there is. It is usually invisible except when a stain is used. Spheres of nucleic acid are present within the cell nucleus and are called *nucleoli* (singular, nucleolus); their size decreases as the cell divides. Outside the nuclear membrane there is a cisterna. Here, nucleic acid from the nucleus is stored before moving to other parts of the cell. The nucleus is the centre of chemical activity in the cell, and plays a major part in determining its size, shape and function. Resting nuclei, not dividing, are described as being at *interphase;* in this phase, the nuclear protoplasm is made up of chromatin threads, but when cell division starts these threads aggregate to form chromosomes.

If cell division is studied using an ordinary microscope, a small body near the nucleus divides in early prophase (see pp 32–3) with the two bodies formed moving to opposite poles of the cell. By electron microscopy these are seen to be cylindrical in shape; the individual elements are the centrioles and the 'duet' is a *diplosome.*

Secretion vesicles occur in the cytoplasm. They originate in the Golgi bodies and transport secreted substances to the periphery of the cell.

The mitochondria are concerned with cell respiration and Krebs' cycle reactions (p 125). There are usually thousands of mitochondria in a cell.

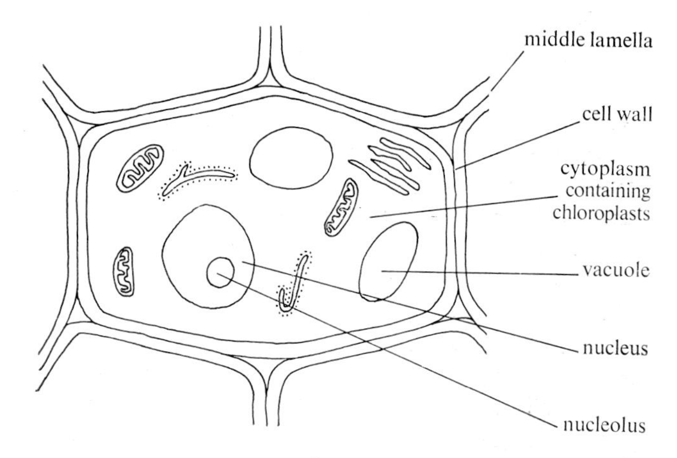

Fig 2.3 **Generalised plant cell**

The plant cell (Fig 2.3) differs from the animal one in two important ways, first being surrounded by a *rigid wall* and second, containing *chloroplasts.* As cells divide the new walls that are laid down commence with the *middle lamella* which chiefly comprises pectates of calcium and magnesium. Upon this is deposited cellulose in *microfibrils,* within which there are well defined crystalline areas called *micelles.* This constitutes the *primary* cell wall, and further depositions of cellulose as macrofibrils, added when the cell has ceased expanding, form the *secondary* cell wall. In the bark of trees there is a layer of suberin between the primary and secondary walls. Chloroplasts contain a *stroma* which contains the enzymes used in the photosynthetic dark reaction, together with DNA and RNA. Within this matrix there are concentrated regions called *grana* which are made up of stacks of chlorophyll-containing discs. Plant cells in stems are of three types:

1 Supporting: to hold up the stem. They have thick walls and contain lignin. When cells are encased with lignin their contents die. Thus, supporting cells are those in which the material is dead and only a cell wall remains, but they may also have a circulatory function.

2 Protective: to protect the delicate parts of the stem. They have fairly thick walls and are found in the outer layers of the stem.

3 Circulatory: to provide paths for circulation of water and food. They are long and must have holes in their walls so that their contents can get to nearby cells.*

Plasmodesmata are interesting regions (strands) in plant cells, being regions of continuity between the cytoplasm of adjacent cells. These strands of cytoplasm pass from cell to cell via the primary cell wall: once the secondary cell wall has been laid down the strands are broken.

Table 2.1 **Comparison of animal and plant cells**

animal cells	*plant cells*
usually smaller, less distinct outline, no distinct wall	usually larger, distinct outline definite wall (made of cellulose)
almost wholly made of cytoplasm	only thin lining of cytoplasm
if they have vacuoles they are small and temporary only (excretory and secretory)	large central vacuole

2.2 Protein synthesis

In the nucleus of the cell there are *chromosomes* made up of DNA arranged in tightly coiled spirals or helices. *Assembling* of amino-acids to give proteins occurs in the cytoplasm and, as DNA is in the nucleus, a particular amino-acid cannot react directly with particular groups of bases in the DNA. It can be shown that *protein synthesis* occurs at the ribosomes.

The synthesis of a particular protein needs the correct assembly of amino-acids in a unique order. The information which specifies the actual acids and the order that is required is carried in the coding of the bases in the particular part of the DNA molecule (Fig 2.4) concerned with that protein. Here, bases present in the nucleus, ribose and phosphate (V) groups assemble to synthesise *messenger* RNA (the template for protein synthesis). RNA, as we saw in §1.12, is similar to DNA, but with only one strand. The double helix of DNA unwinds; the bases line up opposite their partners along the DNA strand, *i.e.* adenine opposite uracil, guanine opposite cytosine (Fig 2.5). This process is called *transcription.*

The message (base sequence) of one DNA strand is copied by an RNA strand with the help of the enzyme RNA polymerase. The RNA strand then leaves the DNA strand and proceeds to the ribosome on the endoplasmic reticulum. In higher organisms, a form of DNA called *information* DNA passes out of the nucleus into the cytoplasm where messenger RNA is synthesised directly. The ribosome consists of protein (30–50%) plus messenger RNA.

Specific amino-acids are 'picked up' and carried to their correct sites on the template situated on the ribosomes by means of adaptor molecules of transfer RNA. These are synthesised in the nucleus and are present in the vesicles of the endoplasmic reticulum. Attachment of an amino-acid is aided by an enzyme called an *amino-acid deactivator*, and there is a different one for each different amino-acid. The whole complex, called an amino-acyl-transfer RNA complex and formed by use of ATP, now moves to the ribosome, and the transfer RNA becomes

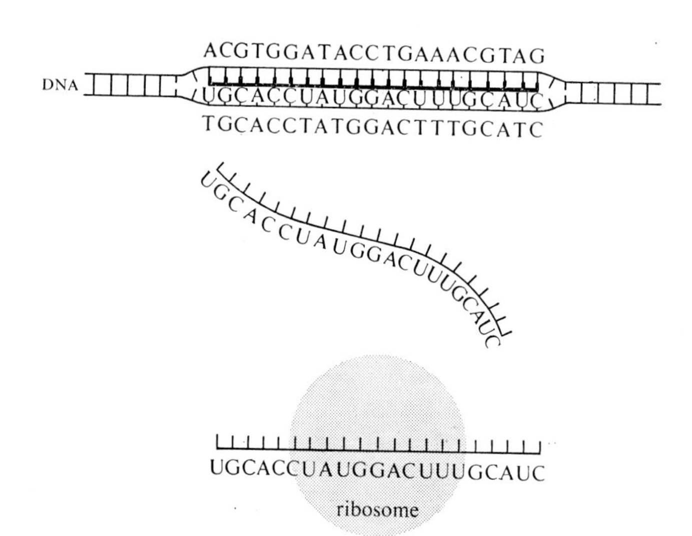

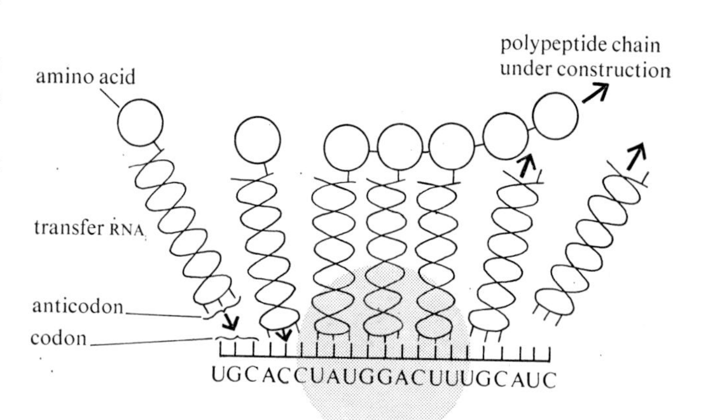

Fig 2.4 **Protein synthesis (a)**

aligned according to the *triplet codes* (the fit of sets of three bases). An enzyme is needed to link the *m*-RNA to the *t*-RNA. Juxtaposed amino-acids now condense together forming part of a polypeptide chain, which becomes detached from the ribosome using ATP. The *t*-RNA is now available to collect more amino-acids from the endoplasmic reticulum. Finally, the polypeptides are folded into their final shape.

The sequence of events can be summarised as follows:

1 Amino-acids enter the cell and each different one is picked up by its own type of *t*-RNA.

* Xylem cells are circulatory, lignified and dead, so categories 1, 2 and 3 are not distinct.

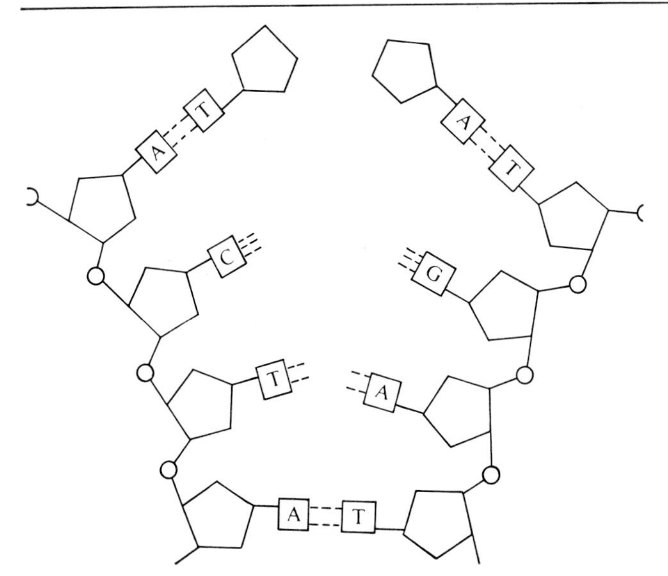

Fig 2.5 **Protein synthesis (b)**

2 The *t*-RNA molecules have, at their ends to which amino-acids are not attached, three bases in a definite sequence which will correspond to three bases positioned on the ribosome.

3 The *t*-RNA moves to the ribosome, and link up of bases takes place according to the triplet code.

4 Polypeptide chains are released, *t*-RNA returns to pick up more amino-acids, and the polypeptide chains become folded.

Let us suppose that the triplet of bases (triplet code) to which each amino-acid corresponds consisted of fewer bases. Now if this group consisted of only one base, then only four amino-acids could be coded. With a theoretical two bases, sixteen (4^2) pairs are possible.

	C	G	A	T
C	1	2	3	4
G	8	7	6	5
A	9	10	11	12
T	16	15	14	13

There are *twenty* amino-acids to encode and so a group of two bases is again insufficient. So the group must be a minimum of three bases giving 64 (4^3) possible options; this is supported by experimental evidence. The groups of three bases (or 'code words') are called *codons*. With 64 different codons available to code 20 amino-acids, it might be thought that there could be more than one codon per amino-acid, but this is not found to be the case. One amino-acid can be coded by related codons, *e.g.* alanine is coded by the codons GCU, GCA, GCG. Notice that the first two bases here are common to all three codons. Some amino-acids are coded by only one codon (*e.g.* tryptophan, methionine) while others are coded by as many as six (*e.g.* serine, arginine). Some codons do not appear to code any amino acids (nonsense codons), and act as full stops in the sequence of acids in the genetic code, *e.g.* UGA, UAA. The same sequence of bases always stands for the same amino-acid, no matter what is the nature of the organism being studied (those investigated have included *Escherischia coli*, clawed toad, yeast, man).

Suppose some synthetic RNA was made from a solution containing 70% adenine (A) and 30% uracil (U), and that when this RNA operated in a cell-free system, it was found that the proteins formed contained amino-acids in this ratio:

5·5 times as much X as Y
2·5 times as much X as Z
12·5 times as much P as Y

i.e. Y:Z:X:P as 1:2·2:5·5:12·5

Can we find out which triplets of bases are giving rise to X, Y, Z and P? Random assortment, by chance, into triplets would give the following probabilities for combinations (bearing in mind that in every ten parts of solution there are seven of adenine and three of uracil) in the proportions:

3 adenines (AAA)	$0{\cdot}7 \times 0{\cdot}7 \times 0{\cdot}7 = 0{\cdot}343$
2 adenines, 1 uracil (AAU, AUA or UAA)	$0{\cdot}7 \times 0{\cdot}7 \times 0{\cdot}3 = 0{\cdot}147$
2 uracils, 1 adenine (UUA, UAU or AUU)	$0{\cdot}3 \times 0{\cdot}3 \times 0{\cdot}7 = 0{\cdot}063$
3 uracils (UUU)	$0{\cdot}3 \times 0{\cdot}3 \times 0{\cdot}3 = 0{\cdot}027$

In the final mixture we would then have a ratio of triplets as follows:

AAU . UUA
AAA: AUA : UAU : UUU
UAA AUU

as 0·343:0·147:0·063:0·027
or 12·7:5·4:2·3:1
which, allowing for experimental error is the same as
P:X:Z:Y
It is clear, therefore, that
X is coded by adenine and one uracil
Y is coded by three uracil
Z is coded by one adenine and two uracil
P is coded by three adenine
Thus, by determining the order of the triplets of bases in the nucleic acids, the order of assembly of the amino-acids to form proteins is also determined. The whole assembly of base groups, which determines the overall pattern of amino-acids, is called the *genetic code*.

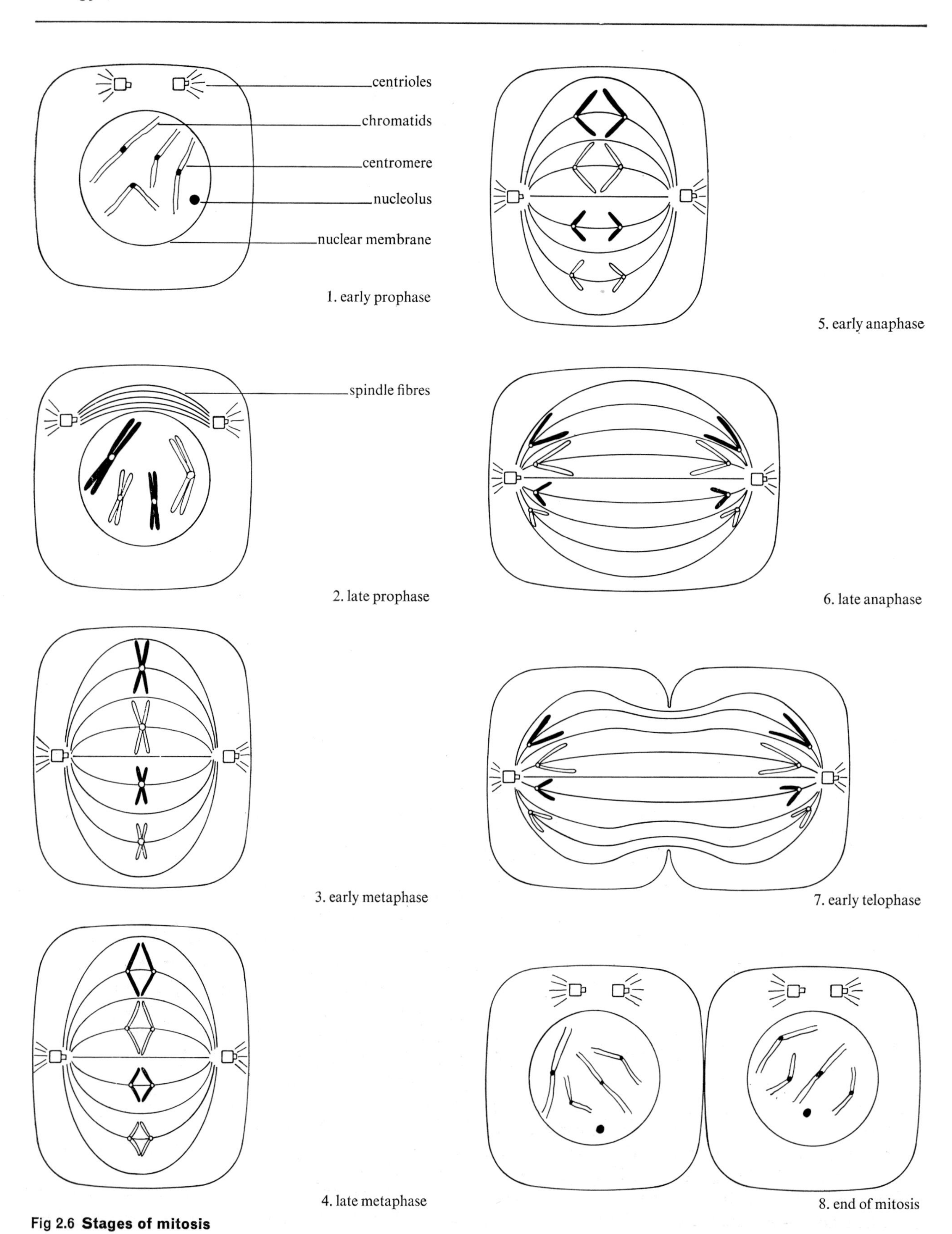

Fig 2.6 **Stages of mitosis**

2.3 Cell Division

Examination of growing organisms reveals that their number of cells is increasing; root tips are convenient for studying cell multiplication. During cell division, the double helix form of DNA first uncoils and then new bases, sugars and phosphate groups condense on each strand to give new strands. This has been demonstrated by tracer techniques *e.g.* using bacteria in which the nucleotides of their DNA had been labelled with ^{15}N. After replication, it is clear that identical sets of genetic information can be passed to each *daughter cell.*

MITOSIS The division of the nucleus that takes place before the division of the cell is called mitosis. The nucleus has to divide so as to provide the two daughter cells with all the information required for them to make all they will need. In mitosis, each daughter cell gets an identical dose of DNA. There are four stages, *prophase, metaphase, anaphase* and *telophase* (Fig 2.6).

1 *Prophase* The chromosomes, viewed with the aid of Foulgen's stain, appear as double coiled threads, each one called a *chromatid* and joined at a *centromere.* The chromosomes become much shorter, thicken, and move to the edge of the nucleus. The nucleolus decreases in size since its nucleic acid content is being passed to the chromosomes. The membrane around the nucleus begins to break. A small body lying outside the nucleus called the *centriole* divides into two, the two halves move apart to different sides of the nucleus, and in so doing give rise to *spindles* of fibres; these are fine tubes of contractile material. After the nuclear membrane has vanished the fibres become the dominant feature of the cell. This is a long stage.
2 *Metaphase* The chromosomes become attached to the spindle fibres at the centromere region. The centromere splits, forming fibres and these fibres elongate while the spindle fibres contract; the daughter chromatids start to separate. This stage is brief.
3 *Anaphase* The separating of the chromatids which began in metaphase continues. This stage is rapid.
4 *Telophase* The divided chromosomes reach their own poles of the cell (and become longer and thinner) and a nucleolus is formed. The nuclear membrane reforms, the spindle fibres reconstitute the centriole and the cell divides.

You can study mitosis using a root tip of garlic. There are 46 chromosomes per cell in garlic root tips.

Mitosis is the means by which all the cells of a multicellular organism (Fig 2.7) are derived from the original *zygote*, the cell formed by fusion of the sex cells (gametes) in sexual reproduction. It is also the means by which organisms reproduce asexually; all asexually produced offspring have the same genetic information as their parents (*e.g.* vegetative reproduction; taking cuttings, potato tubers, strawberry runners, daffodil bulbs, etc.). Examples of asexual reproduction include flowering plants; lower plants (*e.g.* gemmae of liverworts); spores of fungi (*e.g. Mucor*); budding of *Hydra*; parthenogenesis of aphids; production of drone bees from the unfertilised eggs of a queen bee.

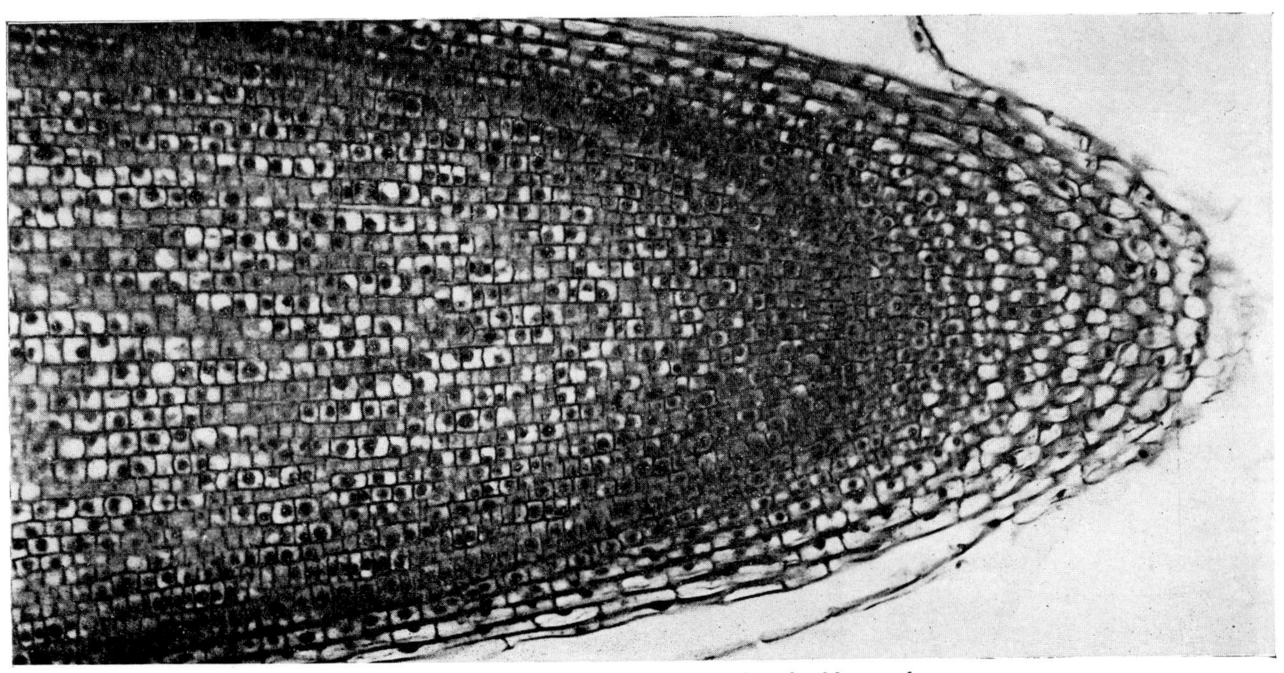

Fig 2.7 **Vicia faba root tip l.s; the large chromosomes are particularly clear in this species**

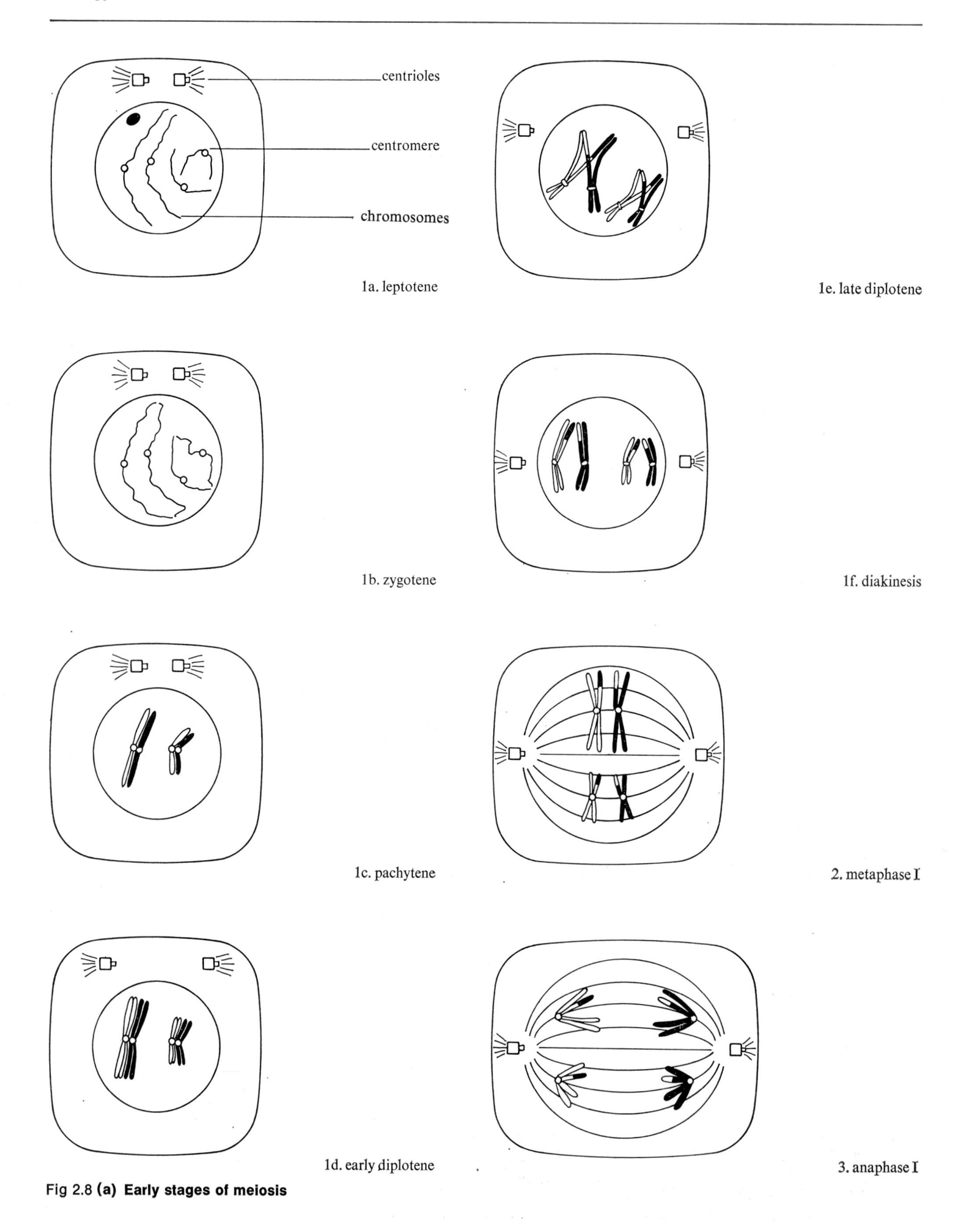

Fig 2.8 **(a) Early stages of meiosis**

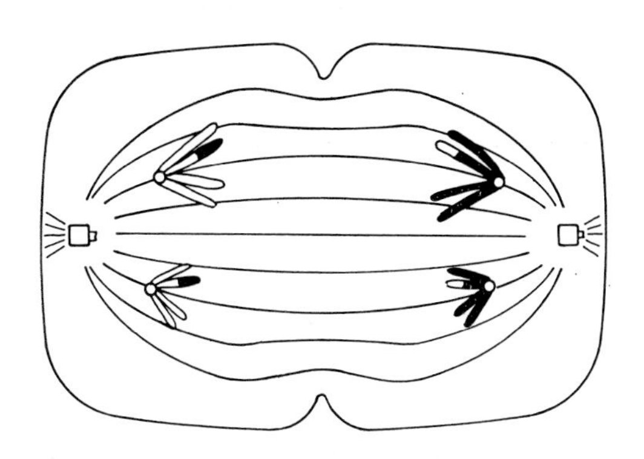
4. telophase I

MEIOSIS We have seen that when a zygote undergoes mitosis to produce first an embryo and eventually an adult, the number of chromosomes remain the same, like those in every other member of the species. If gametes were produced like body cells, by mitosis, then the zygote after fusion of the gametes might be expected to have twice as many chromosomes as its parents. Halving of the number of chromosomes takes place in *meiosis*. To observe this type of cell division we must study cells from organs which form gametes, the gonads (Fig 2.9) of animals and the anthers and carpels of flowering plants.

The stages of meiosis (Fig 2.8) are as follows:

1 Prophase I, which can be subdivided into

a Leptotene: separate homologous (partner) chromosomes are present, the chromosomes being elongated.

b Zygotene: homologous chromosomes unite (to start with at their ends and then over their whole length).

c Pachytene: the chromosome pairs become thicker, each member consisting of two chromatids.

d Diplotene: the homologous chromosomes begin to separate except that they remain fixed together at several points, called *chiasmata*. Part of a chromatid from one partner becomes united with part from the other chromatid, a process called *crossing over*, and it is here that an exchange of genetic information occurs.

e Diakinesis: This is when the pairs of partner chromosomes separate even further.

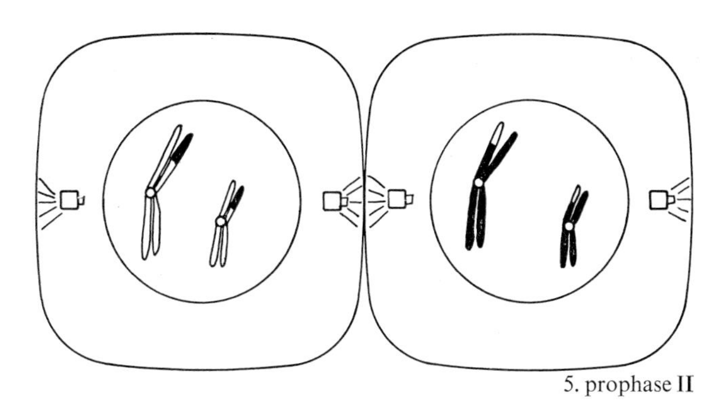
5. prophase II

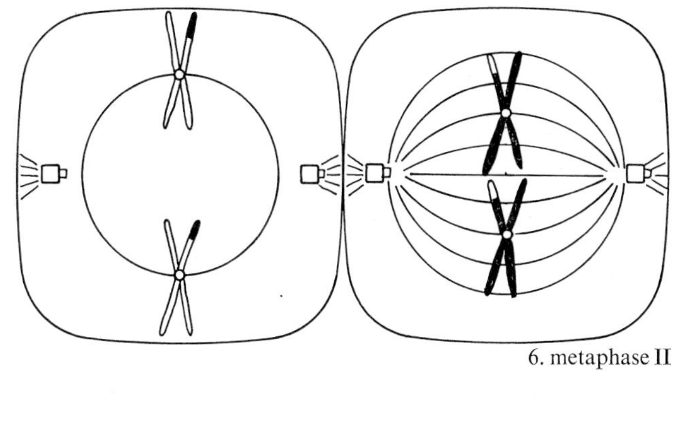
6. metaphase II

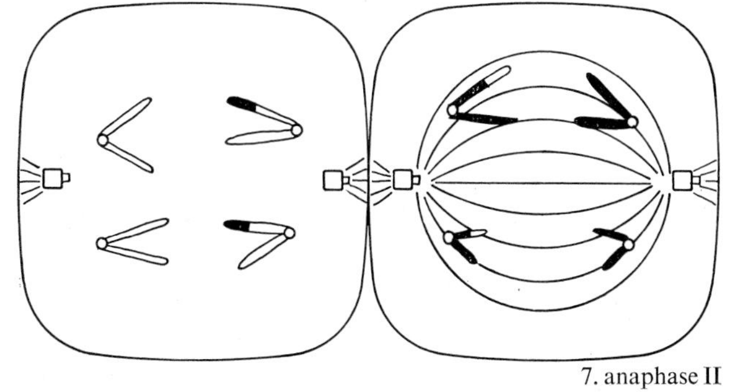
7. anaphase II

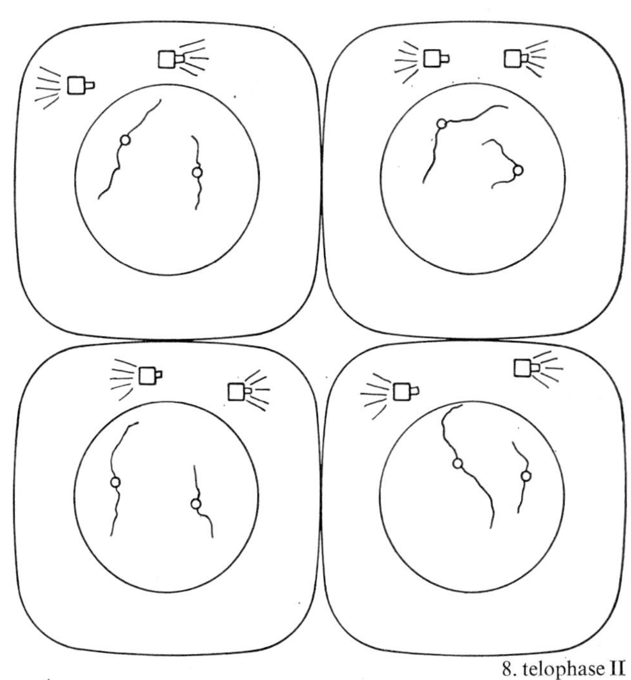
8. telophase II

Fig 2.8 **(b) Late stages of meiosis**

2 *Metaphase I* The chromosome pairs move to the equatorial region of the cell and the centromeres of each pair adhere to the split centrioles. As in mitosis, centromere fibres increase in size and the spindle fibres contract. The centromeres start to separate and the nuclear membrane is broken.

3 *Anaphase I* The pairs of chromosomes are pulled apart, rupture taking place in the region of the chiasmata. The new chromosomes which move to opposite poles derive genetic material from each of the original partner chromosomes.

4 *Telophase I* Here, the sets of chromosomes are at their own poles of the cell and become longer and thinner. The spindle fibres are broken and the nuclear membrane reforms.

5 *Prophase II* This is like prophase of mitosis, and is short, but owing to the earlier stage of meiosis, each pair of homologous chromosomes can be thought of as a single chromosome having some material coming from each member of the original pair.

6 *Metaphase II* similar to metaphase in mitosis.

7 *Anaphase II* similar to anaphase in mitosis.

8 *Telophase II* similar to telophase in mitosis.

Thus, each of the above two daughter cells undergoes division giving four cells from the original one cell.

When chromosomes exist in pairs (though not necessarily associated with each other), as in the cells of the body, the cells are called *diploid.* If the chromosomes are not present in pairs the cells are called *haploid.* In meiosis one diploid cell gives rise to four haploid cells.

Gametes are haploid and when they unite they form a zygote which is diploid. The gonad cells, formed after successive mitotic divisions of the zygote, are diploid; but when they undergo meiosis their chromosome number is halved, giving rise to haploid gametes again.

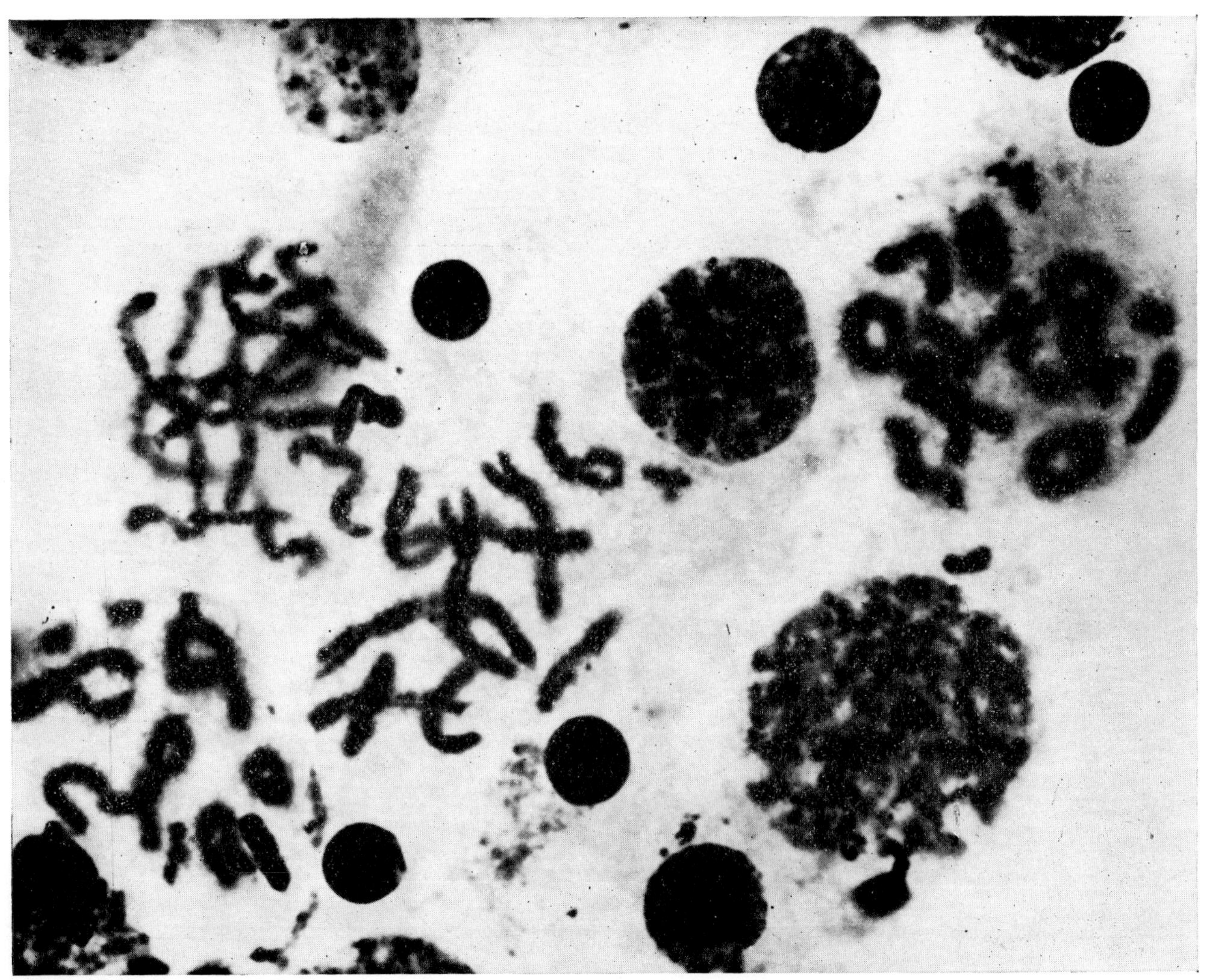

Fig 2.9 **Locust testis squash; useful for study especially up to the metaphase stage of meiosis**

Thus in diploid body cells, one of each pair of chromosomes originates from the male parent, and its partner chromosome comes from the female parent. At meiosis each gamete nucleus will obtain a mixed set of chromosomes, some from the father and some from the mother. The more chromosomes the organism has, the more combinations of male and female chromosomes there can be in the gametes. For man, the haploid number is 23, giving 2^{23} combinations.

Practical work

MITOSIS Using a microscope (high power and oil immersion) examine cells in a slide of the growing point of a shoot. Look to see if you can see cells undergoing cell-division.

GAMETOGENESIS Examine a prepared slide of a section of a vertebrate testis. Look for the stages of spermatogenesis.

2.4 Enzymes in cells

The enzymes in the cell allow reactions to take place there at relatively low temperatures. Heat is often needed to start chemical reactions but this is impossible in a cell since it would cause it to die.

An enzyme molecule has active sites on its surface where the molecules of *substrate* can come into positions favourable to their rapid reaction. If the substrate molecules would react in the absence of enzyme with an energy of activation E, then with an enzyme the molecules react with an energy less than E. This relationship is illustrated in Fig 2.10. Energy of activation is the energy barrier which has to be overcome before the reactions take place.

An enzyme suitable for simple study is salivary α-amylase on the substrate starch, which is hydrolysed to maltose and then glucose. The amount of starch in the reaction mixture is followed colorimetrically using the fact that it gives a blue colour with iodine.

Enzymes consist of two parts, a protein molecule and a smaller part called a *co-factor* or *co-enzyme* which determines the actual reaction catalysed. Inorganic ions are often co-factors, acting by joining the enzyme to the substrate during the reaction, *e.g.* Fe^{2+} for oxidases, K^+ for transferases, Zn^{2+}, Mn^{2+} and Mg^{2+} for hydrolases and additives. Some vitamins can act as co-enzymes *e.g.* Vitamin B_2 for oxidising enzymes such as flavoproteins. Enzymes have the following characteristics. They are specific for their own substrate (closely related as regards molecular geometry), and are affected by its concentration and purity. Unlike catalytic reactions, the concentration of enzyme affects the rate of the reaction with the substrate, but more important is that concentration of substrate which gives half the minimum rate of reaction, given a fixed enzyme concentration; this is the Michaelis constant which may be compared with half-life in radioactivity. The importance of the straight part of the curve obtained when rate is plotted against substrate concentration for a definite enzyme concentration shows that rate is directly proportional to substrate concentration over this region. When other stimulatory or inhibitory substances are present, reactions may be quicker or slower.

Enzymes are affected by temperature and pH. Reaction rates increase as temperature is raised to an optimum but from about 303K their performance falls off rapidly until at about 310K the enzyme molecule is broken down (*denatured*). As proteins are amphoteric (that is, they react both as bases and acids) electrolytes, pH affects their behaviour, and so there is also an optimum pH for each enzyme.

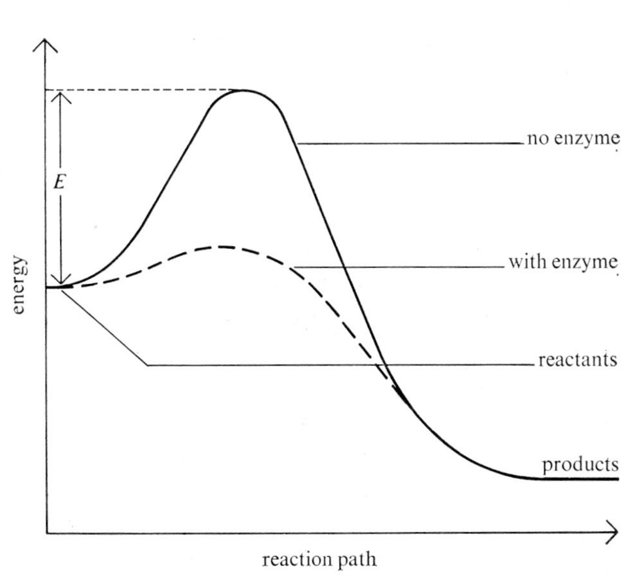

Fig 2.10 **Course of a cell reaction; with enzyme present and absent**

Enzymes can be categorised into the following types:

HYDROLASES These carry out *hydrolysis*, which can be represented as

$$X\text{–}Y + H\text{–}OH \underset{\text{condensation}}{\overset{\text{hydrolysis}}{\rightleftharpoons}} X\text{–}H + Y\text{–}OH$$

They can be subdivided into

1 Carbohydrases which hydrolyse carbohydrates. Examples are sucrase (in yeast) which hydrolyses sucrose to glucose and fructose; cellulase (in the digestive tract of ruminants *e.g.* cows) which hydrolyses cellulose to cellobiose; and amylase (in saliva) which hydrolyses starch to maltose. *Exohydrolases* are those which act by stepwise cleavage of the substrate molecules, whereas *endohydrolases* act by random cleavage.

2 *Proteases* which hydrolyse proteins as shown below.

```
      R      O   |  H      R      O   |  H      R
      |      ||  |  |      |      ||  |  |      |
---- C ---- C ---+- N ---- C ---- C ---+- N ---- C ----
      |      OH  |  H      |      OH  |  H      |
      H          ↓         H          ↓         H
```

An example is pepsin which hydrolyses proteins in digestion.

3 *Lipases* which hydrolyse lipids, *e.g.* they hydrolyse fats to glycerol and a carboxylic acid.

4 *Nucleases* which hydrolyse nucleic acids, and the bases, ribose or deoxyribose sugars and phosphates (V) are formed.

5 *Ligases* These are enzymes that operate in reactions involving the linking of two molecules and the breakdown of ATP to ADP, a thiokinase being an example.

OXIDOREDUCTASES These aid *redox* reactions. Redox means reduction and oxidation and the two must go together. Reduction means gain of electrons by an atom or ion while oxidation means loss of electrons by an atom or ion.

$$X + ne^- \rightleftharpoons x^{n-} \text{ is reduction } (e^- = \text{electron})$$

$$M - ne^- \rightleftharpoons M^{n+} \text{ is oxidation}$$

These electronic definitions are to be preferred to the elementary 'gain of oxygen (or loss of hydrogen) for oxidation, gain of hydrogen (or loss of oxygen) for reduction'. Oxidases catalyse the direct oxidation of their substrates by atmospheric oxygen. They occur in most cells and are very common in plants – think of polyphenol oxidase in apples; when you peel an apple, the 1,2-dihydroxybenzene (catechol) present is oxidised to a brown compound in the presence of polyphenol oxidase which can begin to work when the surface of the apple under the peel is exposed to the oxygen in the air.

In the cell the normal oxidation can be represented as

$$\underset{\text{reduced form}}{AH_2} + \underset{\text{oxidised form}}{B} \rightleftharpoons \underset{\text{oxidised form}}{A} + \underset{\text{reduced form}}{BH_2}$$

or as an electron transfer reaction of the type

$$\underset{\text{reduced form}}{X^{n+}} - me^- \rightleftharpoons \underset{\text{oxidised form}}{X^{(n+m)+}}$$

Reactions involving reduction of nicotinamide-adenine-dinucleotide (NAD) to $NADH_2$ during glycolysis fall into this class (Fig 2.11). Enzymes which carry out the first of the above reactions are referred to as *dehydrogenases*. During glycolysis and the Krebs' cycle reactions in respiration, a series of this type of enzyme act on intermediate compounds [oxidising (dehydrogenating)] them while NAD is reduced. $NADH_2$ has then to be oxidised, and it is during this pattern of reactions that ATP *generation* takes place (Fig 2.12).

$$NADH_2 + \tfrac{1}{2}O_2 \rightarrow NAD + H_2O \quad \Delta G \text{ is very negative (large energy release)}$$

The intermediate compounds are arranged in such a sequence that, as hydrogen atoms pass from one to the other, a small quantity of energy is released and can be removed as ATP. This sequence is called a *hydrogen* or *electron transport chain*. Between NAD and oxygen there are five intermediate carriers (at least) which are alternately oxidised and reduced. These are mainly electron transport carriers rather than hydrogen transport ones and they are called *cytochromes*, iron-containing coloured substances closely related from the structural point of view to haemoglobin. During the redox, the iron ion in each cytochrome is alternately in its +3 and +2 oxidation state [Fe(III) = oxidised and Fe(II) = reduced; only in the last stage does oxidation of iron(II) to iron(III) involve molecular oxygen, this being subsequently reduced to H_2O].

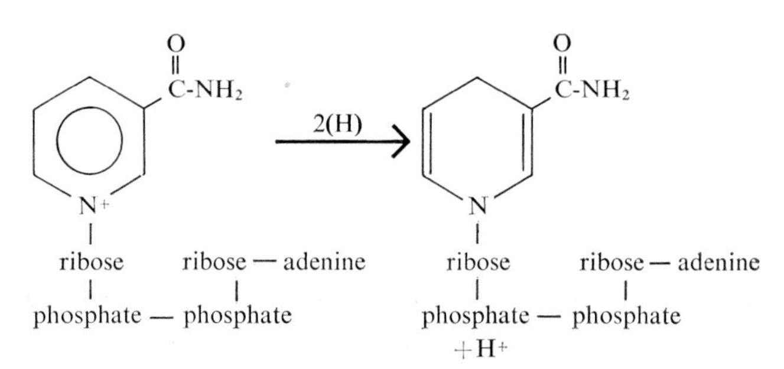

Fig 2.11 **Reduction of NAD to $NADH_2$**

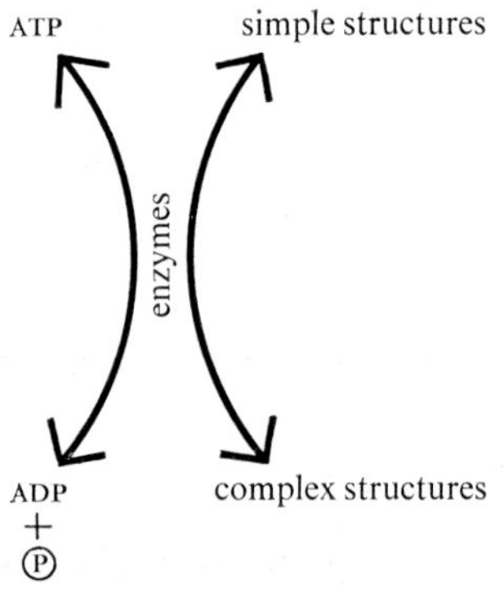

Fig 2.12 **Enzyme action**

Table 2.2 **Other enzymes**

enzyme	*action*
carbonic anhydrase	splits H_2CO_3 (carbonic acid) into H_2O and CO_2
decarboxylase	in animals converts amino-acids to amines and CO_2, acting on tyrosine, histidine and tryptophane
luciferase	acts on luciferin in luminescent organisms such as the firefly
rennin	coagulates milk, converting caseinogen to casein which in the presence of calcium ions forms insoluble calcium paracaseinate
thrombase	aids blood clotting, fibrogen converted to fibrin

TRANSFERASES catalyse the transfer of definite chemical groupings from one molecule to another. Examples are *transaminases* which transfer amino-groups–NH_2 or *phosphorylases* which transfer a phosphate(V) group –PO_4. In anaerobic glycolysis, a phosphorylation takes place initially converting glucose to glucose-6-phosphate, the enzyme involved being hexokinase (see p 124).

OTHER ENZYMES of interest, more difficult to classify, are tabulated in Table 2.2.

Practical work

Study of the salivary α-amylase on starch; the starch gives a blue colour with iodine and the maltose does not. Study the effect of temperature and pH on the activity.

2.5 Types of cells

Unicellular organisms are very tiny; some are just observable with the naked eye, but most can be seen only by microscopy. They do not attain a large size because they need to maintain a large area of surface membrane compared with the volume of the organism. Remember that the biological world is organised round the surface area/volume ratio. The large area of surface allows *diffusion* to be rapid; diffusion is the passage of molecules from one zone to another by virtue of their motion, causing an intermingling of the molecules; remember that molecules will collide with one another and also with molecules and other particles in the suspension medium, so their path will zig-zag. Also, unicellular organisms need to maintain a minimum volume of nuclear material compared with the volume of the cytoplasm.

There are unicellular species of both animals and plants. Examples are *Amoeba, Monocystis* and *Paramecium* (all animals) and *Euglena* (classed as a plant with animal tendencies) and *Chlamydomonas*. Multicellular organisms can range in size from the blue whale (30 metres long and weighing perhaps 90 tonnes) to the small wheel animals (rotifers) which are only about 300 μm long). Man is a complex multicellular animal and we shall be dealing with him predominantly in the following chapters.

In unicellular organisms the one cell performs all functions. In multicellular organisms different cells are adapted for different activities, becoming specialised as they develop from embryonic cells. The specialised cells are grouped in tissues, and the tissues into organs. The organism functions by means of the co-ordinated activities of the organs. In both animal and plant multicellular organisms, there is division of responsibility amongst specialised tissues for *reproduction, nutrition, conduction, support, co-ordination, excretion, protection* and *locomotion*. Each of these is described in the appropriate section of this book. There are also cells called omnipotent cells like plant cambium that give rise to different tissues.

A summary of cells in animals and plants is given in Table 2.3.

Table 2.3 **Summary of cells in animals and plants**

function	*plant*	*animal*
protective	epidermis	skin
internal surface	parenchyma	endothelium
secretion	gland cells	gland cells
food transport	xylem vessels	via blood
	xylem tracheids	
	phloem sieve tubes	
contraction	movements caused by changing turgor	muscle
impulse conduction	none	nerve
mechanical	collenchyma (living support cells)	connective tissue
	schlerenchyma (hard lignified cells)	cartilage
	xylem vessels	bone
	cell sap of parenchyma	

Xylem cells have large diameter and tube-like appearance. Vessels are made by an end to end arrangement of cells, which have end-walls with pores. Unlike pits (seen in longitudinal walls of vessels) the pores are holes whereas pits are just thin areas of cell wall.

Phloem sieve tubes are the main conducting structures of the phloem in angiosperms. They consist of longitudinally arranged cells. Sieve tube members have end-walls with sieve plates.

2.6 Orders of magnitude of cells

Table 2.4 gives the approximate size of a number of cells. Remember that

1 to be seen with the naked eye a cell must have dimensions at least of the order of 10^5 nm.

2 an optical microscope can be used to study cells with dimensions down to 2×10^2 nm.

3 an electron microscope can be used to study cells with dimensions down to 1 nm.

4 below 1 nm, *i.e.* at atomic and molecular levels, investigations are carried out using X-ray diffraction techniques.

Table 2.4 **Orders of magnitude of cells**

name of cell	*diameter (or dimensions) in nm*
yolk of ostrich's egg	8×10^7
yolk of hen's egg	$2{\cdot}5 \times 10^7$
parenchyma cells of plum	$(1 \times 1 \times 1) \times 10^6$
mature human ova cell	$(1{\cdot}2 \times 1{\cdot}2 \times 1){\cdot}2) \times 10^5$
sclerenchyma fibre cell of *Pinus*	$(4 \times 4 \times 400 \times 10^4$
human liver cell	2×10^4
onion meristem cell (root)	$(1{\cdot}7 \times 1{\cdot}7 \times 1{\cdot}7) \times 10^4$
human intestine muscle cell	$(6 \times 6 \times 200) \times 10^3$
human red blood cell	$(7{\cdot}5 \times 7{\cdot}5 \times 1) \times 10^3$
human nerve cell (neuron)	$(5 - 130) \times 10^3$ body; 10^9 length
typhus rickettsia	$(5 \times 2) \times 10^2$
pneumococcus bacteria	$(1 \times 2) \times 10^2$
bacteriophage virus	$60 - 80$
foot and mouth virus	21
ribosome of cell	15

3 Life Forms

3.1 Classification

Although every living thing belongs to a *species* a universal definition of a species is not possible. In practice a species is usually recognised by the similarity in the appearance of its members and the fact that they can interbreed successfully in nature. Domesticated animals and plants are bad examples because breeders have exaggerated the differences. Members of one species cannot produce fertile offspring with members of another species.*

Related species are grouped in *genera* (singular *genus*). Under the binomial system of nomenclature, first adopted by Linnaeus in 1735 and now generally used, the name of each animal or plant is that of its genus and species. Latin is used to avoid ambiguity and has the great advantage that the names are common to biologists of all languages.

Related genera are grouped into *families*, families into *orders*, orders into *classes*, classes into *phyla* (singular phylum) and phyla into *kingdoms*. Using the system a man would be classified as follows:

kingdom	ANIMAL
phylum	CHORDATA
sub-phylum	VERTEBRATA
class	MAMMALIA
order	PRIMATES
family	HOMINIDAE
genus	*Homo*
species	*sapiens*

Name: *Homo sapiens*

A beech tree would be similarly classified as follows:

kingdom	PLANT
phylum	SPERMATOPHYTA
class	ANGIOSPERMAE
order	FAGALES
family	FAGACEAE
genus	*Fagus*
species	*sylvatica*

Name: *Fagus sylvatica*

* Fertilisation is sometimes possible, *e.g.* between the horse and the donkey.The offspring (mule) is infertile.

A sparrow would go like this:

kingdom	ANIMAL
phylum	CHORDATA
sub-phylum	VERTEBRATA
class	AVES
sub-class	NEORNITHES
order	CARINATAE (PASSERIFORMES)
family	PASSERIDAE
genus	*Passer*
species	*domesticus*

Name: *Passer domesticus*

3.2 Animals and plants

It must be stressed that although there often are differences between plants and animals, it is often difficult at this level of study and beyond to differentiate between them. The main differences between the two Kingdoms are as follows.

ANIMALS Their nutrition is *holozoic* (p 131), that is, characteristic of animals, in that they need organic materials and use mouths to acquire complex organic substances; they have compact bodies, are motile, and respond quickly to stimuli; they have no rigid cell walls and cells without large vacuoles; they frequently store glycogen, but have no chlorophyll.

PLANTS Their nutrition is *holophytic* (p 131) or characteristic of plants; they manufacture their food from carbon(IV) oxide, water and mineral salts in the presence of chlorophyll and energy – this being called photosynthesis; they are sessile or fixed but move parts of their bodies which are branching and are less sensitive than animals, responding slowly to stimuli; they frequently store starch; their cells have rigid walls of cellulose and have large fluid-containing vacuoles. Fungi, however, which are plants, have no chlorophyll, do not photosynthesise, but take in their food in liquid form, store glycogen, often have cells with walls of chitin and have thread-like cells.

ANIMAL PHYLA

3.3 Protozoa

These are unicellular organisms which normally have blunt processes called pseudopodia or a whip-like thread called a flagellum, or cilia, each made of protoplasm. They may have more than one nucleus and reproduce by simple fission. They include the following classes.

RHIZOPODA These are usually free-living and moving, have pseudopodia with which they ingest their food but no flagella. They include *Amoeba* (Figs 3.1, 3.3) which lives in moist places such as soil, ponds, ditches.

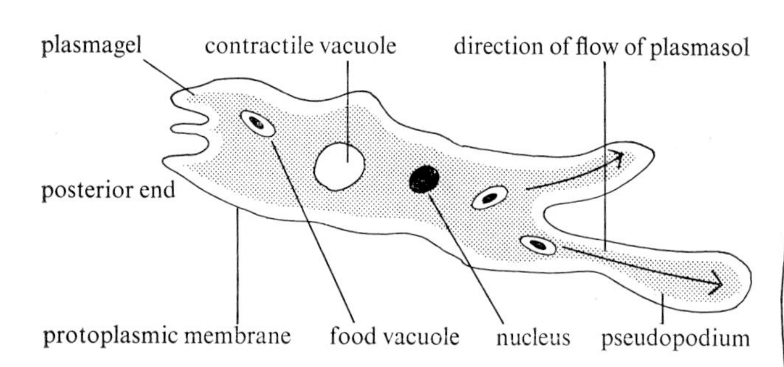

movement

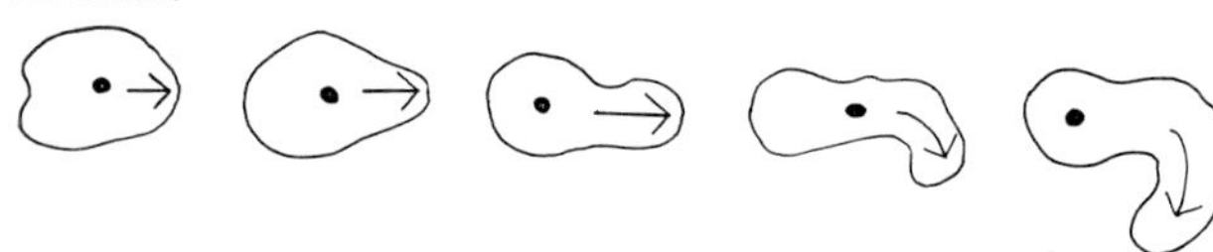

nutrition

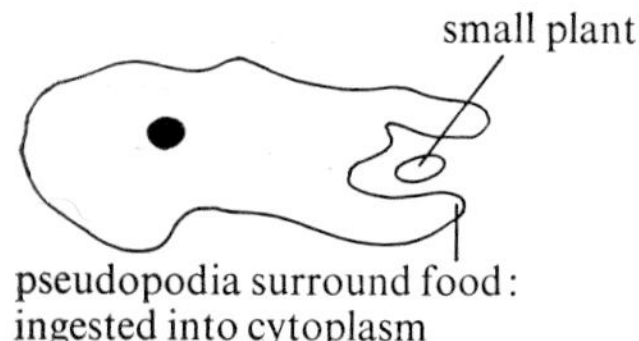

pseudopodia surround food: ingested into cytoplasm to give food vacuole

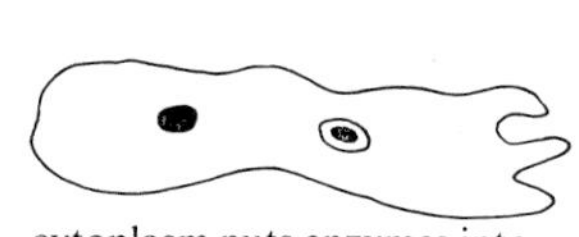

cytoplasm puts enzymes into vacuole to digest part of plant: undigested part egested

binary fission

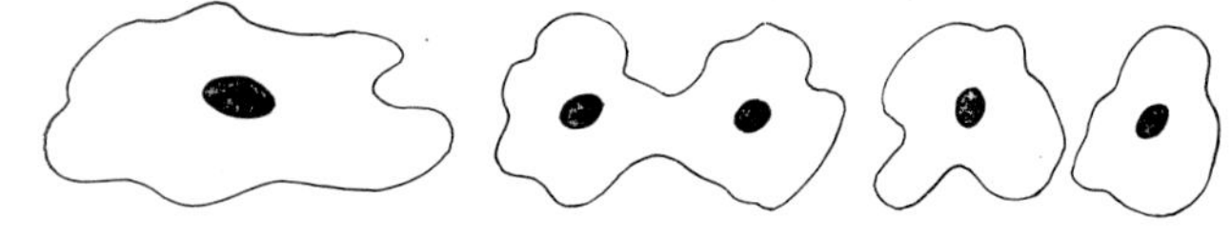

life cycle

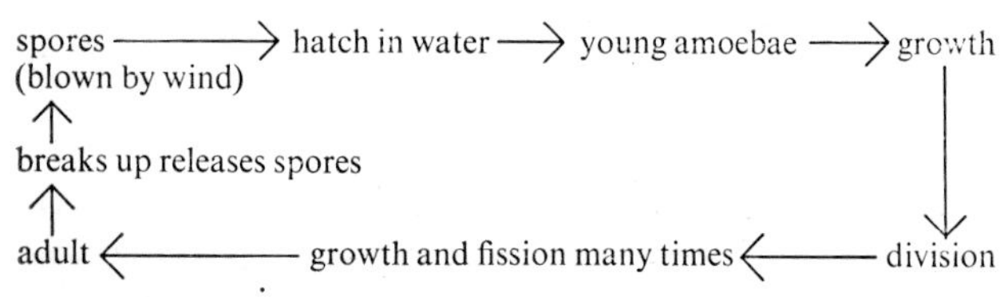

Fig 3.1 **Amoeba**

MASTIGOPHORA These have one or more flagella (and hence are sometimes called Flagellata). They are divided into two sub-classes:

1 PHYTOMASTIGOPHORA These are plant-like flagellates with a chloroplast. They include *Euglena* (Fig 3.2).

2 ZOOMASTIGOPHORA These are animal-like flagellates showing holozoic nutrition. They include *Trypanosoma* which causes sleeping sickness, and *Trichonympha* which lives in termites and helps them digest wood.

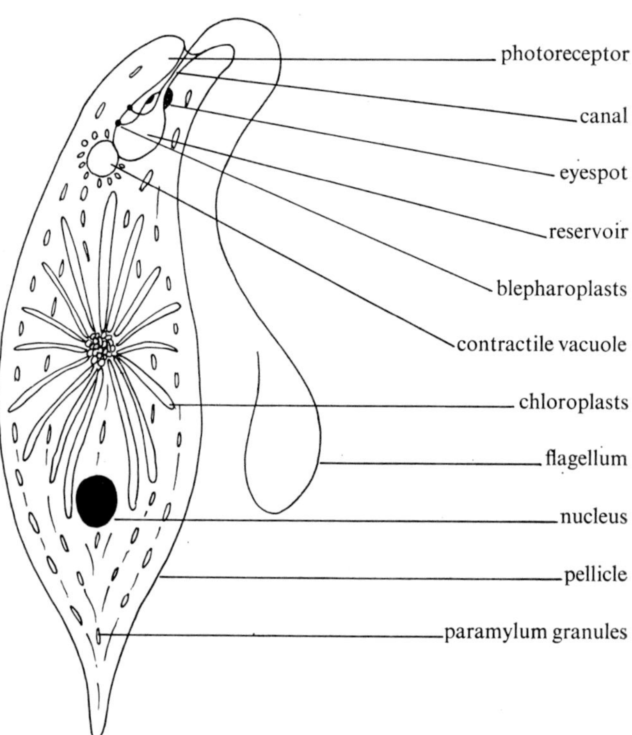

Fig 3.2 **Euglena, plant or animal?**

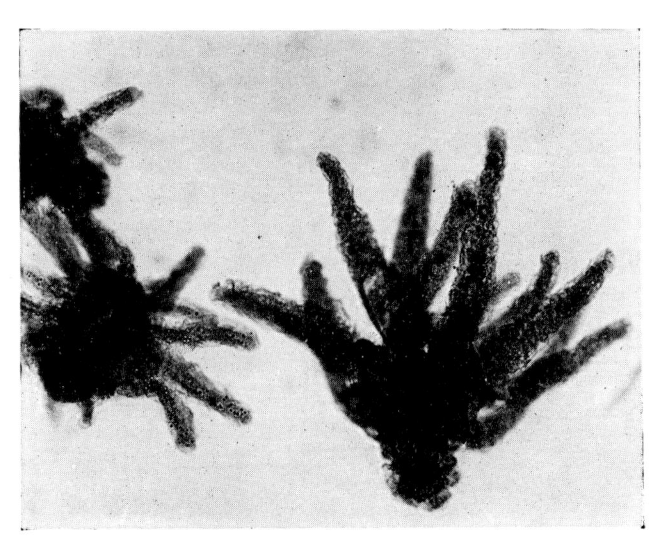

Fig 3.3 **Amoeba proteus**

CILIOPHORA These have cilia for movement/feeding, two nuclei and are complex protozoa. They include *Paramecium* (Fig 3.4) and *Vorticella*. *Paramecium* can reproduce asexually by fission. Sexually the micronucleus forms four haploid nuclei and exchange of these with other individuals can occur at conjugation.

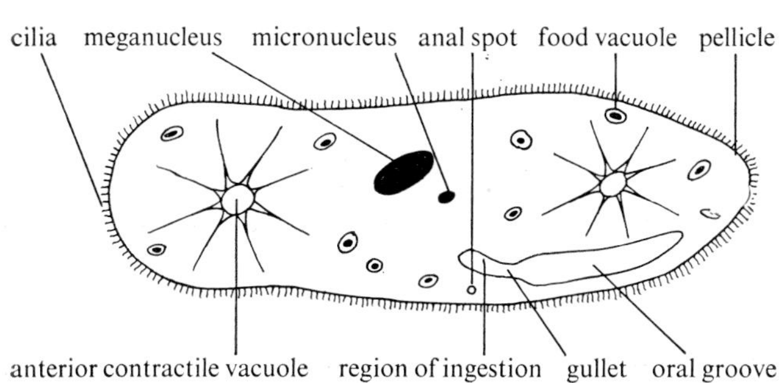

Fig 3.4 **Paramecium**

SPOROZOA These are spore-producing parasites which absorb soluble food all over the body surface and reproduce both sexually and asexually. They include *Plasmodium vivax* (Fig 3.5) (malarial parasite) and *Monocystis* which lives in the seminal vesicle of *Lumbricus terrestris*, the earthworm.

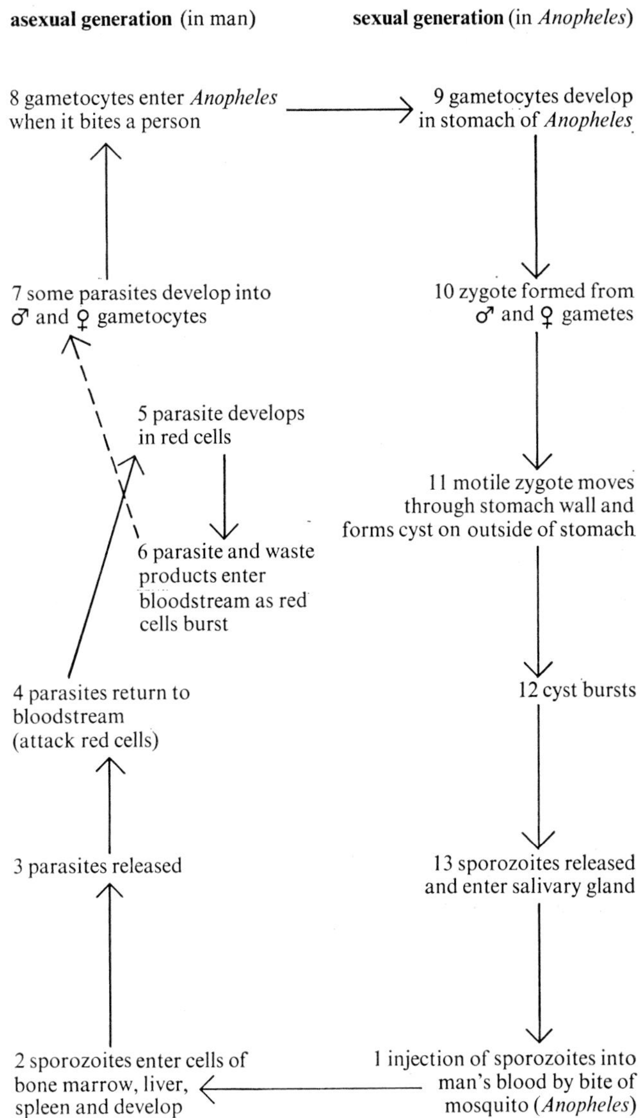

Fig 3.5 **Life cycle of malarial parasite (Plasmodium vivax)**

Practical work

AMOEBA What practical work can be done with *Amoeba*? Its movements can be studied by placing droplets of muddy pond water on a microscope slide. Its reactions to objects placed in its way (*e.g.* sand grains) should be studied. The specimens should be stained before studying the internal anatomy. This is done on the slide. The specimen is first fixed in Bouin and the slide warmed. The Bouin is replaced, successively, with 70% ethanol, iron(III) alum, distilled water, and iron haematoxylin. Each specimen is then black. The stain is replaced with iron(III) alum until only the nuclei are black. Tap water is then drawn through, followed by an increasing gradient of ethanol, then dimethyl benzene (xylene). The specimen is set in Canada Balsam, examined and drawn. (Successive fluids can be drawn between slide and cover slip by filter paper, each new fluid being introduced from the other side of a cover slip.)

PARAMECIUM These can be obtained by adding hay to pond water. This provides food for the bacteria on which *Paramecium* thrives. It can be stained as for *Amoeba* and its movements should also be studied.

TRYPANOSOMA Examine prepared slides – oil immersion – and draw what you observe.

MONOCYSTIS Take the seminal vesicles of *Lumbricus terrestris* (earthworm) and make a smear on a microscope slide of the contents. Immediately this is dry, add 95% ethanol, stain with haematoxylin and clear. Mount in Canada Balsam, examine and draw.

Note: Cilia and flagella. Practice staining with iodine by adding a drop of the solution to a culture fluid. This is a temporary stain.

3.4 Porifera

These are the sponges which although consisting of many cells cannot be considered as co-ordinated multicellular individuals, rather as *colonies* of single cells. There is no

Fig 3.6 **Hydra**

trace of a nervous system, and what co-ordination there is is by direct mechanical action or chemical activators.

3.5 Coelenterata

These have two layers of cells (inner endoderm and outer ectoderm, separated by jelly-like mesogloea). They are referred to as diploblastic. They have cells called *nematoblasts*, which secrete structures called *nematocysts* which can trap food and sting. There are *medusoid* (jellyfish-like) and *hydroid* (animal-like) forms, the former being motile and the latter normally sessile.

There are four classes, Hydrozoa, Scyphozoa, Anthozoa, Ctenophora.

HYDROZOA Examples are fresh-water *Hydra* (Fig 3.6) and *Obelia* (*Campanularia*) which grows on shells or seaweed on sea shores (rocky type) and can be seen in large fixed colonies. Normally there are both medusoid and hydroid forms, although *Hydra* does not have a free-living medusoid form. *Hydra* is a common animal in ponds in late spring and summer. The greater part of the ectoderm is composed of muscle-tail cells, their inner ends being drawn out into threads which are parallel to the tentacle or body. The protoplasm of these cells contracts making the tentacle or body shorter. The nutritive cells have muscle-tails arranged so as to reduce the diameter of the body or tentacles, making the organs longer when they contract. The *Hydra* waves its body and tentacles all the time and can make limited movements. The body and tentacles are stretched out in the water and can be made to contract by touching the animal. *Hydra* catches, for example, small water fleas with its tentacles. Nematocysts, called thread capsules, with barbs pierce the prey and a paralysing fluid stuns it. Other tentacles draw the prey into the digestive cavity and there enzymes aid digestion, undigested matter being ejected from the mouth. New individuals can develop from buds on the body but sexual reproduction is also possible (particularly at the onset of winter). The embryo *Hydra* exists in a horny case in the mud until spring, when it emerges.

Obelia (*Campanularia*) (Figs 3.7, 3.8), in its hydroid form, exhibits asexual budding and can form regions which produce medusoid forms. The latter can swim, and have sex cells which produce larvae (planulae) which in turn start new hydroid colonies.

A little thought shows that the medusoid form of coelenterate can be derived from the hydroid form. If the latter is progressively flattened, the enteron is reduced to a small cavity from which come the gullet and mouth. If the mesoglea is thickened and the tentacles increased, the medusoid form is obtained. The Portuguese Man-of-War has stinging capsules for protection and to aid capture of prey.

SCYPHOZOA Here the medusoid form is dominant. Here belong the 'jelly fish', e.g. *Chrysaura isosceles* (brown jelly fish) and *Aurelia* (found on sea shores). Medusoid forms are produced from a sessile hydroid form by budding.

ANTHOZOA Here the hydroid form is dominant and there is no medusoid at all. Included are the corals and the sea anemones including *Metridium* and *Tealia felina* (dahlia anemone). Corals secrete a calcium carbonate exo-skeleton at their base and they are generally to be found in warm sea regions, *e.g.* Great Barrier Reef.

The hydrozoans, most of which have no skeleton, are seldom found as fossils but the anthozoans are geologically important, *e.g.* the rugose and tabulate corals of the Paleozoic era.

Evidence for the theory that higher animals have arisen from coelenterate-like ancestors includes the fact that coelenterates have several types of cells, and also that the mesoglea resembles mesoderm layers in higher animals. The two layered body of the coelenterate is similar to the gastrula stage through which many higher animals pass during development of their embryos.

CTENOPHORA These include the sea gooseberries which are solitary, transparent animals, globular and which move by cilia. Only one genus has a planula larva (Venus' girdle). They have no nematocysts.

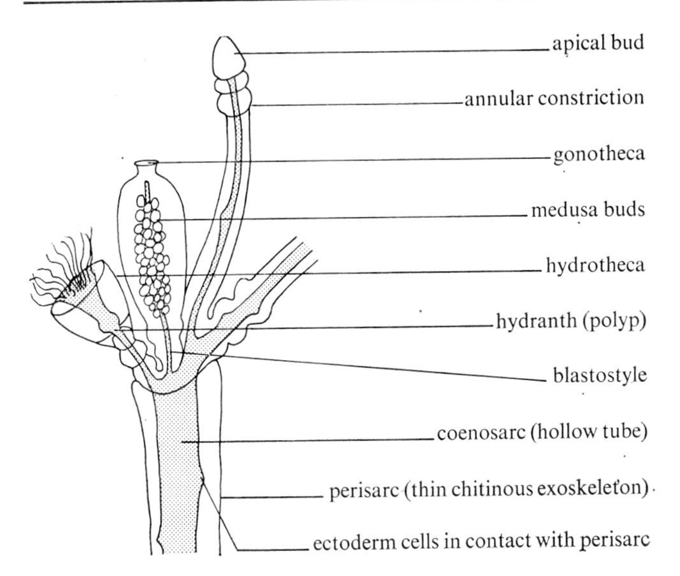

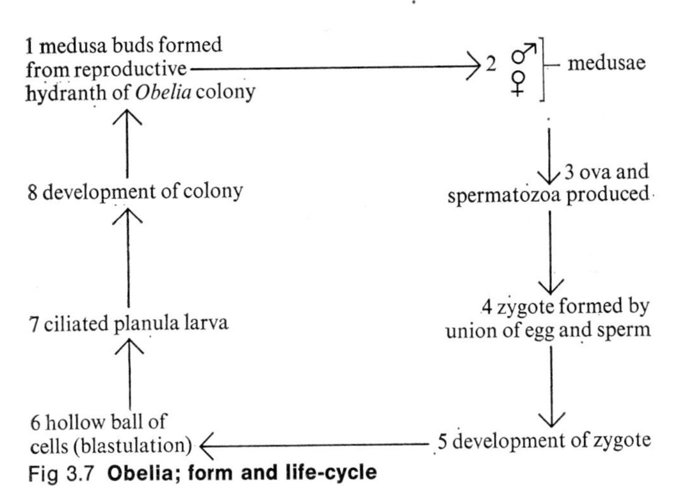

Fig 3.7 **Obelia; form and life-cycle**

Fig 3.8 **Obelia hydroids**

Practical work

What practical work can be done? *Hydra* is visible to the naked eye – staining can be done in a watch-glass or slide, fluids being introduced and removed with a very fine capillary tube. *Hydra* is fixed by use of mercury(II) ethanoate (acetate) solution. The poisonous solution is replaced by 70% ethanol, containing excess iodine. This is immediately withdrawn and replaced by borax carmine. Absolute ethanol, followed by dimethylbenzene (xylene), is introduced and mounting is done in Canada Balsam.

If *Hydra* is mounted in 1% sodium chloride solution, a cover-slip added, and this is tapped sharply, the nematocysts can be seen to explode. *Obelia* can be similarly examined. Draw an entire colony (*e.g.* from fronds of seaweed), one hydranth, and a blastostyle.

3.6 **Platyhelminths**

These *flatworms* are bilaterally symmetrical and triploblastic, having three layers, the ectoderm, mesoderm and endoderm. There is no coelom (cavity) and so they are acoelomate animals. The bodies are flat and a single opening to the gut (as for coelenterates) is both the mouth and anus. The body is unsegmented. Flame-cells (Fig 3.10) serve in excretion and osmotic pressure regulation. Respiration is simple, by diffusion.

The platyhelminths normally have both male and female sex organs (called hermaphrodite) but there is no self-fertilisation; they lay eggs with a yolk.

There are three classes, turbellarians (free-living marine) trematodes (parasites) and cestodes (parasites).

TURBELLARIA The best examples are provided by the planarians (Fig 3.9), found in ditches and muddy water. They are carnivorous. They have fairly elaborate sense organs (more so than the parasites) and their bodies are ciliate.

TREMATODA An example is *Fasciola*, (Fig 3.11), the sheep liver fluke. This has a complex life cycle (Fig 3.12). In the Chinese liver fluke, man replaces sheep and freshwater snail replaces snail. Other examples include *Polystoma* (in bladder of frog) and *Schistosoma* (in abdominal veins of man).

The Trematoda are thin with leaf-like bodies, have thick cuticles, and have suckers. There are no cilia.

CESTODA These include the tapeworms which are parasites. They have thick cuticles. There is no digestive tract or cilia. They fix themselves to the host by suckers or hooks (or both) and have very high powers of reproduction. Examples are *Dipylidium caninum* (dog tapeworm) and *Taenia solium* (Fig 3.13) (man tapeworm derived from measly pork).

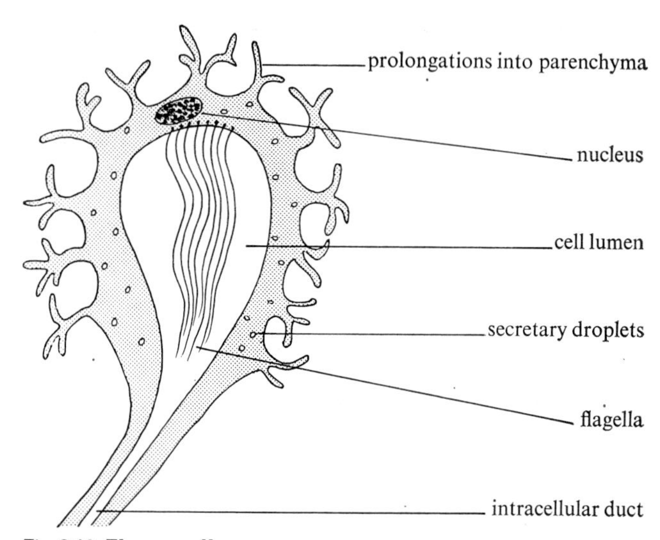

Fig 3.10 **Flame cell**

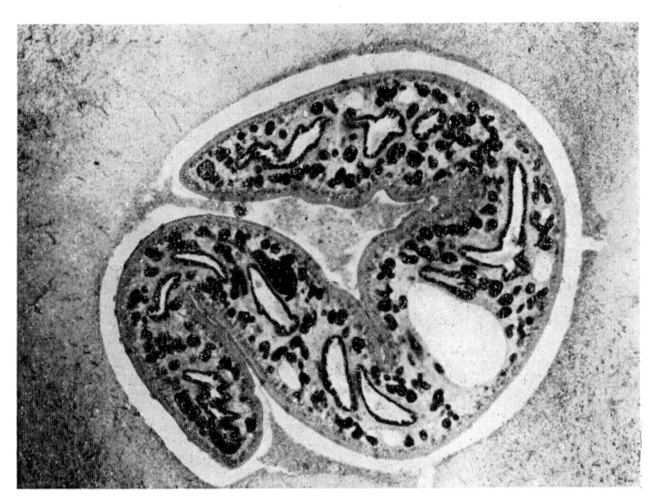

Fig 3.11 **Fasciola hepatica in situ**

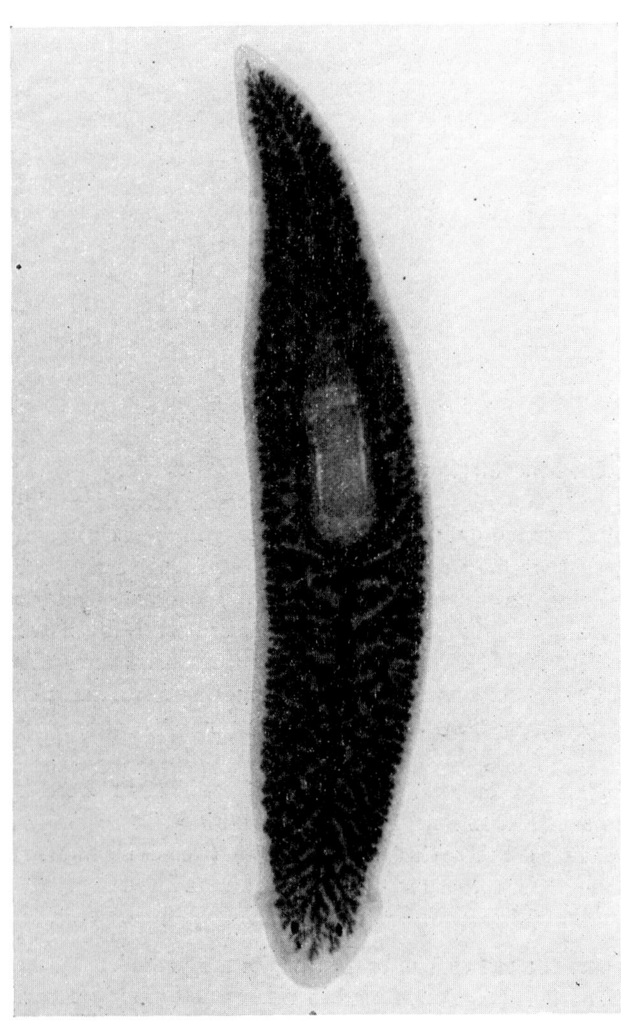

Fig 3.9 **Planarian**

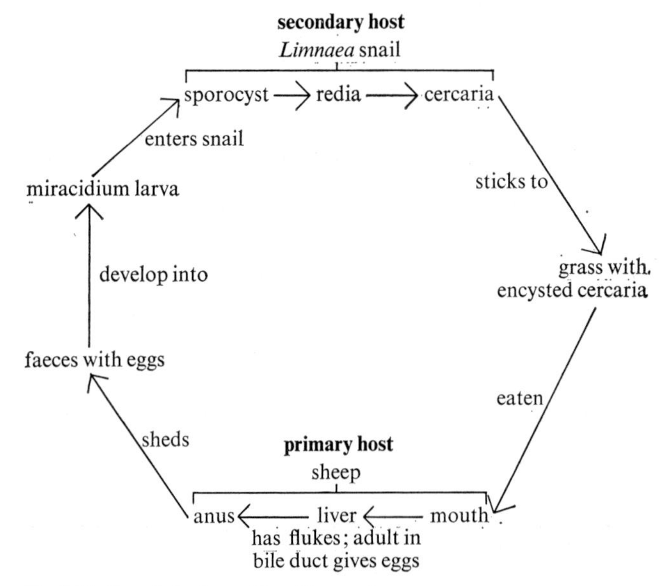

Fig 3.12 **Fasciola hepatica life-cycle**

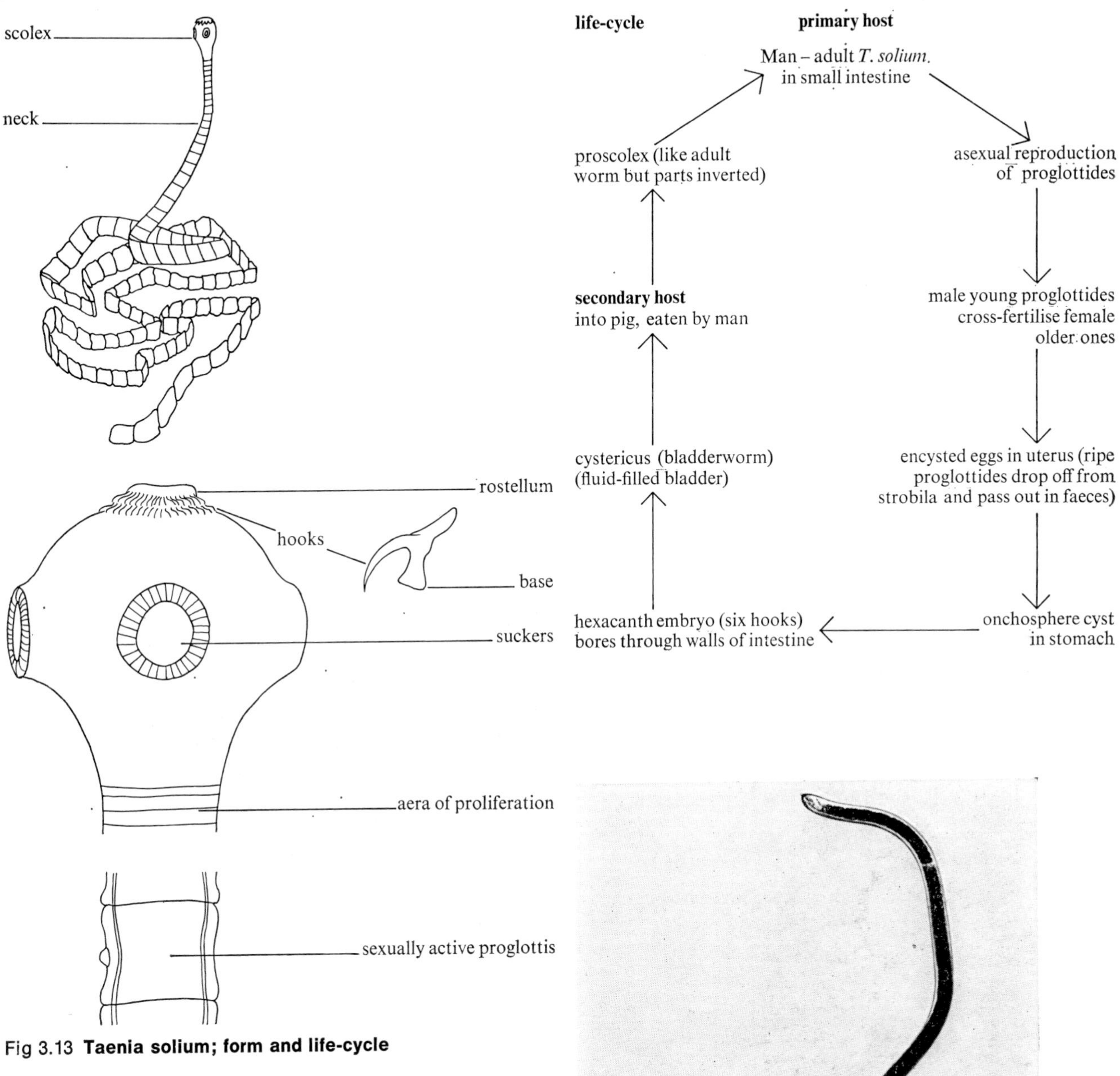

Fig 3.13 **Taenia solium; form and life-cycle**

Practical work

Examine specimens of *Taenia* and *Fasciola* and examine prepared slides of both. Draw what you observe, *e.g.* the layout of the reproductive organs.

3.7 Nematoda

Nematodes (Fig 3.14) are bilaterally symmetrical and triploblastic. They can be parasites or live free. They have a thick elastic protein cuticle, a digestive tract with mouth and anus, and bodies which are long and pointed at the ends. There are no cilia. The adults have a body cavity.

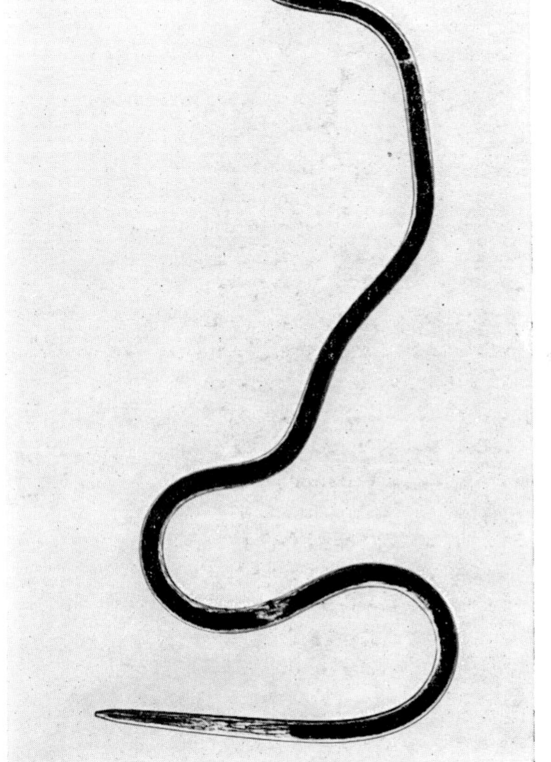

Fig 3.14 **Nematode**

There are two separate sexes. The excretory system consists of two intracellular tube-like vessels. The nervous system is not greatly developed. Examples are *Ascaris lumbricoides* (pig nematode), *Ancylostoma* (hookworm), *Oxyuris vermicularis* (children's threadworm), *Xiphenema* (plant virus vector), *Heterodera* (root eelworm), *Filaria* (causes elephantiasis, transmitted by mosquito).

3.8 Rotifera

These include the wheel-like animals, minute water creatures, which are triploblastic. They have a complex digestive tract with mouth and anus, a ciliated anterior disc for locomotion and collection of food, a very simple nervous system, separate sexes (with parthenogenetic female producing male-producing animals), a complex excretory system of canals with flame-cells at the end. Examples are *Callidina* (roof-gutter rotifer), *Hydatina* (pond and dirty puddle rotifer), and *Pedalion*.

3.9 Polyzoa

These include the sea mats, most of which are marine and live in colonies. They are triploblastic, coelomate, non-segmented creatures, having a very simple nervous system (one ganglion, plus nerves), an excretory system like the rotifera, and a U-shaped digestive tract. They are hermaphrodite. An example is *Flustra*.

Fig 3.15 **Nereis (King Ragworm) showing vicious jaws**

3.10 Brachiopoda

They are the lamp shells. They are triploblastic, non-segmented, coelomate and are inside a bivalve shell attached to the substratum. They are marine. They have a ciliated region round the mouth. There are two outlets in the coelom to function as genital and excretory passages. The blood system is simple. The nervous system is simple, with two ganglia. There are two separate sexes. Examples include *Terebratula* (in deep water around the British Isles), *Lingula* (tropical seas – Hawaii). The brachiopods constitute a most important group of fossil invertebrates. The bivalve shell is either calcareous (made of calcium carbonate) or a mixture of calcareous and organic material.

3.11 Annelida

These are worms with metamerically segmented bodies; they are triploblastic, coelomate and bilaterally symmetrical. The phylum is large and very successful. The body segments are similar and at the head end several segments can function together. The body wall is made of a chitinous cuticle and there are larger rods (of chitin) called chaetae. The body also has a gland-containing epidermis and has circular and longitudinal muscles. A ventral nervous cord is connected by commissures to pairs of cerebral ganglia in the pharyngeal region. The excretory organs are nephridia (p 174). Feeding and sensory structures are at the head end where several segments act together.

There are four classes, Polychaeta, Oligochaeta, Hirudinea and Archiannelida.

POLYCHAETA These include the marine bristle-worms. The body is oval and segmented and each segment bears a bristly parapodium which is a projection from the body wall. The head carries a pair of tentacles and a pair of palps and four simple eyes. The larva is called a *trochosphere* or *trochophore*.

Polychaete means with many bristles. Examples would include the ragworm *Nereis* (Fig 3.15), a swimming, carnivorous animal, the lugworm *Arenicola*, a burrowing, mud-eating worm and the filter-feeding tube-dweller, *Chaetopterus*.

OLIGOCHAETA These have few bristles and include the earthworm and a few freshwater worms. They have no

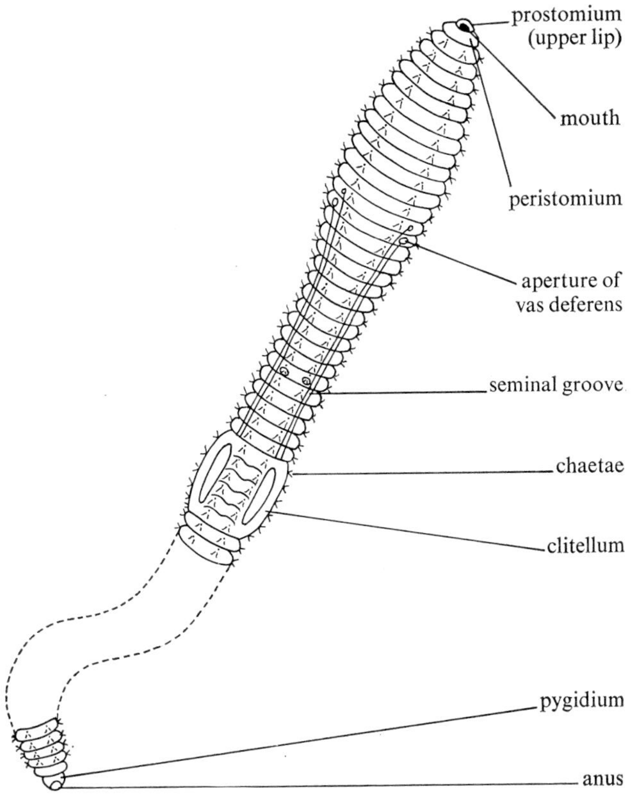

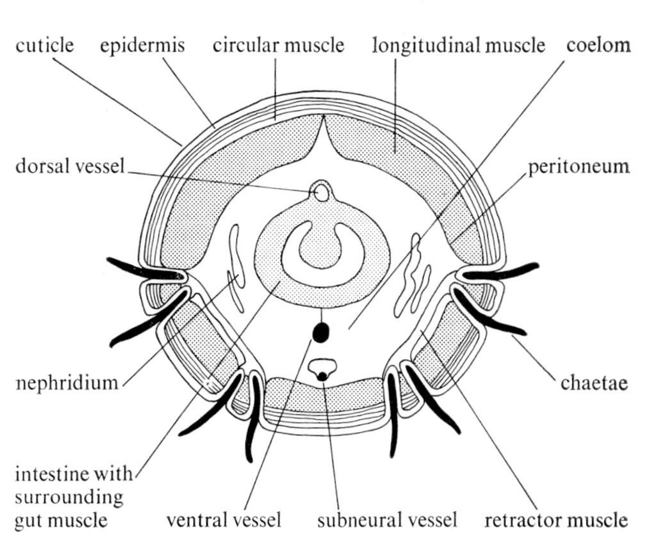

Fig 3.16 **Structure of the common earthworm**

parapodia. They are hermaphrodite and have gonads in definite anterior segments; the products of the gonads pass into the coelom. During copulation there is cross-fertilisation. The eggs and sperms are surrounded by a cocoon, having a protective layer and being filled with albumen. A single worm emerges from the cocoon; there is no larval stage. Examples are *Lumbricus* (Fig 3.16) the common earthworm, and *Tubifex* the freshwater bloodworm.

HIRUDINEA These are the leeches with short bodies with few segments and anterior and posterior suchers; there are no parapodia and few bristles. Some are carnivorous and some exo-parasitic. The coelom is filled with mesenchyme and consists of two long tubes. There is no larval form, the newly hatched young resembling the adult. They are hermaphrodite with eggs laid in cocoons. Examples are *Haemopsis* (horse leech) and *Hirudo* (medicinal leech).

ARCHIANNELIDA These are small marine worms with no parapodia or bristles, derived from polychaetes, *e.g. Histriobdella*, a parasite of lobster eggs.

Practical work

What practical work can be done on the earthworm? Several specimens can be dissected and a transverse section drawn. A nephridium can be removed, mounted in salt solution and examined (high power). A nephrostome can be mounted as described for *Hydra* (p 45). An ovary can be removed and stained in borax carmine. A smear of the inside of a vesicula seminalis should reveal stages of spermatogenesis.

3.12 Arthropoda (jointed limbed)

These make up over 80% of the animal kingdom. They are metamerically segmented, bilaterally symmetrical, triploblastic and coelomate. Their main feature is a chitinous exo-skeleton. There are chitinous membranes between the segments and jointed limbs (with some functioning as jaws). There is much evidence of cephalisation, the coelom is not pronounced (small cavities in gonads and excretory organs) and there are no nephridia. Normally there are no cilia. There is a dorsal contractile heart, lying in a haemocoelic cavity; the blood flows in this cavity. The nervous system is well developed, having paired dorsal cerebral ganglia and ventral nervous cord.

The phylum can be divided into six classes.

ONYCHOPHORA This class includes *Peripatus* which has cilia in the excretory tubes and parts of genital organs. It is terrestrial, breathes by tracheae, and has a head of three segments and one pair of jaws. Each remaining segment has a pair of appendages. It has a thin cuticle, and soft muscular body wall.

TRILOBITA Numerous fossil forms occur. They are oval-shaped and flat. All were marine creatures and became extinct at the end of the Paleozoic era. They have a skeleton divided into three lobes. In life, double appendages existed on each thoracic segment. The head (five pairs of appendages) had structures which bore compound eyes. Most were mud eaters. Examples include *Orria elegans* and *Pseudokainella keideli.*

CRUSTACEA These are marine arthropods, breathing by gills. They have a thick exo-skeleton, two pairs of antennae and three pairs of head appendages. There is a diverse specialisation of limbs. Examples include *Daphnia* (water flea, Brachiopoda), *Cyclops* (Fig 3.17) (Copepoda, with large single eye and large antennae, swimming type), *Balanus* (barnacle, Cirripedia, sessile filter-feeder, plates of calcium carbonate in skin), *Astacus* (crayfish, Malacostraca, decapoda), *Gammarus* (Malacostraca, Amphipoda), *Argulus* (fish louse, Malacostraca, Isopoda), and *Oniscus* (wood louse, Isopoda, Malacostraca, non-aquatic).

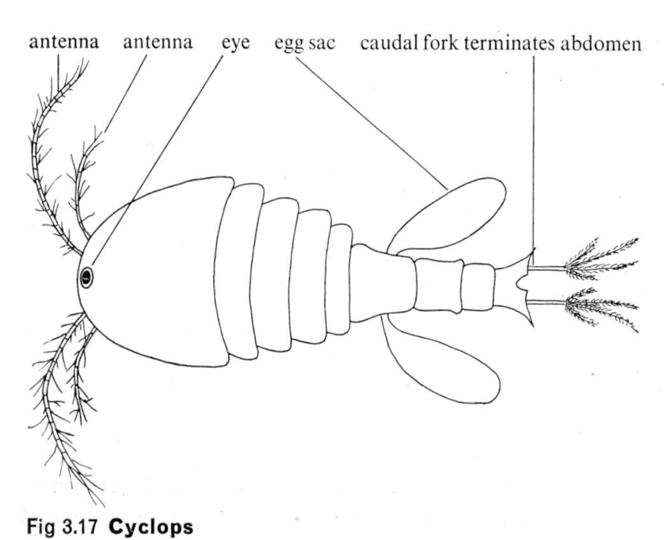

Fig 3.17 **Cyclops**

INSECTA These have the body divided into three parts (Fig 3.18) the head, thorax and abdomen. The head has six segments with one pair of antennae. Respiration is through spiracles in the body wall which communicate with internal tracheae. There are three thoracic segments, each having a pair of legs, and there is usually a pair of wings attached to each of the two rear segments. There are eleven abdominal segments without appendages. Primitive types of insect are wingless. In such orders as

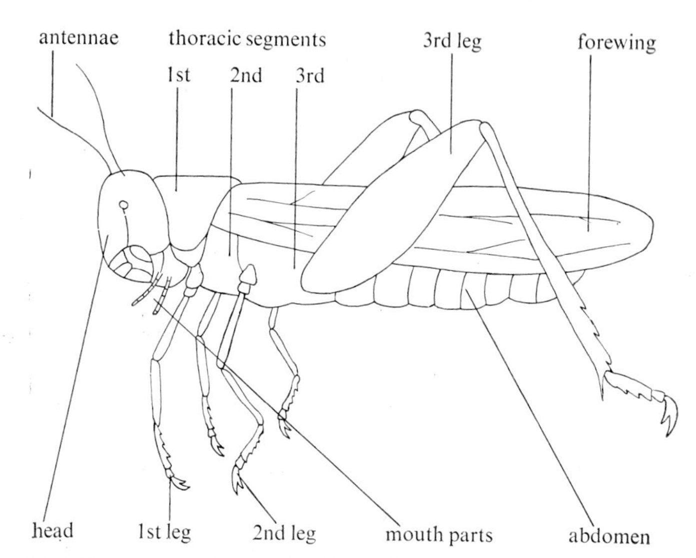

Fig 3.18 **The locust, a typical insect**

the Orthoptera (*e.g.* locust, cockroach) there is an incomplete metamorphosis; the nymphs resemble the adult insect, with wings developing externally. The newly hatched nymph is called the first instar; locust nymphs pass through five instars (stages) before becoming adult. The moulting of the skin of one stage, out of which the next stage emerges, is called ecdysis. The development of the locust proceeds as follows: 1st instar black, 9 mm long through instars 2, 3, 4 to instar 5 (32 mm), all of these being orange and black. Eggs are laid in warm damp sand in a frothy liquid, the female pushing her abdomen into the sand.

The more advanced insect orders show complete metamorphosis from the final larval stage through an

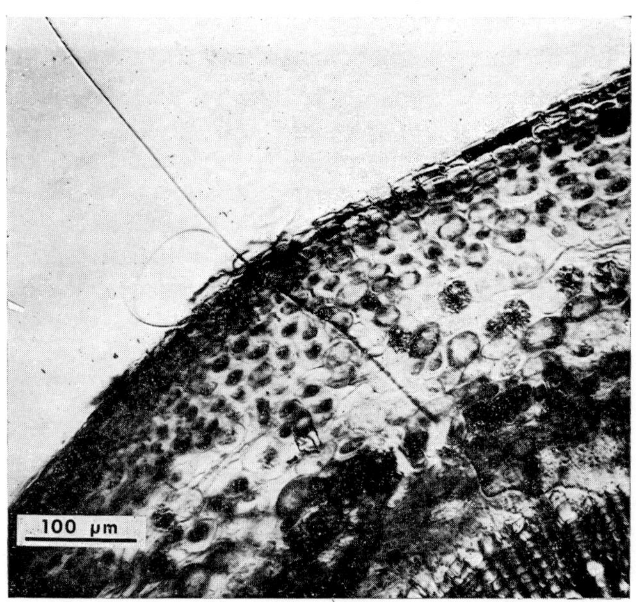

Fig 3.19 **Aphid feeding on stem phloem via stylets**

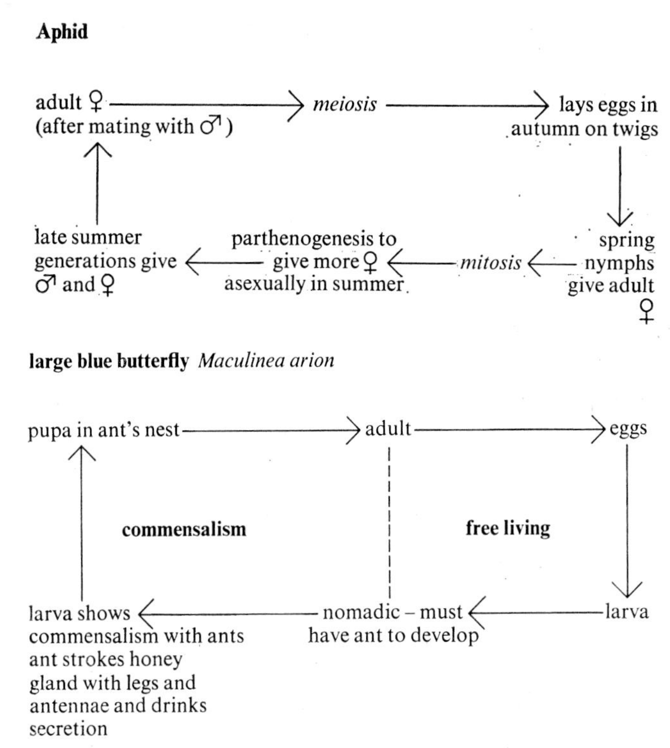

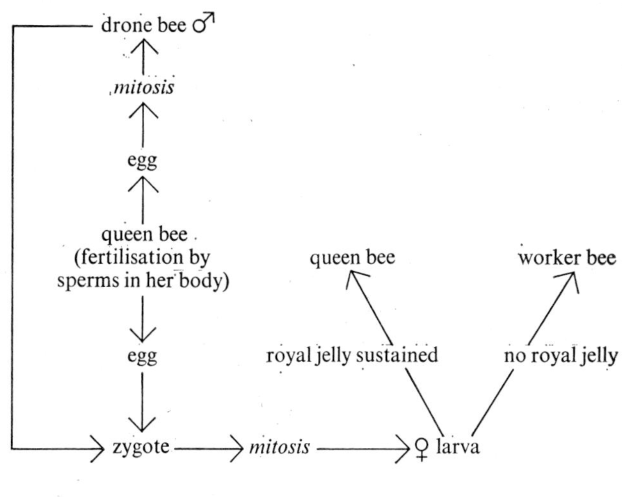

Fig 3.20 **Some insect life-cycles**

extra stage called the pupa to the adult imago. Contrasts are provided by (Fig. 3.20) Aphids, *Maculinea* (large blue butterfly), *Coccinella* (ladybird beetle), *Apis* (bee) and *Musca* (housefly). There are about 850 thousand known species of insect, and they are the most ubiquitous of animals. Most live on land. They have a wide range of feeding habits (Fig 3.19) and some are parasites (*e.g.* mosquitoes, fleas, flies). The tough exoskeleton made of chitin and the ability of many species to fly affords protection against predators. There are few land and fresh water environments which they have not colonised and their rapid life cycle, great fecundity and close adaptation to different sorts of food enables them to multiply quickly.

ARACHNIDA These have bodies which are divided into two zones, the prosoma and the opisthosoma. The former has six segments and the latter thirteen. The first prosoma segment bears chelicerae, the second bears pedipalps and the remaining four have legs (8 in all). Respiration is usually by tracheae. Examples are *Epeira* (common web-spider), *Lycosa* (wolf-spider), *Boophilus* (cow tick), *Limulus* (king crab), *Scorpio* (scorpion), *Tetranychus* (red mite) and *Letradectus* (black widow spider) (Fig 3.21).

MYRIAPODA These have the body divided into similar segments with similar pairs of legs on nearly all segments. These are terrestrial and breathe by tracheal systems. The head has one pair of jaws. Examples are *Lithobius* (centipede) and *Iulus* (millipede, two pairs of legs on each segment).

Practical work

What practical work can be done with, say, the locust or cockroach? A specimen can be dissected – fixed in molten wax with its dorsal surface protruding, as well as the head. The salivary glands are removed and stained as described for *Hydra*. The head and neck are removed, boiled in sodium hydroxide solution (why?) and the mouth-parts removed. Each part is dipped into 70% ethanol, then absolute ethanol, then dimethylbenzene (xylene), and mounted in Canada Balsam. A portion of the trachea should be mounted in propantriol (glycerol) in a similar manner. The external features and the life cycle should be studied.

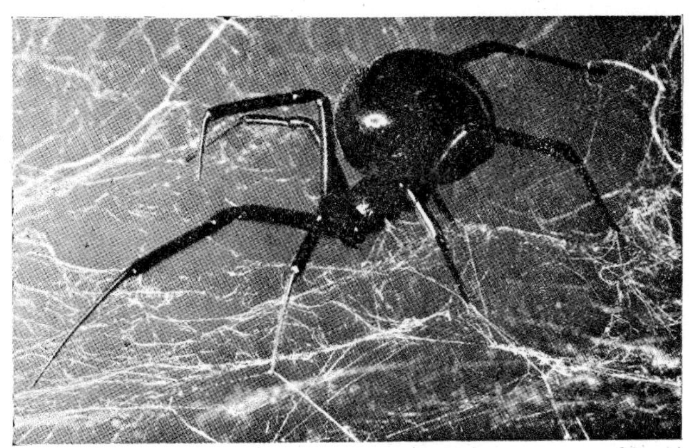

Fig 3.21 **Black widow spider; an arachnid**

3.13 **Mollusca**

Most molluscs are marine and the earliest ones appeared in the Cambrian era. There are about 80 thousand species living, and 125 thousand have been described. Fossils are important. The body is divided into a head, a muscular foot and a visceral hump. They are unsegmented, bilaterally symmetrical, coelomate and triploblastic. The skin is soft but that covering the visceral hump forms a mantle which secretes a shell. There is a heart and blood system. The nervous system has ganglia joined by commissures. The larva is often a trochosphere type. The mantle cavity contains gills and operates as a lung in air breathing forms. The following classes exist:

AMPHINEURA These are marine, with a long body and a shell of eight plates. They have an anterior mouth and posterior tail. They have no eyes, no tentacles and the ganglia of the central nervous system are very poorly developed. An example is *Chiton* (coat of mail shells).

GASTROPODA These are terrestrial, marine and freshwater, and have a head with eyes and tentacles. They have an anterior anus and a coiled shell. An example is *Helix* (land snail (Fig 3.22), hermaphrodite, adapts to herbivorous nature).

SCAPHOPODA These have a long body in a tube-like shell open at both ends, and a foot for burrowing in the sand. An example is *Dentalium* (elephant's tusk shells).

LAMELLIBRANCHIATA These have bivalved bilaterally symmetrical shells, extended plate-like gills for filter feeding, and a reduced head and foot. Examples are *Mytilus* (mussel), *Ostrea* (oyster), *Pecten* (scallop).

CEPHALOPODA These are bilaterally symmetrical, with complex eyes and a well developed brain. The head is surrounded by tentacles probably derived from the foot. They have a siphon through which water can be expelled, giving jet propulsion. Examples of cephalopods are *Octopus*, *Nautilus*, *Sepia* (cuttlefish) and *Loligo* (squid). In most cephalopods the shell is either absent completely (*Octopus*) or has become internal (squid). In *Nautilus*, however, it is extremely well developed, consisting of many chambers.

3.14 **Echinodermata**

These are radially symmetrical (adult), triploblastic, coelomate, with the larval form segmented but the adult not so; the larva shows bilateral symmetry. There is a calcareous exo-skeleton. The coelom has a water vascular system in contact with the surroundings and dilates the tube feet of the echinoderms. The nervous system is reduced, there are separate sexes and there is no specialised excretory organ.
The following classes exist.

ASTEROIDEA These are free-living carnivores with a flat star shaped body. They have five arms (with pouches) and a ventral mouth. Examples are *Asterias* and *Asterina* (starfishes).

ECHINOIDEA These are free-living, but slow moving, herbivores, with a globular discoid body. Examples are *Echinus* and *Echinocardium* (sea urchins).

CRINOIDEA These are primitive, spineless, sessile, with the young attached by a stalk. They have branched arms with ciliated food paths ending at the mouth; they are filter feeders. Examples are *Antedon* and *Actinometra* (sea lilies and feather stars).

OPHIUROIDEA These are free-living with spines; they are star shaped, but with no pouches. An example is *Ophiura* (brittle star).

HOLOTHUROIDEA These are free-living, cucumber shaped with no spines. An example is *Cucumaria* (sea cucumber).

3.15 **Chordata**

The feature that all chordates have in common (and which is used to define the phylum) is that at some stage in their development they have a *notochord*, or dorsal stiffening rod. (Figs 3.23, 3.24). In many cases the notochord is only present during one stage of development,

Fig 3.22 **Helix aspersa; the garden snail**

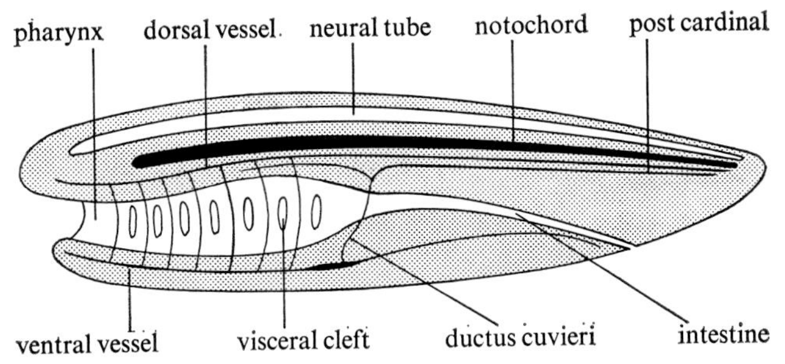

Fig 3.23 **Generalised chordate structure**

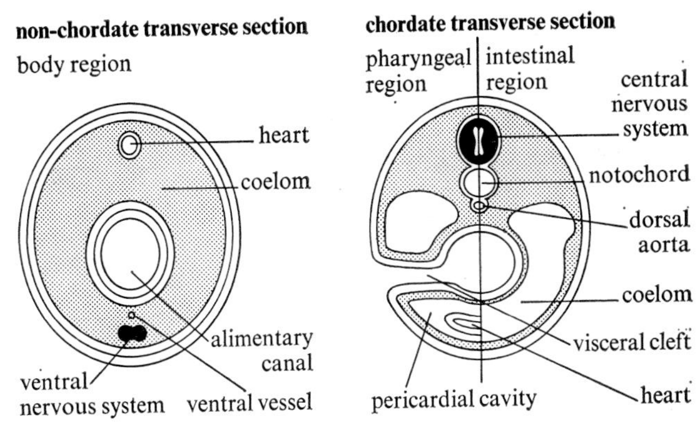

Fig 3.24 **Comparison between chordate and non-chordate structure**

and becomes much modified. Typically the chordates also have a *tubular dorsal central nervous system*, a *closed blood system* in which blood flows forward ventrally and backward dorsally, *pharyngeal slits* (gill slits) connecting the pharynx with the outside of the animal) and a *post-anal segmented tail.*

There are four sub-phyla; Hemichordata, Urochordata, Cephalochordata and Vertebrata. The first three are often grouped together as the protochordata, or invertebrate chordates.

HEMICHORDATA Apart from the notochord which is in any case confined to the proboscis before the mouth the hemichordates show few of the features of the chordates. They have much in common with echinoderms, having a similar embryonic development. There is no tubular central nervous system, no post-anal tail and the blood flow is in the reverse direction. The body can be divided into three distinct regions, the collar, the trunk and the proboscis. Examples include *Balanoglossus* and *Dolichoglossus* (marine burrowing worms).

UROCHORDATA Here the chordate characteristics reveal themselves in the larvae only. The notochord which is confined to the tail disappears during metamorphosis and a large perforated pharynx develops (by multiplication of gill slits); this becomes used for feeding via cilia. There is no coelom and the sexes may be combined or separate. An example is *Cionia* (shallow water sea squirt).

CEPHALOCHORDATA These are small and fish-like. The notochord lies along the length of the body, there is no heart, the head is not predominant and there are no paired limbs. Nitrogenous excretion is by nephridia. An example is *Amphioxus* (lancelet).

VERTEBRATA These are chordates with well developed head and brain, and having a backbone. The notochord is replaced in the adult by an internal skeleton of bone or cartilage. Nitrogenous waste is excreted via the kidneys. The vertebrates have a muscular ventral heart, a tail after the anus, and usually two pairs of limbs. There are seven classes, Agnatha, Chondrichthyes, Osteichthyes, Amphibia, Reptilia, Aves and Mammalia.

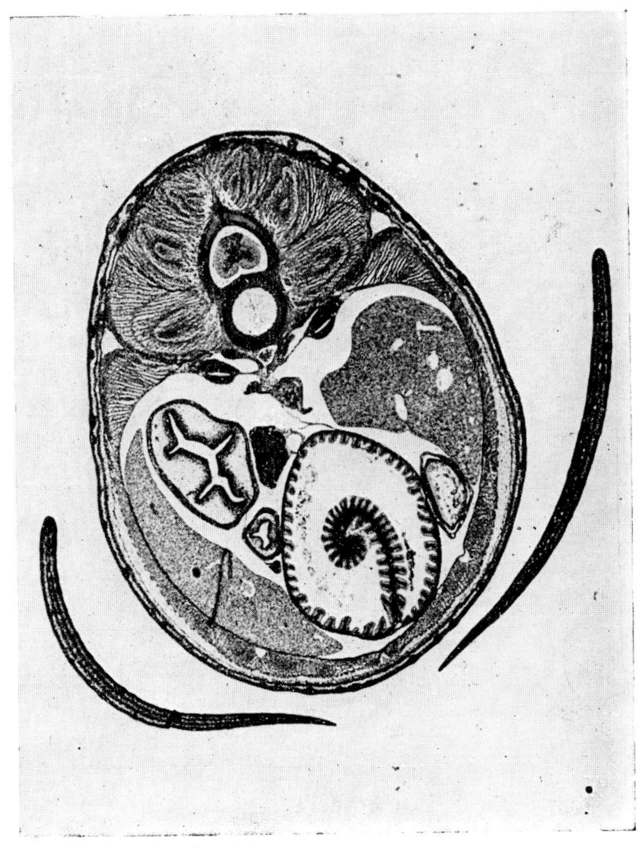

Fig 3.25 **T.s. of a dogfish**

1 AGNATHA (Cyclostomata) These have no jaws (a feature which distinguishes them from the other vertebrates), round mouths and the notochord is present in every stage of development. They are primitive having a much less well developed brain than the other vertebrates, and are mostly represented by fossils from the Silurian period. An example is the lamprey, with slimy skin and no placoid scales.

2 CHONDRICHTHYES These are marine fish with a skull and backbone consisting entirely of cartilage. In the order Selachi (no operculum covering gills, many replaceable teeth) are *Scyliorhinus* (dogfish, Fig 3.25) and *Selache* (basking shark); in the order Bradyodonti (gills with operculum, few teeth) is *Chimaera.*

Practical work

What practical work can be done with the dogfish? The external features should be examined. The dissections should include the alimentary canal, the urino-genital, venous, afferent arterial, efferent arterial and nervous systems. The skeleton can be drawn after removing the flesh. Scales can be removed by boiling the skin in sodium hydroxide solution; they can be mounted directly into Canada Balsam.

3 OSTEICHTHYES These are fishes with a bony skeleton and scales. They have an air bladder as an outgrowth of the gut. In the order Teleostei (homocercal tail) are *Salmo* (salmon) and *Gadus* (cod): in the order Dipnoi – the lungfishes – (gills and lungs, upper jaw and cranium fused, diphycercal tail, teeth joined as plates) are *Ceratodus* and *Protopterus* (mud fish of Africa); in the order Coelacanthini (hollow spines, diphycercal tail) are *Latimeria* (living) and *Macropoma* (extinct). It is interesting to note that the lungfishes which flourished in the Devonian lakes are probably ancestors of the amphibians.

4 AMPHIBIA In this class legs replace fins and the skin is soft and usually not scaly. The skin is permeable, so Amphibia cannot survive in salt water. Breeding takes place in the water giving a larva which undergoes metamorphosis to the adult. The adults have pentadactyl limbs and a middle ear, but no outer ear. There are gills in the larval form and lungs in the adult. Amphibians were the first tetrapods, were dominant in the Carboniferous period, and gave rise to reptiles in the Permian.

In the order Stegocephalia (with bony exoskeleton and large endoskeleton) there are the extinct Permian amphibians. In the order Urodela (tail plus short limbs, adult often has gills) is *Triton* (newt). In the order Anura (adults with neither gills nor tails, powerful hind limbs) are *Rana* (frog) and *Bufo* (toad). In the order Gymnophiona (burrowing animals, no limbs or tails, often with large eggs and no larva) is *Coecilia.*

Practical work

The frog, or a similar animal like the claw toothed toad, should be dissected to study the alimentary canal, urino-genital and vascular systems, skeleton etc., as it demonstrates all the features of vertebrate anatomy. Try and study the life-cycle as well.

5 REPTILIA In this class the characteristics are a dry impermeable skin, horny scales or plates, pentadactyl limbs and eggs with shells; there is no larval stage, an amnion develops from the embryo within the egg, reproduction takes place on land. Respiration is by means of lungs.

Reptiles (dinosaurs) completely dominated the Mesozoic period, but most (and all the largest) are now extinct. Modern reptiles include tortoises, turtles, lizards, snakes, crocodiles and alligators.

6 AVES In this class, the birds, the characteristics are a warm-blooded nature, pentadactyl limbs (the front pair as wings), feathers, lungs for respiration, large eggs filled with yolk encased in a calcareous shell, no larva. The heart has ventricles and auricles divided; the aortic arch is present on the right side only. There is a large brain, complex behaviour including courtship, nesting and care of the young. There is the development of an amnion from the embryo within the egg. Birds can be terrestrial or aquatic.

In the order Ratitae there are all the large running birds such as *Dromaeus* (emu) and *Apteryx* (kiwi). In the order Carinatae there are the flying birds such as *Turdus* (thrush), *Columba* (pigeon) and *Carduelis* (linnet).

7 MAMMALIA This class (Fig 3.26) are characteristically warm-blooded, covered with hair, terrestrial, but also successful in water. The members have pentadactyl limbs, a diaphragm to help respiration (lungs), and a divided heart, with left systemic arch. They show variety in teeth and have three ossicles (malleus, incus, stapes) in the ear. They have a unique and extremely advanced method of reproduction and rearing, with retention of the foetus in the uterus, placental nutrition until birth, and feeding on milk from the females mammary glands after birth.

The class is divided into three sub-classes:

PROTOTHERIA Of the mammals the monotremes are the most like their reptilian ancestors. They lay large-yolked eggs, and lack proper teeth. Examples are the duck-billed platypus and the spiny anteater.

METATHERIA These are the marsupials (from the Latin *marsupium*, a pouch on the belly of the female).

Fossil evidence suggests that they were once widely distributed but today the opossum is the only common form living outside Australasia. The young are born in an extremely immature state, and climb into the pouch where they are nourished. Young marsupials are no more than animated embryos, unlike their parents in appearance. Examples include kangaroos, wombats, phalangers, bandicoots and opossums. Marsupials were probably exterminated in areas other than Australasia because they could not compete with other species, especially the placental mammals. There is no fossil record of there ever having been any placental mammals in Australasia (except those taken there recently by man), so it is likely that it was cut off from the rest of the world before they had got there (see Chapter 6), reducing competition for the marsupials. Opossums survived elsewhere because of their ability to climb trees.

EUTHERIA These are the placental mammals, making up 95% of living mammals. They are distinguished from other species by the presence of a placenta in development of the embryo. The young develop within the body of the female.

The main orders of the placental mammals are as follows:

Insectivora (shrews, moles)
Rodentia (mice, beavers)
Lagomorpha (rabbits, hares)
Carnivora (dogs, cats, weasels, badgers, otters)
Artiodactyla (cows, goats, deer, sheep, pigs – even-toed ungulates)
Perissodactyla (horses, zebras – odd-toed ungulates)
Subungulata (elephants)
Sirenia (sea-cows)
Cetacea (whales, dolphins)
Chiroptera (bats)
Primates sub-order Lemuroidea (lemurs)
sub-order Anthropoidea (man, monkeys, apes)

The more advanced the order the more the brain is developed, giving man (it is hoped) the power to reason and learn.

Practical Work

The rabbit should be dissected, after examination of the external features. The order should be: alimentary canal (abdomen), urino-genital and vascular systems, followed by a study of the neck structure. The skeleton should be studied and the separate bones drawn. The skull of a dog should also be studied. The life history should be studied.

(a)

(b)

Fig 3.26 **Mammalia**

(a) great grey kangaroo and baby, a marsupial which lives in Australia

(b) bottle-nose dolphin, a placental mammal

PLANT PHYLA

3.16 Thallophyta – Bacteria

These are very small, single-celled, rarely more than 5 μm in length. They have no cellulose walls or chloroplasts and no large vacuoles. Their cell walls are of protein and fatty materials and, although DNA is present, it is not there as a nucleus. Granules of glycogen, fats, etc. may also be present. They are classified as spherical (*cocci*), rod-shaped (*bacilli*) and spiral (*spirilla* or *spirochaetes*). Some have flagella, the lashing motion of which aids motion. They reproduce very rapidly by simple fission giving huge populations. Some can exchange genetic information by conjugation. Some form spores, a resting stage in which the protoplasm is surrounded by a tough wall. These spores are dispersed by air currents and are often very resistant to extremes of temperature. Most are stable at low temperatures and most can be killed above 323K. In sterilisation to remove bacteria, the object is heated in an autoclave or treated chemically.

Most bacteria are *saprophytes* (live on dead matter – see §9.9) and can digest materials by their wide range of enzymes, although some cause important animal and plant diseases.

Bacteria can live in soil or water or decaying matter. They bring about decay of dead materials and so prevent accumulation of leaves, etc. *Rhizobium* bacteria fix nitrogen by symbiotic association with leguminous plants (peas, beans, lucerne); *Clostridium* and *Azobacter*, which live freely, also fix nitrogen. *Nitrosomonas* and *Nitrobacter* bacteria effect the oxidation of ammonia to nitrate(III) and of nitrate(III) to nitrate(V) respectively, § 9.4. Bacteria in the large intestine of man play a part in the synthesis of the vitamins K and B_2. Not many animals can digest cellulose by a direct enzyme, and those which feed on vegetation having cells with cellulose walls do so by utilising the enzymes of bacteria and simple animals, living symbiotically in the gut (rumen of cows and sheep, caecum and appendix of rabbit).

Some bacteria are toxic, producing in their host proteins called toxins *e.g.* tetanus toxin and botulin toxin. Some diseases such as typhoid, cholera and diptheria are produced by bacteria. These diseases can be 'caught' by eating food containing bacteria or spores, by inhaling moisture containing them or by contact with an infected person. Invasion of harmful bacteria can be counteracted by leucocytes or by anti-toxins in the blood. If an animal recovers from a bacterial disease it acquires immunity to further attack if antibodies are formed.

Prevention of infection can be achieved by cooking food, disposal of sewage, use of antiseptics, personal hygiene, sterilisation and isolation of persons with bacterial diseases. Plant diseases (*e.g.* fire blight of pear, *Erwinia amylovora*) are usually controlled by prevention; there are no effective chemical controls. Among other groups, iron-bacteria oxidise iron(II) to iron(III) and sulphur-bacteria (*Thiobacillus thio-oxidans*) oxidise a mixture of sulphur and water to sulphuric(VI) acid.

Bacteria are observed by culturing them on media containing nutrients and agar jelly in petri-dishes or on slopes in test tubes.

3.17 Thallophyta – Algae

It is considered that the algae have descended from the earliest forms of life on earth. They are mostly aquatic (except *Pleurococcus* on tree bark), all with photosynthetic pigments and have large chloroplasts. Their cells have cellulose walls and vacuoles (as for other plants) and they can be unicellular, extending to multicellular and filamentous. Asexual reproduction involves a free-swimming stage (zoospore). They show a variety of types in life cycle, e.g. *Spirogyra* with haploid dominant and the seaweed *Fucus* with diploid dominant. Sexual reproduction is usually well defined and varies from simple *isogamy* in which there is conjugation of identical gametes; if the similar gametes are of different sizes – micro/macro-gametes it is callde *anisogamy;* and if male and female cells unite it is *oogamy*.

There is no regular alternation of generations as shown in higher plants and even when the life history exhibits two forms of plant (gametophyte and sporophyte) these forms do not usually alternate regularly.

The sea-water absorbs the component wavelengths of white light in various amounts, according to depth; red is absorbed near the surface and green penetrates down to lower reaches. This has forced algae (which, remember, get energy from light) to become adapted to their environment; green algae near the surface can utilise the red wavelength while deeper living red algae utilise the red pigment to change the green wavelength into suitable energy.

The classes of algae are Chlorophyceae, Bacillariophyceae, Dinophyceae, Cyanophyceae, Phaeophyceae, Xanthophyceae, Rhodophyceae and Euglenophyceae.

CHLOROPHYCEAE (green algae) These are primitive algae, mostly freshwater, few marine, occasionally terrestrial. They are unicellular and some can hardly be distinguished from flagellate protozoans. Pigments include chlorophylls, carotenes and xanthophylls, but no biloproteins. The food reserve is starch or fat. There are two or four flagella, smooth or whiplash types. There are pyrenoids in the chloroplasts. There are unicellular sex organs, ranging in type from isogamy to oogamy, always

with motile male gametes. Asexual spores are zoospores.
The orders are as follows:

VOVOCALES, which are freshwater and motile, have a single chloroplast in each cell and include *Chlamydomonas* (Fig 3.27), *Eudorina*, *Volvox*, *Gonium* and *Pandorina*. (Note that one of the following is essential to the more 'animal type' of alga: swimming through water, allowing water to pass over organism, floating on top of water).

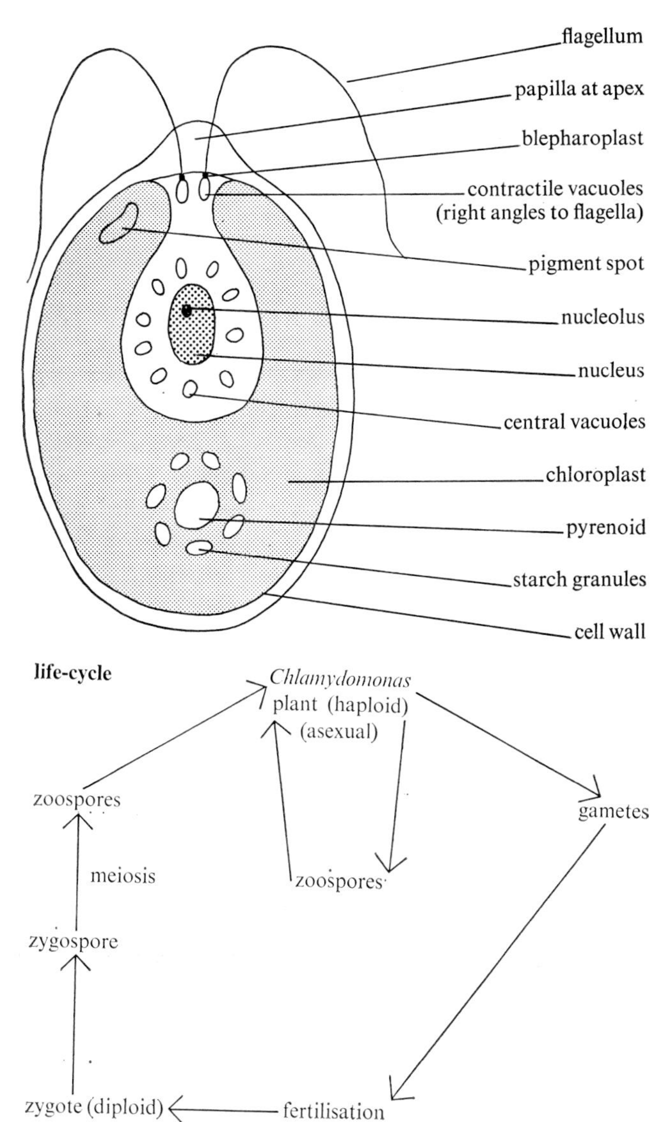

Fig 3.27 **Chlamydomonas; form and life-cycle**

CHLOROCOCCALES, which are mainly freshwater (plankton), non-motile, single cells or colonials, never filamentous and include *Chlorella* and *Pediastrum*.

ULOTRICHALES, mainly freshwater, body with filament of haploid cells, fixed to substratum in early stages, and includes *Ulothrix*.

ULVALES, marine, filamentous, body expands to sheet thallus by cells dividing longitudinally, anchored to substratum and includes *Ulva* (sea lettuce) and *Enteromorpha* (filamentous), both of which are seaweeds found in the upper shore zones.

OEDOGONIALES, freshwater, anchored to substratum when young, filamentous, multiflagellate zoospores for asexual reproduction, and includes *Bulbochaete*.

CONJUGALES, freshwater, free floating single cells or filament like haploid body, distinctive chloroplast structure (spirals, lobes, stars), only fragmentation as means of asexual reproduction, sexual reproduction isogenous with amoeboid gametes, conjugation by tube like connections between pairing cells, and including *Spirogyra* (Fig 3.28) and *Zygnema*.

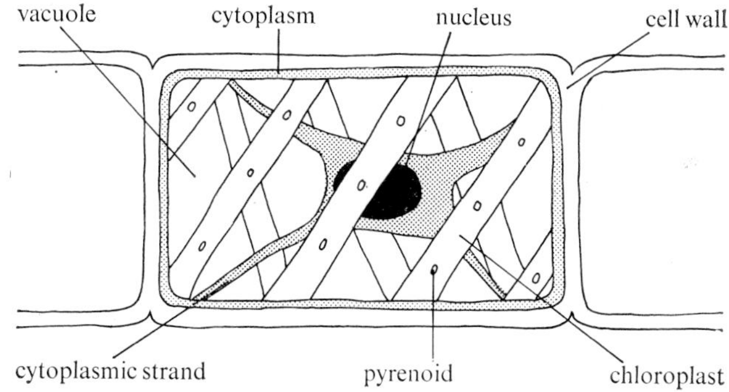

Fig 3.28 **A Spirogyra cell**

CHLADOPHORALES, freshwater and marine, branched filament like body with haploid or diploid cells having many nuclei, asexual reproduction by four flagellate zoospores, sexual reproduction isogamous or oogamous, shows alternation of generations with meiosis at zoospore formation, and including *Chaetomorpha*.

There are many other orders known but they are less important.

The green algae range in number of cells from *Chlamydomonas* (1), through *Pandorina* (16) to *Volvox* (about 20 thousand). Reproduction ranges from isogamy (*Chlamydomonas* – most), through anisogamy (*Chlamydomonas braunii*) to oogamy (*Volvox*, *Eudorina*).

Special mention must be made of *Pleurococcus naegelii* the terrestrial green alga covering tree trunks, palings, etc. The cells can withstand much desiccation without damage. *Pleurococcus* consists of spherical green cells, sometimes single and sometimes aggregated into groups or colonies. The protoplasm has chloroplast and nucleus, but no pyrenoids. Reproduction is by cell division and separation, and if moist conditions prevail, the cells may not separate fully, which leads to groups or colonies.

BACILLARIOPHYCEAE (diatoms) These have single cells or chains of cells. Each cell has two interlocking valves and is silicified and sculptured. The cell wall contains some pectinous material. They are marine or freshwater and in fact occur near any damp surface; they constitute most of the plankton and are free floating. They have chlorophylls and xanthophylls, including the brown fucoxanthin. Their storage food supplies are oils and the glucosan polysaccharide (starch-type but with β-1-3 links) chrysolaminarin. Asexual reproduction is by cell division in which generations get smaller and smaller. The smallest cells reproduce sexually (isogamy or oogamy). The zygote forms an enlarged auxospore which then gives a full size vegetative cell. The cell wall, after death, forms diatomaceous earth (called kieselguhr, used in gas/liquid chromatography and thin-layer chromatography). Examples (Fig 3.29) include *Pinnularia, Asterionella, Tabellara, Amphura and Cymbella.*

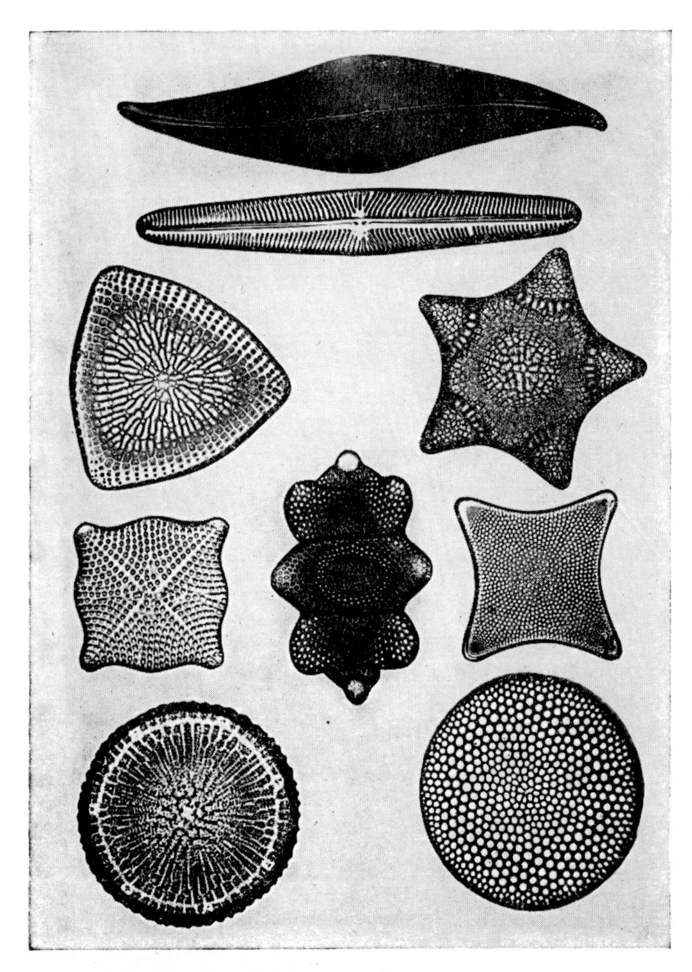

Fig 3.29 **A selection of diatoms**

DINOPHYCEAE (dinoflagellates) Occur in fresh water and marine plankton. Some are parasites. They are very numerous in upper waters of the sea (10^6 per dm^3). The cell has a definite shape due to plates of cellulose. Most are rounded, not pointed. They have two unequal cilia in grooves between the plates. Some occur in masses or as short filaments. Some lack chlorophyll and feed by surrounding other organisms. It is clear that some have animal and plant like characters.

PHAEOPHYCEAE (brown algae) These are marine seaweeds of the littoral shore zone. Pigments present include chlorophylls, xanthophylls, fucoxanthin (brown) and diatoxanthin, but no biloproteins. Storage of food is as mannitol (the alcohol from the aldohexose mannose) and laminarin (polyglucosan, β-1-3 links). The cell wall has a layer of cellulose and an outer layer of alginic and fucinic acids. The thallus is formed of tissues for different functions. There are motile reproductive stages, asexual and sexual, which are pear-shaped; they have two flagella. There are alternations of generations (haploid/diploid) in all except one sub-class. The brown algae show clearly the zonal colonisation of a seashore with a strict pattern in which each species exists in a zone best suited to itself. The male cells (antherozoids) and female cells (oospheres) are shot out from the ripe conceptacles, and form an oospore which floats out on the sea; if it reaches the correct zone, it grows; otherwise it dies.

In the sub-class Isogeneratae (e.g. *Dictyota*) there is isomorphic alternation of generations. In the sub-class Heterogeneratae (e.g. *Laminaria*) there is heteromorphic alternation of generations but in the sub-class Cyclosporeae no alternation is found.

The order Fucales is marine, found on rocky coasts, usually between high and low tide levels. The diploid thallus is large and well differentiated into holdfast (to anchor to the substratum), stipe and blade. There is no asexual reproduction. Sexual reproduction involves oogamy with the sex organs in conceptacles. Examples are *Pelvetia* and *Fucus* (bladder wrack) (Fig 3.30).

XANTHOPHYCEAE (yellow-green algae) These are fresh water algae; they store oil, not starch, and have chlorophyll. An example is *Vaucheria.*

RHODOPHYCEAE (red algae) These contain the pigment phycoerythrin. The thallus is very large. The haploid phase dominates. An example is the red seaweed *Batrachospermum* which occurs in calcareous unpolluted rivers on stones.

EUGLENOPHYCEAE These are the algae, small in number, which are sometimes classed as flagellate protozoans. *Euglena* (Fig 3.2) (see Protozoa) is an example. Some forms (colourless) feed heterotrophically.

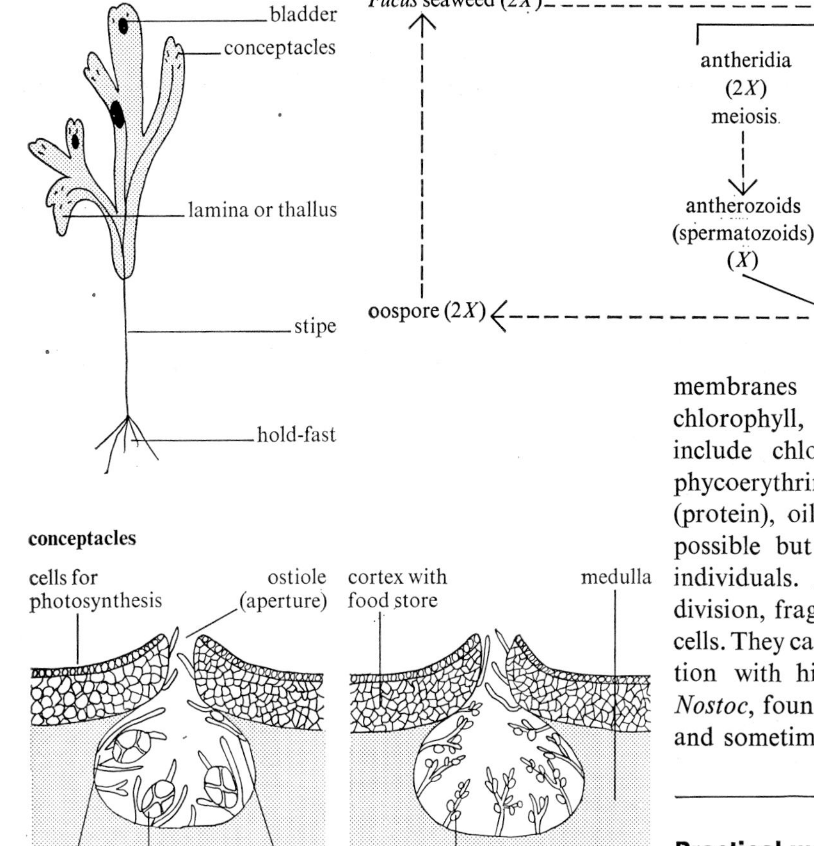

Fig 3.30 **Fucus**

The algae show a steady evolution, starting with the simple *Chlamydomonas*, through varying degrees of complexity (illustrated by *Pandorina* and *Eudorina*) to give the complex *Volvox*. Loss of cell motility (*Chlorella*) and stringing together of cells like sausages gives filamentous types (e.g. *Spirogyra*) and to complex branching filament structures (e.g. *Cladophora*).

Algae demonstrate clearly a series of stepping stones and a jumping off point from which more complex plant life must have evolved. They provide food for the larger animals in the sea and (*Pleurococcus*) for some land animals as well. Alginates are increasingly used in food processing. Natural gas is extracted from plankton deposits formed on the sea bed in the geological past.

CYANOPHYCEAE (blue-green algae) These are related to bacteria. They are marine, fresh water and occur on damp surfaces (*e.g.* soil). They are unicellular or have simple filaments, usually in a jelly-like matrix. The cells have a distinct wall with a mucilaginous sheath outside this. No membranes separate the nuclear substances, such as chlorophyll, from the cytoplasm. Pigments present include chlorophylls, biloproteins, phycocyanin and phycoerythrin. Storage compounds are cyanophycin (protein), oil and glycogen. No sexual reproduction is possible but there may be genetic exchange between individuals. Asexual reproduction takes place by cell division, fragmentation, non-motile spores or vegetative cells. They can colonise new ground, frequently in association with higher plants. Examples are *Anaboena* and *Nostoc*, found in paddy fields in Asia. Both fix nitrogen and sometimes live as symbionts. Both are windborne.

Practical work

What practical work can be done? A complete thallus of *Fucus* should be sketched. A transverse section of the frond should be examined (low power) see palisade layer, cortex and medulla. Some portions of a transverse section of the frond should be examined (high power); look for the conducting cells, 'fibres', etc. Longitudinal sections through the apex of a frond should be examined to see the meristem, apical cell, etc. Sections through male and female conceptacles should be examined (low power) – antheridia and oogonia under high power.

3.18 Thallophyta – Fungi

Fungi and yeasts are important economically – as saprophytes, disease agents, through spoilage of food, in the production of antibiotics (penicillin) and in brewing. Although lower forms are widespread in aquatic conditions, fungi are mainly terrestrial. They have no chlorophyll and are either *saprophytic* or *parasitic*. The vegetative body is made up of filaments called *hyphae*. A conglomerate of such threads gives the much-branched thallus called a *mycelium*. This may take a compact form if the threads are much interwoven as in toadstools and bracket fungi and in the underground *rhizomorphs* ('bootlaces') of the parasitic fungus *Armillaria* which attacks trees. Specialised hyphae terminate in *haustoria* in parasitic fungi (to penetrate the host) while root-like *rhizoids* are formed by saprophytes. The cell wall is of a special type of cellulose called fungal cellulose, or of chitin. Any colour is due to pigments in the cell wall. Fungi form spores (Fig 3.31), which are often held by a sporangiophore to aid dispersal, or conidia in more advanced types. Many reproduce asexually as well as by true sexual fusion of gametes formed either in special organs by fusion of the contents of two specialised hyphae or by fusion of two nuclei in a vegetative cell. Reserve food is held as glycogen or in oil drops.

The fungi are divided into three classes Phycomycetes, Ascomycetes and Basidiomycetes.

PHYCOMYCETES These are the most primitive with a well developed, non-septate, mycelium. They reproduce asexually by removal of spores from the end of hyphae. Two types of spore are produced, flagellate, swimming zoospores and wind dispersed conidiospores resistant to desiccation. The Phycomycetes can be subdivided into:

1 ZYGOMYCETES which reproduce sexually by zygospores (Fig 3.31) formed from two similar hyphae during conjugation (isogamy). An example is the mould *Mucor*.

2 OOMYCETES reproducing sexually by small male antheridia from which a fertilisation tube grows out to the larger female archegonium; a single male nucleus enters the oosphere to give an encysted zoospore or oospore. Examples are *Phytophthora infestans* (potato blight) (Fig 3.32), *Pythium de baryanum* ('damping off' on seedlings, *e.g.* cress (Fig 3.33)) and *Albugo candida* (downy mildew of Cruciferae).

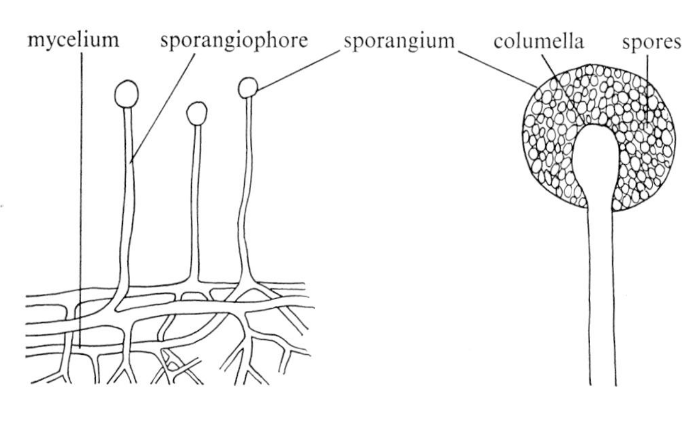

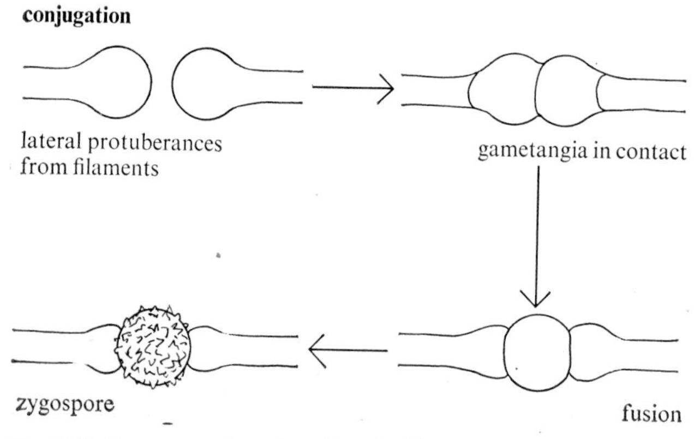

Fig 3.31 **Spores and conjugation in Mucor**

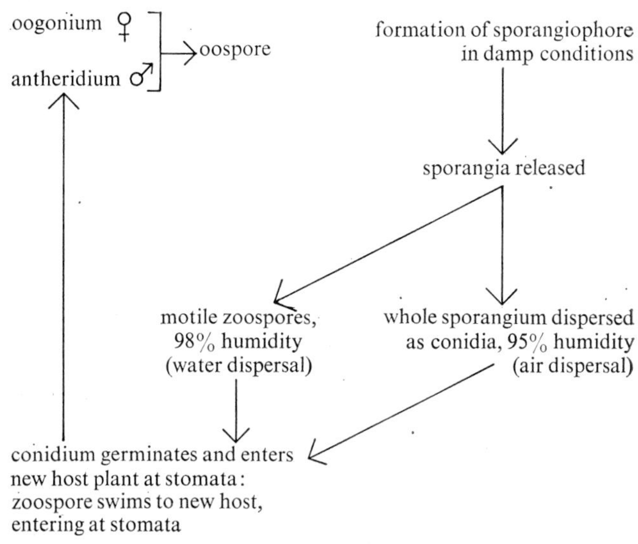

Fig 3.32 **Phytophthora infestans**

Phytophthora infestans causes rapid destruction of potato leaves and tubers. Control is by spraying with copper based fungicides when warning of impending infection has been given. Aerial surveys, spore traps, and forecasts based on weather data are used to provide advice to farmers. Crop rotation, pre-planting fungicidal treatment, destruction of infected stems and tubers, and the use of resistant varieties are other measures that assist control.

Practical work

What practical work can be done on *Mucor*? If a piece of bread is soaked in water and left under a glass cover for a few days hyphae of *Mucor* should be seen. Hyphae of the *Mucor* can be examined (low and high power). Draw hyphae and developing sporangia. Sexual organs can be drawn from a prepared slide.

ASCOMYCETES The mycelium is septate and the spores (ascospores) are formed in a sac called the ascus (flask-

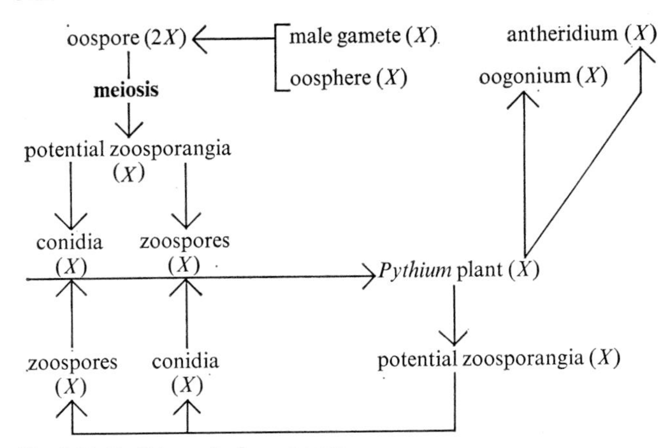

Fig 3.33 **Pythium de baryanum**

shaped). Examples are *Saccharomyces* (yeast) *Penicillium*, *Aspergillus* and *Erysiphe*.

Yeast brings about alcoholic fermentation in sugary solutions. Under natural conditions, yeast spores from other colonies or from the soil blow onto injured fruits and colonise the exuding juice. This contains oligo- and polysaccharides which the carbohydrases in the yeasts convert to di- and monosaccharides. As death of the injured cells take place, the yeast proliferates in them. The respiration can be aerobic or anaerobic:

$$C_6H_{12}O_6 \rightarrow 2\,CH_3COCOOH \quad \triangle G - ve \text{ (energy liberated)}$$

glucose — 2-oxopropanoic (pyruvic) acid

$$2\,CH_3COCOOH \xrightarrow{\text{decarboxylase}} 2\,CH_3CHO + 2CO_2$$

ethanal (acetaldehyde)

$$2\,CH_3CHO + 2\,NADH_2 \rightarrow 2\,CH_3CH_2OH + 2\,NAD$$

ethanol

Nitrogen and mineral ions are also required.

These reactions are made use of in the fermentation of alcoholic drinks. Grape and other fruit juice supplies the sugar and selected strains of *Saccharomyces cerevisiae* convert this to alcohol solutions of the order of 10–14% under anaerobic conditions in wine making.

The common method of reproduction is asexual, by budding off small outgrowths which grow to full sized yeast cells. Sexual reproduction is by ascospores which fuse in pairs to give diploid cells (Fig 3.34). *Penicillium notatum* is famous as the source of the first described antibiotic, penicillin.

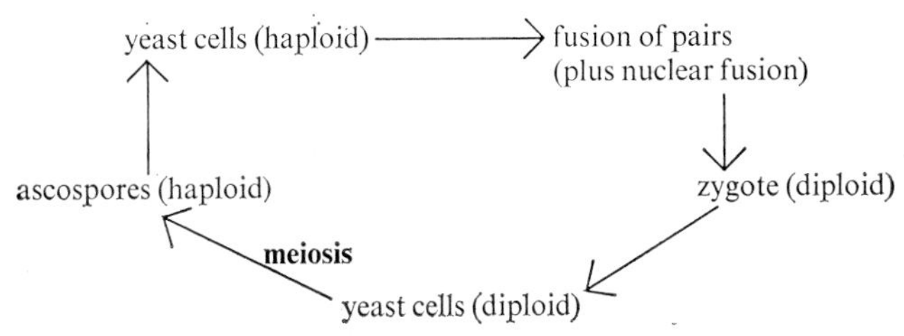

Fig 3.34 **Life-cycle of yeast**

BASIDIOMYCETES The mycelium is septate. Some members (*Puccinia*, wheat rust) are parasitic and others (*Psalliota*, mushroom – Fig 3.35) are saprophytic. *Puccinia* is heteroecious with a phase of the life cycle on wheat and a phase on barberry. Several different types of spore are produced but the characteristic basidiospores are formed on wheat straw or on the large fruiting body of the mushroom. Basidiospores from wheat straw infect barberry and in turn new wheat plants are attacked. Those from mushrooms (which may be observed by placing a mature mushroom on a clean piece of paper in a still atmosphere for a few hours) infect new areas of humus in meadows and pastures. Basidiomycetes often form associations with trees in which they envelop and slightly penetrate the root of the tree; the fungus is nourished by the tree, which appears sometimes to benefit, and only to suffer if it is already in poor health, through the association. This type of fungus growth is called mycorrhizal; it is not confined to Basidiomycetes.

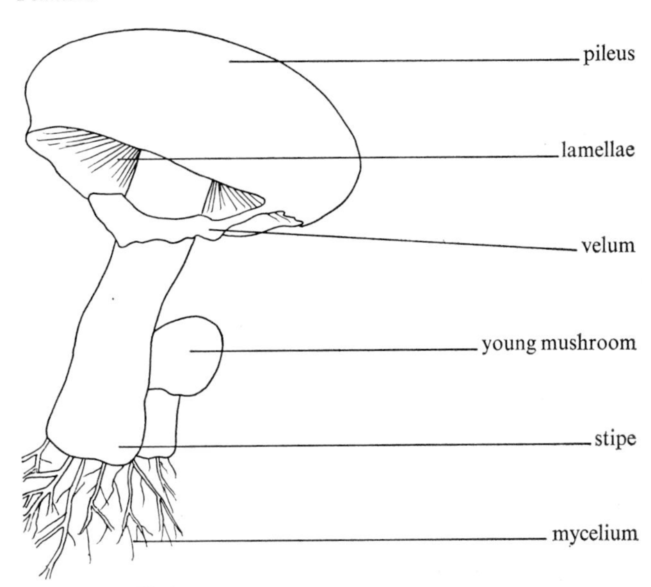

Fig 3.35 **Psalliota**

3.19 Bryophyta

This phylum includes the mosses and liverworts. The vegetative body is attached to the sub-stratum, being differentiated into attaching and absorbing organs, rhizoids and aerial parts which may or may not exhibit a stem/leaf system. They have photosynthetic pigments and large chloroplasts but no cuticle. There is a clear cut heteromorphic alternation of generations with gametophyte and parasitic (or partly so) macroscopic sporophyte generation growing from it (Fig 3.37). Asexual reproduction occurs and spores can develop into a protonema (Fig 3.36). The Bryophyta can be subdivided into:

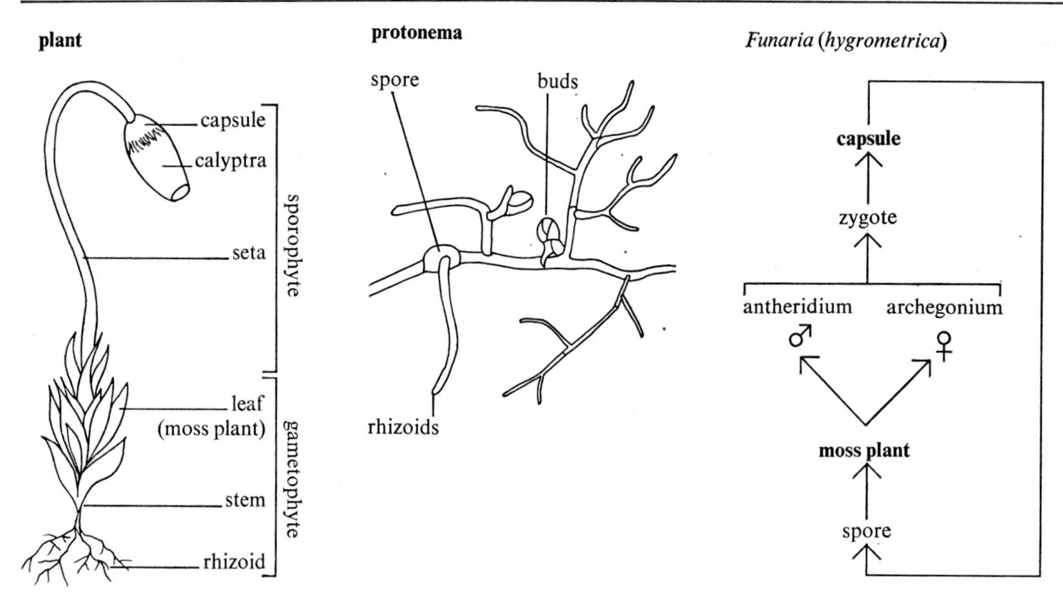

Fif 3.36 **Funaria; form and life cycle**

MOSSES An example is *Funaria* (Fig 3.36), a common moss which grows in tufts or patches on the surface of ground (particularly burnt woodland). The gametophyte is differentiated with many leaves. The antheridia and archegonia are borne at the apices of two branches of one shoot. The sporophyte is persistent and is partially synthetic. The sporogonium consists of capsule, seta and foot. The ripe capsule contains many spores and opens, when dry, to scatter them.

Other mosses are *Polytrichium* and *Sphagnum* (Irish bogmoss).

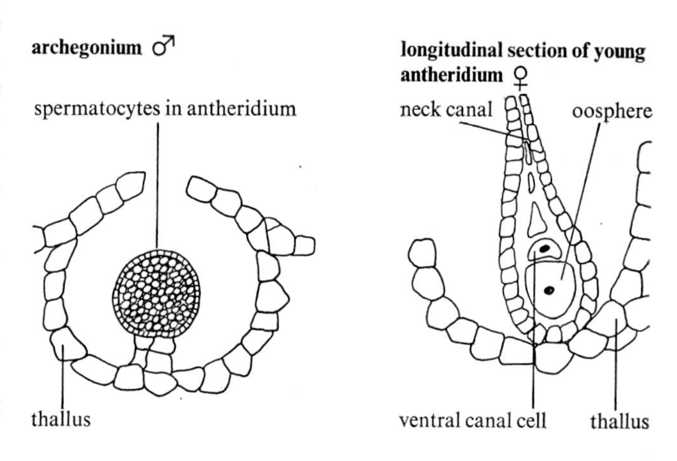

HEPATICAE (liverworts) The prostrate gametophyte is simpler in structure and less differentiated than in mosses. More advanced Hepaticae may develop an axis bearing

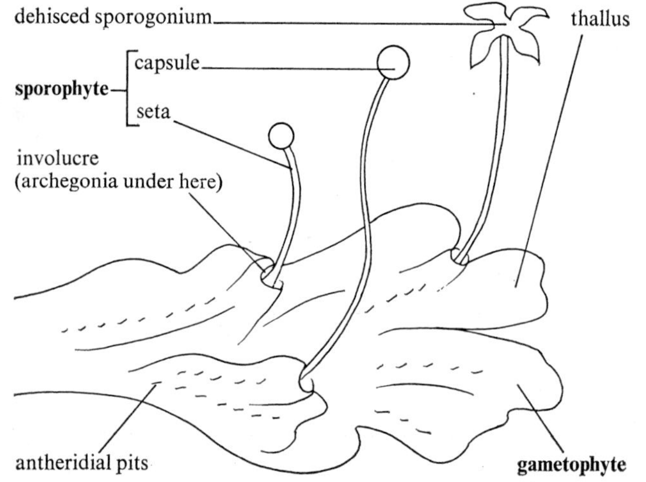

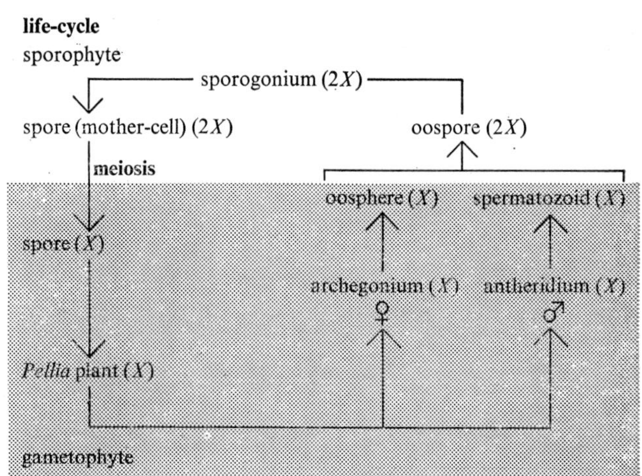

Fig 3.37 **Pellia epiphylla; form and life-cycle**

leaf-like structures. The thallus of the gametophyte (Fig 3.37) grows from meristems and throws out unicellular rhizoids into the substrate. The sporophyte is simple, lasts only a short while, and cannot photosynthesise; it sends out its spores by elators (cells inside the capsule which twist as they become dry.) Spores never develop into a protonema. Liverworts are more dependant than mosses on water. An example is *Pellia epiphylla* which is found on damp ground.

3.20 **Pteridophyta**

Pteridophytes, (these include ferns, horsetails and clubmosses), are not low plants but tend towards complexity. The earliest known examples are fossil remains in Silurian and Devonian rocks. Filicales (ferns) are the most important sub-division. Many need shade and moisture (woods, hedges and banks). There is a well marked differentiation (root, stem, leaf). In most cases the stem is a rhizome, growing horizontally or obliquely upwards through the soil. Roots are fibrous and adventitious, developing from the rhizome surface or leaf-bases. A good example is *Dryopteris filix-mas* (Fig 3.38), the male shield fern. Ferns show alternation of generations with dominant diploid sporophyte and a rather transient thalloid haploid gametophyte. The leaves and stem have a cuticle and there is a stele with xylem and phloem. Pteridophytes need water for reproduction since fertilisation is by a swimming motile antherozoid (ex antheridium) to oospore and archegonium.

Growth is from an apical cell. Pigments are chlorophylls and carotenoids. Ferns were richly represented in the Carboniferous period (coal measures). Most ferns are terrestrial with a few aquatic ones. The horsetails (e.g. *Equisetum*) are related to the ferns.

The Lycopods are another subdivision. A few are living and some are fossil. The sporophyte is differentiated into root, stem and leaf but the leaves are always small compared with stem size (microphyllus). There are homosporous and heterosporous members. Sporangia come in cones or strobili, on the ends of branches. The two most important orders are:

LYCOPODIALES, mostly tropical, with two extant genera, eligulate leaves, homosporous, e.g. *Lycopodium*.

SELAGINELLALES, mostly tropical, one extant genus, ligulate, heterosporous, e.g. *Selaginella*.

Heterospory (Fig 3.38) shows that the higher plants may have evolved from the Pteridophyta.

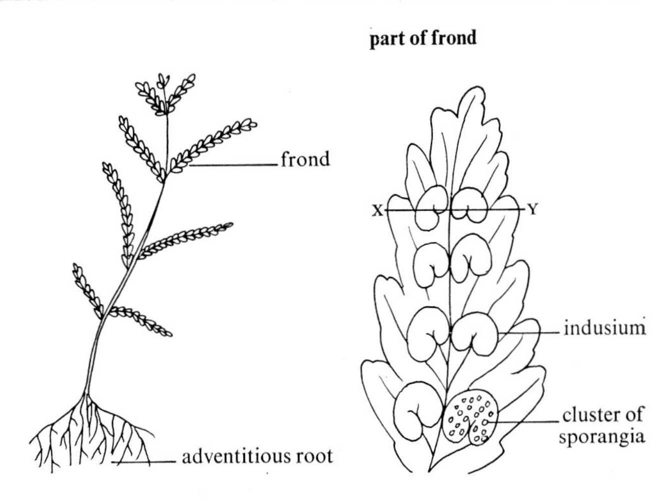

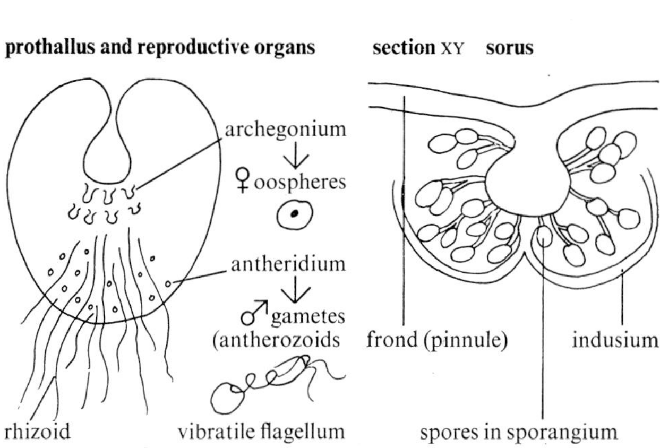

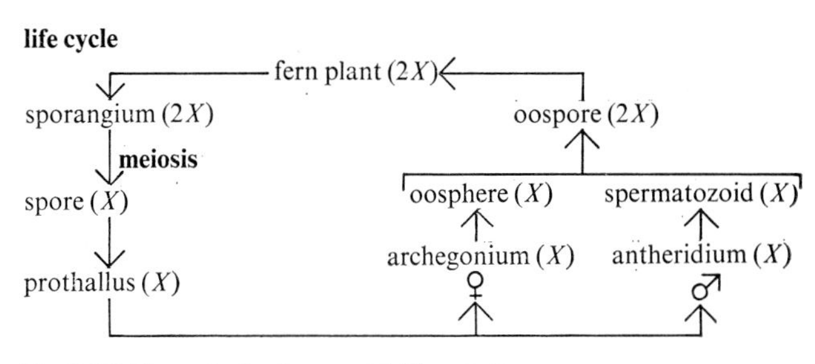

Fig 3.38 **Dryopteris; form and life-cycle**

Practical work

A plant of *Dryopteris* should be examined and sketched. The under-surface of a frond should be examined. A short stem should be boiled in dilute hydrochloric acid so that the remaining tissues can be removed to reveal the conducting vessels. A vertical section through a sorus should be examined and a single sporangium examined under high power. A prothallus should be examined and the positions of antheridia and archegonia noted.

For *Pellia*, transverse sections of the thallus should be drawn; they should be parallel to the major axis. Sporogonia should be examined and so should the contents of a capsule.

3.21 Spermataphyta

These show the highest degree of internal tissue differentiation. The body has root, leaf and stem with clearly defined xylem and phloem. Growth is from an apical meristem. Secondary thickening is common. Pigments are chlorophylls and carotenoids. There is well-defined sexual reproduction and clearly defined alternation of generations. Fertilisation is by means of a pollen tube. The male gametophyte escapes as a pollen grain (microspore) and is taken, usually by air current or insect transfer (Table 3.1) to the female gametophyte which develops in the sporophyte. In some primitive forms (*Ginkgo*) the male gametes are motile. The reproductive parts are carried in strobili (lower groups) and flowers (higher groups). After fertilisation, the resulting structure is called a seed. Seeds can be borne as they are or in a closed structure developed from the sporophyte to form a true fruit.

Table 3.1 **Wind and insect pollination**

	insect pollination	*wind pollination*
flowers	conspicuous, often large	small, inconspicuous
petals	coloured, often red or white (visible to bees)	often green (grasses: green glumes)
scent	often	rarely
nectaries	usually	rarely
anthers	small, firmly attached	large, loosely attached
filaments	short	long to catch the wind
pollen	sticky, moderate amount	lots of it
stigma	short, sticky	large, feather like, pollen traps
selfing	avoided by protandry (pollen ripe before stigma) or protogyny (opposite) or genetic self-incompatibility	often avoided by genetic self-incompatability, and by separate male flowers earlier on individual plants than the female ones
examples	buttercup deadnettle lupin	grasses willows

The microspores (pollen grains) are borne in large numbers in microsporangia (pollen-sacs) associated with microsporophylls (stamens). The seed (integumented megasporangium) has a single megaspore (ovule) and within this a vestigial female gametophyte generation is formed. Two male nuclei (or motile antherozoids) are transferred through the micropyle (pore) into the ovule. The sporophyte nourishes the seed during development. Spermatophytes are mainly terrestrial, though some are aquatic. Fossils are common; they go back to the Devonian period (Paleozoic era). Fossil remains are found alongside pteridophytes in Paleozoic rocks.

The main divisions are gymnospermae and angiospermae.

GYMNOSPERMAE These are mainly trees and shrubs with relatively few different species. The seeds are borne naked, with no carpel. There are no vessels in the xylem. Each megasporangium consists of a nucellus (body of ovule) invested by one integument, except for the micropyle. The main orders are:

1 CYCADALES, a small group, mostly in warmer climates, usually with large male and female cones (*Cycas*, *Zamia*).

2 CONIFERALES, which provide much of the world's timber, are mostly in the Northern Hemisphere; they are large trees with needle-like leaves, e.g. *Pinus* (pine), *Larix* (larch), *Picea* (spruce), *Abies* (fir), and *Sequoia* which includes the world's largest trees; broader leaved species like *Araucaria* (monkey-puzzle) and smaller species with flat leaves, e.g. *Juniperus* (Juniper), *Cupressus* (Cypress).

3 TAXALES, dioecious, with no female cone, ovules borne singly, within a fleshy aril e.g. *Taxus* (Yew).

4 GINKGOALES, e.g. *Ginkgo biloba*, maidenhair tree with motile male gamete.

5 GNETALES with enclosed ovules, e.g. *Gnetum*.

ANGIOSPERMAE These dominate the world's vegetation in most parts and show a great diversity of form (Fig 3.39). They have true xylem vessels. The most characteristic feature is the flower with whorls of sterile and fertile parts (Fig 3.40). They are usually hermaphrodite. The endosperm is unique, formed within the ovule as a result

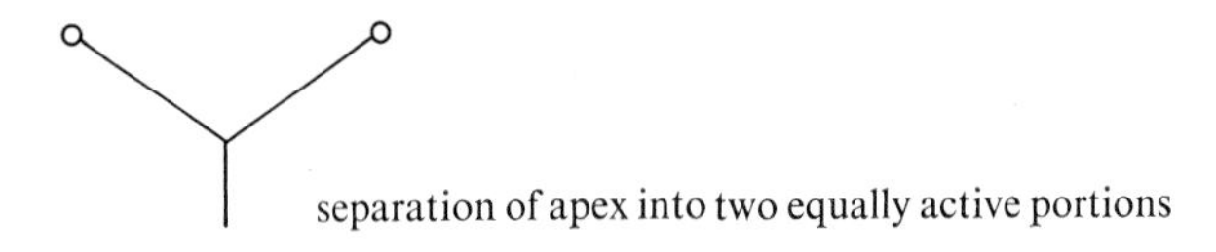

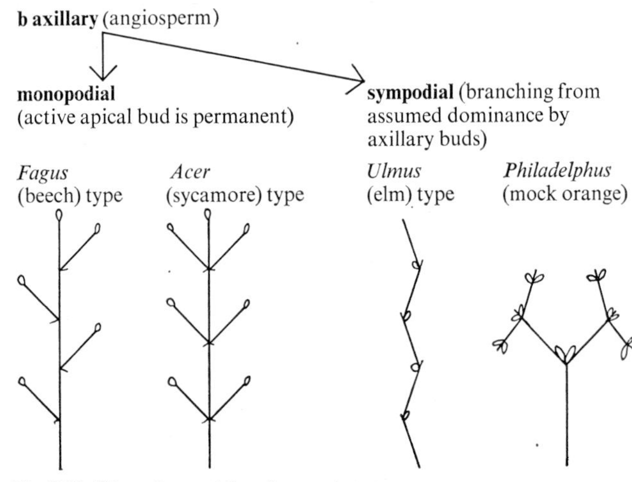

Fig 3.39 **Stem branching in angiospermae**

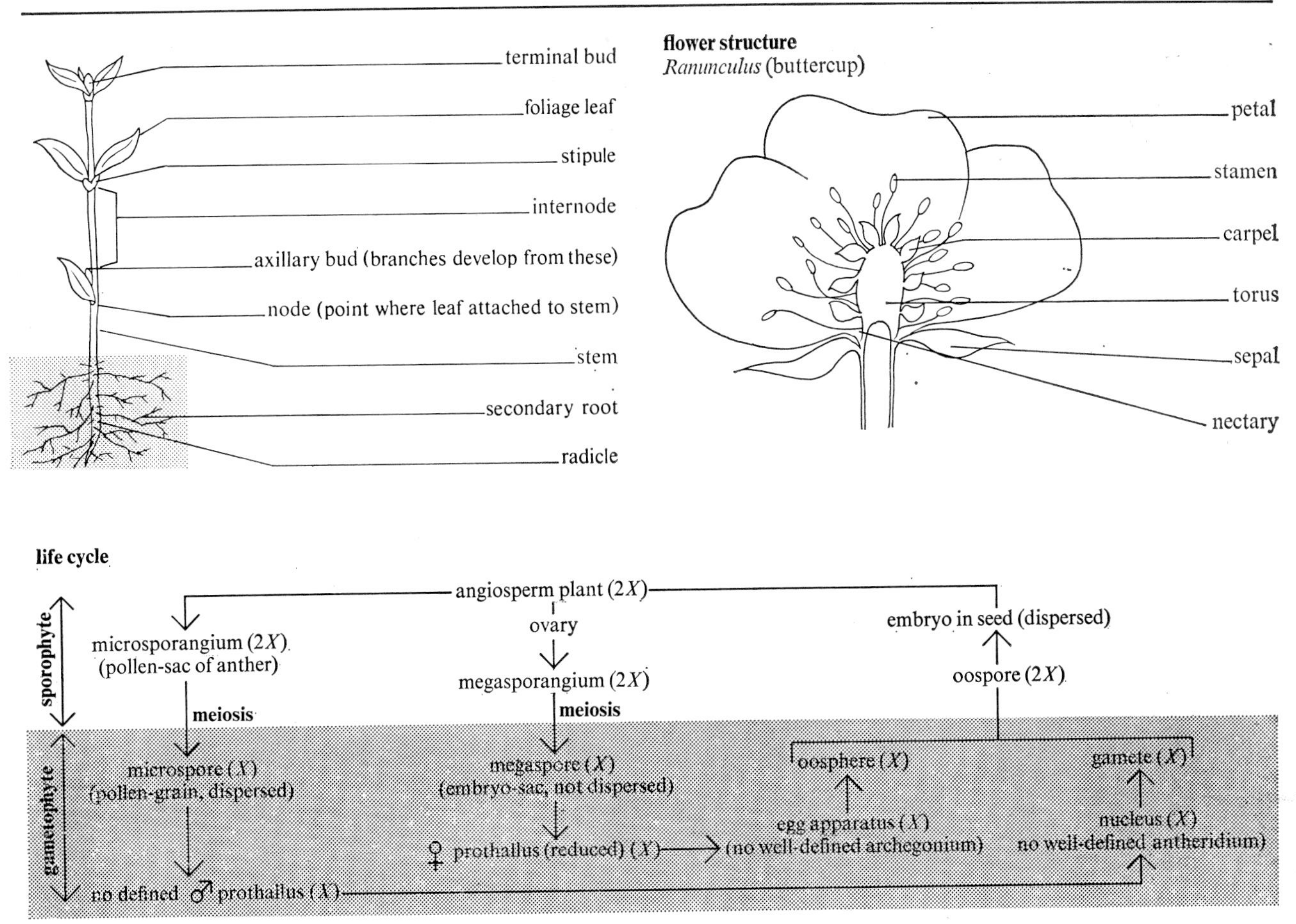

Fig 3.40 Generalised angiospermae

of a triple fusion in which both male and female cells are involved. Each seed (megasporangium) consists of a nucellus enclosed in special carpels, forming an ovary which ripens into a true fruit.

There are two sub-classes, monocotyledons and dicotyledons.

1 MONOCOTYLEDONS Most of these are herbaceous, none are true trees but some are palm-like. They have adventitious root systems and the leaves usually have parallel veins. The stem contains irregularly arranged vascular bundles. The flower parts are usually in multiples of three. The seed contains a *single* cotyledon (special storage leaf) – hence the name monocotyledon. If the perianth (the outer series of floral organs, outside the stamens and carpels) is rudimentary or absent, the families concerned are Cyperaceae (sedges) and Gramineae (grasses) (Fig 3.41). Those with petaloid perianth are Iridaceae (*e.g.* irises) with inferior ovary, or Liliaceae (*e.g.* lilies) and Juncaceae (rushes) with superior ovary. The ovary is inferior if it is below the point of origin of the petals.

2 DICOTYLEDONS These are herbaceous shrubs and trees. They usually have net-veined leaves. Vascular bundles are arranged in rings. Secondary thickening is common. Flower parts are in fours or fives (or multiples). There are *two* cotyledons in the seed – hence the name dicotyledon. Further subdivisions are:

Apetalae: without petals, *e.g.* Salicaceae (willows).

Sympetalae: petals united, rarely free; ovary inferior, *e.g.* Compositae (daisies); ovary superior, *e.g.* Labiatae (deadnettles), Primulaceae (primroses).

Polypetalae: petals free, rarely united. Polypetalae can be further subdivided according to the insertion of the floral organs on the torus or receptacle. This is nearly always short and is only occasionally elongated between the whorls of floral organs. The torus can assume a variety of shapes, from convex to flattened, hollow or cup-shaped. The shape of torus determines the way in which floral organs are inserted.

In the *hypogynous* (Fig 3.43a) arrangement (*e.g.* buttercup, poppy) the torus is convex and the stamens, petals and sepals inserted in order on the side of the torus below the gynoecium. In the apocarpous pistil type, the carpels are free, *e.g.* Ranunculaceae, buttercups (Fig. 3.40) and

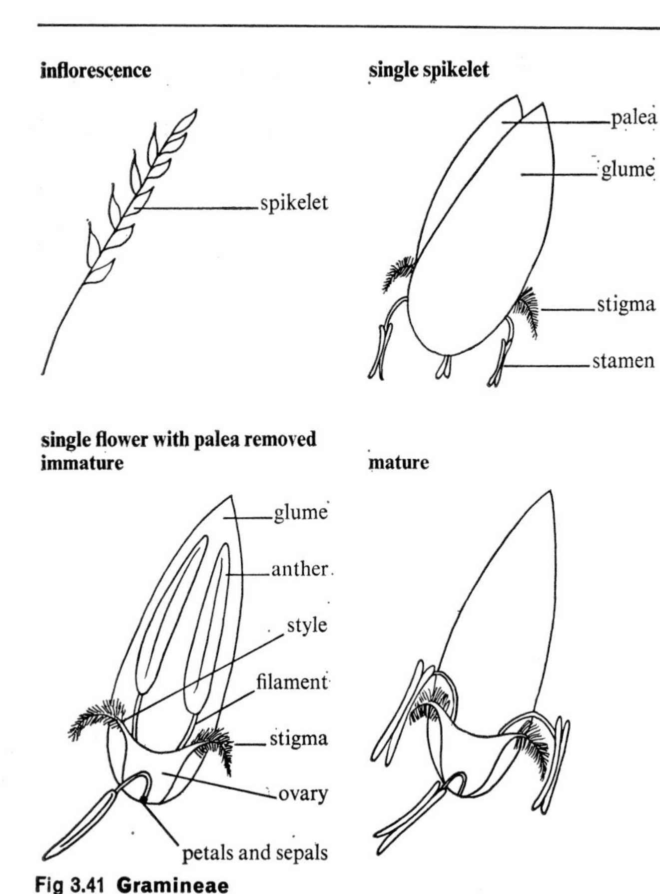

Fig 3.41 **Gramineae**

in the syncarpous pistil type the carpels are united to form a single compound ovary (examples are Cruciferae (Fig 3.42), cabbage or wallflower, Violaceae, violets, and Caryophyllaceae, pinks or campions).

In the *perigynous* arrangement, e.g. *Rubus, Filipendula, Prunus, Rosa* (Fig 3.43b), the torus is a flattened disc, the gynoecium developed in the middle of the disc with the sepals, petals and stamens round the rim (not under the gynoecium but round it).

In the *epigynous* arrangement, e.g. *Pyrus* (Fig 3.43c), the torus is a deep cup with carpels the first adherent, the sepals, petals and stamens being inserted on the gynoecium.

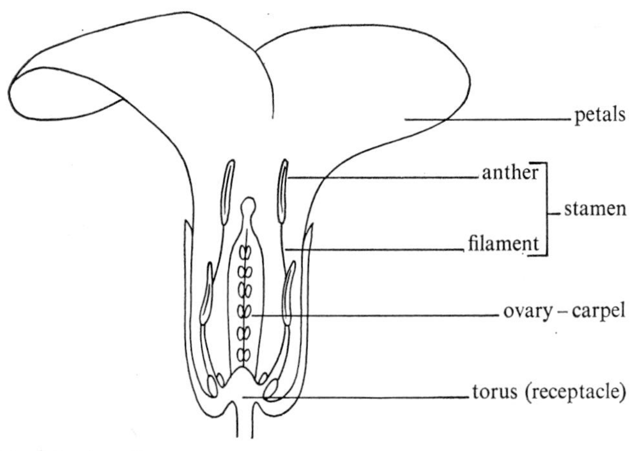

Fig 3.42 **Cruciferae**

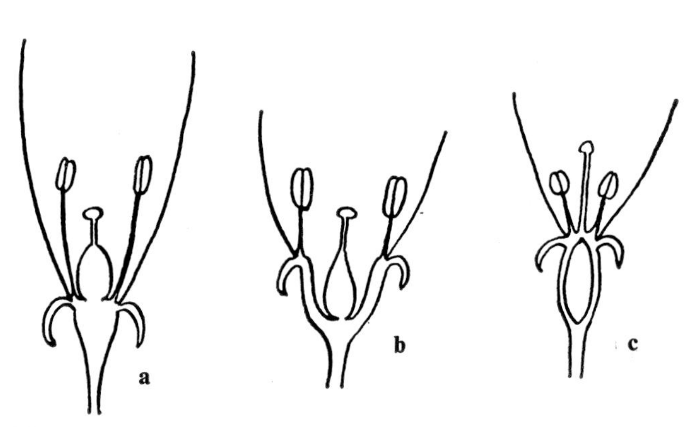

Fig 3.43 **Arrangement of stamens, petals and sepals on the torus**

3.22 Summary of flowers (Figs 3.40 to 3.43)

An *inflorescence* is a group of flowers (or floral region)
An inflorescence is borne on a *peduncle*
An individual flower-stalk is a *pedicel*
A flower borne on a pedicel is *pedicellate*
A flower with no visible pedicel is *sessile*
Parts of the flower are borne on the *floral axis* (*torus* or *receptacle*)
The outer ring is of *sepals* making up the *calyx*
The next is of petals making up the *corolla*
The next is of *stamens* forming the *androecium*
The innermost is of *carpels* forming the *gynoecium*
Stamens consist of *filament*, *connective* and *anther*
At the base of each carpel is an ovary surmounted by a *style* and *stigma*
Flowers at the end of the stem axis are *terminal*
Flowers borne in the axils of leaves are *axillary*
Single flowers are *solitary*, multiple inflorescences are *compound*
A *bract* is a reduced leaf in whose axil a pedicel arises
A *bracteole* is a reduced leaf borne on a pedicel
Calyx and corolla (or those parts present) make up the *perianth.*

The *floral diagram* (Fig 3.44) is a ground plan of the flower showing how the various parts are related to one another and to the axis. Cohesion is union between members of the same series of floral leaves and adhesion means union between members of different series. In the diagram cohesion of parts can be indicated by connecting lines. The *floral formula* enables the botanist to present morphological features of the flower without description provided the floral diagram and a longitudinal section are also given. The symbols ⊕ and ↑ denote respectively,

a *Taraxacum officinale*

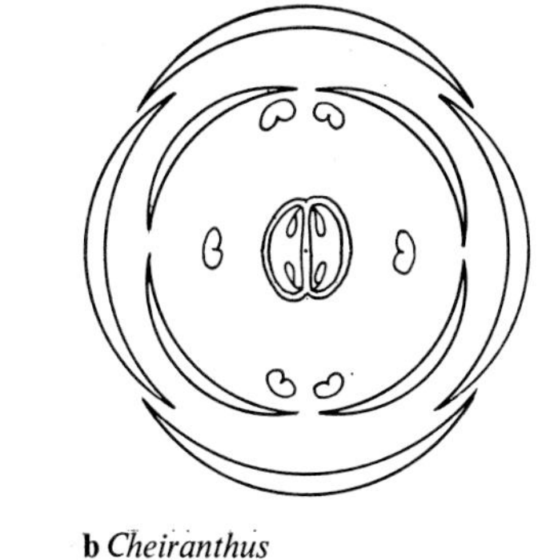

b *Cheiranthus*

Fig 3.44 Floral diagrams

radially and bilaterally symmetrical flowers, the direction of the arrow showing the symmetry plane which divides the flower into equal halves. The symbols ♂, ♀, ☿ denote, respectively, staminate, carpellary and hermaphrodite flowers. The letters K, C, P represent calyx, corolla and perianth, A and G represent the androecium (stamens) and gynoecium (pistil) and the number after each letter represents the number of parts in each series. A bracket enclosing the parts indicates cohesion, and ⌒ means adhesion between the parts of successive whorls. Ovary inferior is indicated by a line *above* the number following G. Ovary superior is indicated in the same way, but with the line *below* the number. Very many parts in each series is written as an infinity sign (∞). For example *Primula* is represented by

$☿ \oplus K(5)\ \overset{\frown}{C(5)\ A5}\ G(\underline{5})$

which means hermaphrodite, radially symmetrical flower, gamosepalous calyx (5 sepals), gamopetalous corolla (5 petals), androecium of 5 free epipetalous stamens, syncarpous pistil (5 carpels) and superior ovary.

Can you interpret the formulae

$☿ \oplus P\ 3+3,\ A\ 3+3,\ \underline{G(3)}$

as applied to *Lilium* and

$☿\ \uparrow\ K0,\ C0,\ A3,\ \underline{G1}$

as applied to perennial rye grass?

Practical work

What practical work should be done on Angiospermae? This can only be sketched in outline here.

STEM Examine various types (protective, climbing, propagative). Examine twigs, shoots, bulbs, corms. Examine a transverse section of a young stem (*e.g.* sunflower, marrow). Make a regional diagram, then a high-power study of a typical vascular bundle. Cut and examine sections of a tulip (or hyacinth) stem as an example of a monocotyledon.) A general stain for stems is safranin and light green; the general ground tissue is then green and the vascular xylem and fibres become red.

LEAF Examine various types and their arrangements. Examine some sweet peas bearing tendrils and expanded stipules. Examine leaf-bases of a rose bush; note the stipules. Examine various foliage buds and a prepared transverse section of a bud. Cut a transverse section of a dicotyledonous leaf. To do this, split vertically a piece (1 cm) of elder pith and insert the leaf in the crack. Cut sections of pith and leaf together and examine in propantriol (glycerol) (low power). Mount a portion of the epidermis of an iris leaf in dilute propantriol (glycerol). Examine the stomata.

FLOWER Examine the entire structure of buttercup and sweet pea, draw and label. Cut a medium longitudinal section through the flower. Construct a floral diagram. Examine the receptacle and gynoecium of a rose and narcissus (cut sections longitudinally). Remember buttercup is hypogynous but rose and apple are perigynous and narcissus is epigynous. Cut transverse sections of ovaries of various flowers – use a hand lens to examine. Examine types of inflorescence. Cut transverse sections of flower buds, *e.g.* buttercup, tulip. Examine pollen grains (low power). Place some grains in dilute (1·5 – 2·5 %) sucrose solution – notice germination in due course and examine pollen tubes; stain in haematoxylin to observe contents of nuclei.

Cut transverse sections of carpels, *e.g.* columbine or other plant having large carpels. Find a section which passes longitudinally through an ovule. Stain in haematoxylin and examine (high power) – look for integuments, nucellus, embryo sac, etc. Consider the ecology of several flowers. Study pollination. Examine fruits and seeds.

ROOT Examine tap root (*e.g.* beetroot, carrot) and fibrous root (*e.g.* nasturtium). Germinate a pea or bean and examine root tips under low power; note the root cap, root hairs, etc.

Examine adventitious roots – split a leaf of begonia and lay on moist soil.

Cut and examine transverse sections of a young root of a monocotyledon (*e.g.* iris) and a dicotyledon (*e.g.* sunflower); stain the thin sections as for the stem.

Examine early stages of thickening by cutting sections of a root of a sunflower some distance behind the root hair region.

Cut a section through the hypocotyl of a bean seedling and examine; draw the stages to show the transition from root to stem.

3.23 Annuals, biennials and perennials

ANNUAL PLANTS These survive as seeds. After germination, formation of flowers and formation of seeds, the plant dies and the life cycle is completed within the year. Examples are groundsel and poppy.

BIENNIAL PLANTS These develop large storage organs in the first year and food from these is used during the second year's growth, when the main flowering, if not the only flowering, occurs. The plant then dies, after formation of seeds. An example is the carrot. Some biennials live for a further year or so.

PERENNIAL PLANTS These may be *herbaceous* or *woody*. Herbaceous plants during the winter remain either as dormant underground forms, with no leaves above ground (crocus, tulip) or have a low tuft of foliage above ground (iris, grasses). The structures below ground (rhizomes, bulbs, corms, etc.) store food for use in the spring. Many herbaceous perennials do not flower for the first year or so. Woody perennials (trees and shrubs) have a branch/trunk system which persists for many years. If the leaves are shed in autumn, they are called *deciduous* (*e.g.* oak, ash, elm) and if they shed their leaves at intervals but retain some throughout the year they are called *evergreen* (*e.g.* holly, most conifers).

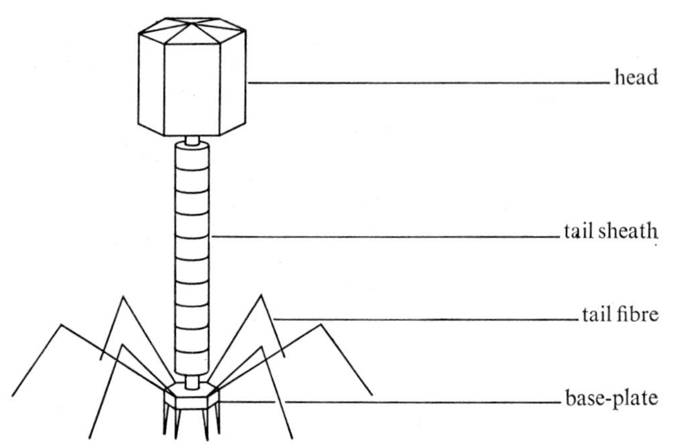

life cycle

host cell attacked ⟶ viral DNA enters ⟶ host cell synthesises new viral DNA and proteins ⟶ new virus built up ⟶ host cell bursts to release virus ⟶ host cell attacked

Fig 3.45 **Bacteriophage; form and life-cycle**

3.24 Viruses

Although viruses are not necessarily plants, they are dealt with here simply for completeness.

Viruses are small, about 10 to 800 nm in length. They can be 'seen' with the electron microscope. Viruses are so simple that they cannot grow and reproduce on their own, only with the aid of a living cell. In general, a virus can only penetrate one type of host cell. Thus T2 virus attacks the bacterium *Escherichia coli*. Viruses which live inside bacterial cells are called *bacteriophages* (*phages*), (Fig 3.46). T2 is made up of about 30 proteins, plus DNA. When infected with virus particles, the cells of the host build up new T2 proteins. The infected cell now bursts to release the new virus, which starts off again on its cycle of infection/growth (Fig 3.45). In carrying out this synthesis *E.coli* uses its own enzymes, or it makes new ones in order to assemble amino acids into the sequence required for T2 proteins and DNA. Since it would normally make its own proteins, the virus has altered its behaviour. Hershey and Chase (1952) showed that viral DNA is more likely than its protein to cause the alteration in behaviour. It is now possible to remove the cell wall from living bacterial cells, then mix DNA extracted from a *phage* with the naked bacteria, and then to notice that the bacteria respond by forming a complete infectious virus. Since no viral protein is present in the mixture, it could never have been responsible for altering the behaviour of the bacteria. The virus so formed is complete in that its DNA is enclosed in its protein sheath.

All viruses are parasitic. They cause the diseases small-pox, chicken-pox, influenza and common cold in man, foot and mouth and swine fever in farm animals, rabies and psittacosis in pets and myxomatosis in wild animals (rabbit) as well as some forms of cancer.

Most cultivated plants are now known to have virus diseases, which include potato mosaic, sugar beet yellows, cucumber mosaic, plum pox, and raspberry mosaic. Viruses do not usually invade seeds (though there are exceptions like lettuce mosaic) so they are often spread by vegetative propagation. Meristems are often virus free and new healthy stocks are sometimes raised by culturing meristems until they are large enough to propagate from. Some hosts are attacked by viruses but the effects are not noticed for a long time because the virus is dormant; the virus of this type is called a *prophage*.

Most viruses are resistant to antibiotics and drugs. The cell itself makes an anti-viral agent called interferon. Outside the host cells, viruses are easily dealt with (heat, short wavelength light, disinfectants, etc.).

Some viruses can become permanently incorporated into the genetics of the host cells (called *reduction*) and some can put into cells genetic information derived from cells they have previously occupied (*transduction*).

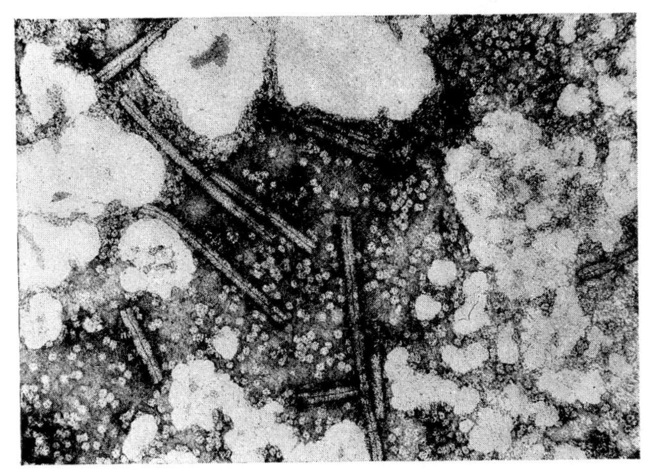

(a) Rods

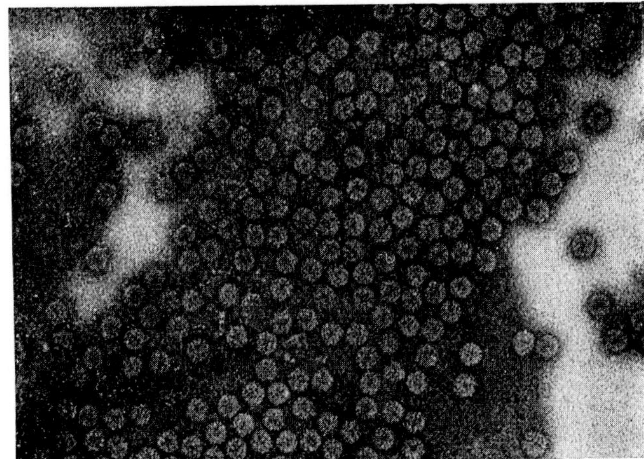

(c) Spheres

(b) Flexuous Rods

(d) Grand Soleil d'Or

Fig 3.46 **The electron micrographs (a), (b), and (c) show the three major shapes of virus particles.**

(d) Illustrates the difference in growth rates when viruses are eliminated from the narcissus variety Grand Soleil d'Or (below) compared to growth when viruses are present (above).

(e) Virus identification. The centre well in the agar-filled plate contains antiserum of cherry leaf roll virus (CLRV). The wells contain prepared extracts as follows 1 and 2– CLRV, 3 and 4– suspected CLRV from elderberry trees, 5 and 6– extract from healthy plants.

The extracts diffuse through the gel. The precipitation between the antibody well and the four virus wells shows that the two viruses are related. The precipitation 'spur' between wells 2 and 3 shows a serological distinction between them. The elderberry virus is a strain of CLRV, similar to, but not identical with the cherry strain.

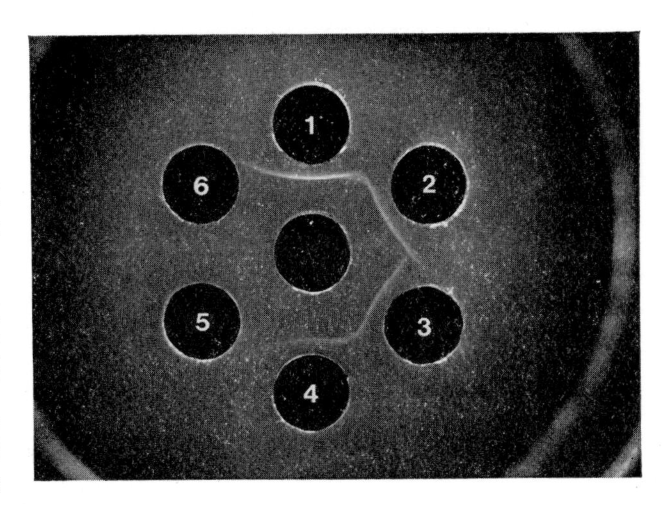

(e) Serology

(a) Sexual reproduction

(b) Vegetative reproduction; higher plants

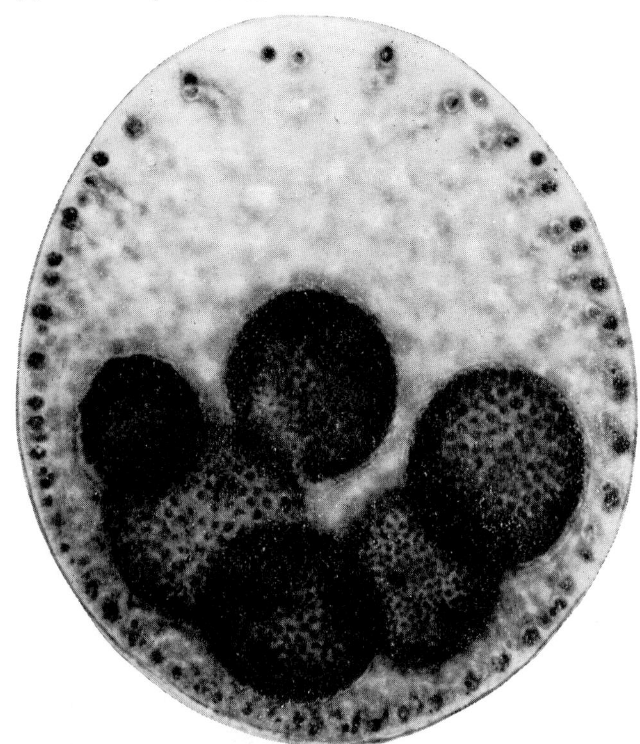

(c) Vegetative reproduction; the algae

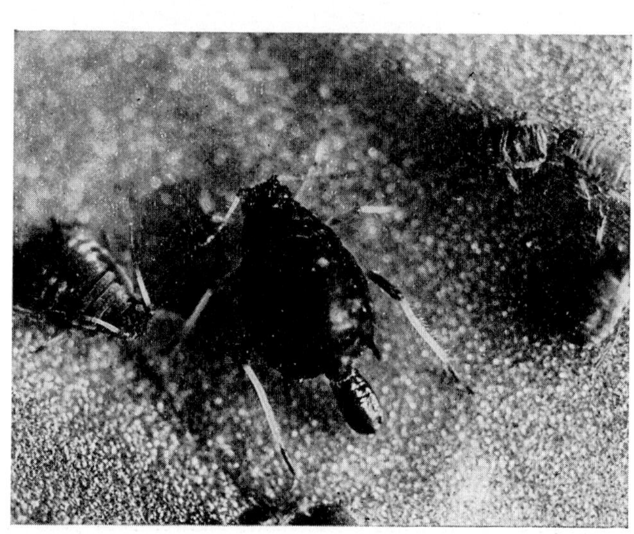

(d) Parthenogenesis

The photographs show different kinds of reproduction (Fig 4.1)

(a) Frogs reproduce sexually. When eggs are laid by the female the male produces seminal fluid containing sperms which fertilise the eggs by penetrating the albumen. Pairing ensures that the eggs are fertilised when they leave the female's body.

(b) This is one type of asexual reproduction. When valuable plant cultivars are selected or bred, vegetative reproduction provides a means of keeping them pure. It also enables plants to be grown to maturity more rapidly than is the case with propagation from seed. Here scientists examine a bundle of well rooted fruit tree root stocks propagated in heated bins.

(c) Volvox tertius provides a good example of vegetative reproduction in the algae. The photograph shows six daughter plants ready for release from the parent.

(d) Aphids are insects that can reproduce by parthenogenesis which is the development of the young from an unfertilised egg. They are also unusual among insects in that they give birth to live young (viviparity).

4 Reproduction and Growth

4.1 Introduction

If a species is to survive it must reproduce; reproduction is thus the key characteristic of living organisms. Deaths caused by *predators* (animals that prey on others for food), by disease and by natural causes must be replaced. Reproduction can be either sexual or asexual.

ASEXUAL REPRODUCTION In asexual reproduction a single individual organism produces offspring *mitotically*, *i.e.* without meiosis. Asexual reproduction is more common among plants, although there are examples in the animal kingdom *e.g.* parthenogenesis in summer generations of wingless aphids.

The characteristic feature of asexual reproduction is that each new individual exactly resembles the parent, since in mitosis there is no reassortment of genetic information (cf. p 33). Obviously new offspring from the same parent exactly resemble one another, and collectively form a *clone*.

Asexual reproduction occurs in different ways:

1 Single cells divide into two (*binary fission*), the method of reproduction of unicellular organisms.

2 Single cells may be budded off from a larger body, as in spore formation.

3 Single cells form eggs (bees), or after embryonic development within the mother (aphids) are born viviparously ('live birth').

4 Larger collections of cells split off from the main body. Although disorganised masses of cells may have some capacity for regeneration, normally specialised parts of organisms do this. In plants this is called *vegetative reproduction.* Some remain attached to the parent and derive nutrition from it while forming roots (*e.g.* runners, Table 4.1), and only become independent when the connecting tissue dies. Others form new individuals when already split off. Often parts of plants that reproduce vegetatively are storage organs. You will be familiar with many of these (Table 4.1 and Fig 4.2) from your elementary studies of biology.

Plants can be artificially propagated by vegetative means. The plant can be divided and the parts planted separately. In this case man sometimes anticipates nature by splitting off a specialised group of cells before it is ready to start an independent existence, keeping it alive in high humidity conditions. Runners, offsets and suckers can be cut off and, if rooted, planted. If not rooted, they may be kept in a closed atmosphere like a garden frame until roots appear.

Cuttings can be taken from suitable shoots, roots and leaves. Shoots *e.g.* blackcurrant (hardwood—winter) or geranium (softwood—spring) are most frequently used and are usually cut off just below a node (high concentration of food and auxin for good rooting). Root cuttings from the thick side roots are used to propagate some plants *e.g.* horse radish and leaf cuttings are used to increase some ornamentals, *e.g.* African violet, or *Begonia rex*. Roses and fruit trees are propagated by grafting buds or stems (scions) on to older plants (root stocks).

Higher plants in nature sometimes propagate both sexually and asexually. Where the environmental conditions are fairly constant, vegetative reproduction is very effective; but where new factors enter the environment sexual reproduction (seedlings) may be an advantage.

Table 4.1

name	*example*
tuberous roots	dahlia, lesser celandine
stem tubers	potato (Fig 4.2) Jerusalem artichoke
rhizomes (food storage organs)	iris, solomon's seal (Fig 4.2)
bulbs (condensed shoots)	tulip, snowdrop
corms (short swollen stem)	crocus, cuckoo-pint
runners (creeping stems root at tip or nodes to form independent plant)	strawberry (Fig 4.2) creeping buttercup, *Saxifraga sarmentosa*
offsets (short runners)	daisy, houseleek
tip-root (tips of arching shoots form roots)	blackberry
suckers (shoots from lateral buds below ground)	elm, plum, raspberry (Fig 4.2) rose
rhizomes (not food storage organs)	lily-of-the-valley, couch grass
bulbils (axillary buds give new plants directly)	*Lilium bulbiferum*
adventitious buds on leaves which develop to new plants	*Bryophyllum*, the fern *Asplenium bulbiferum*, *Rolmeia*

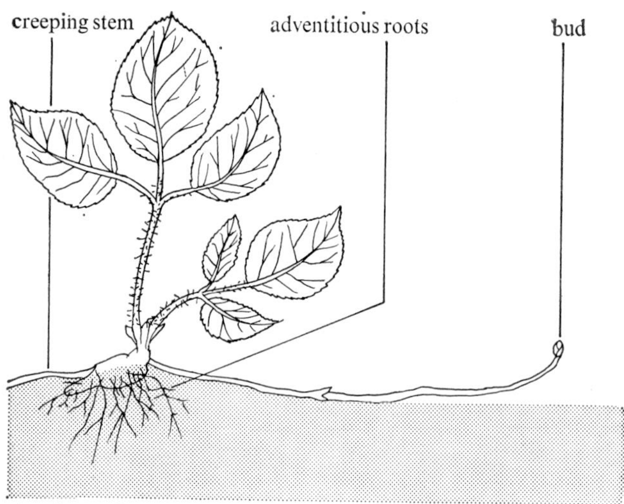

strawberry plant and runner

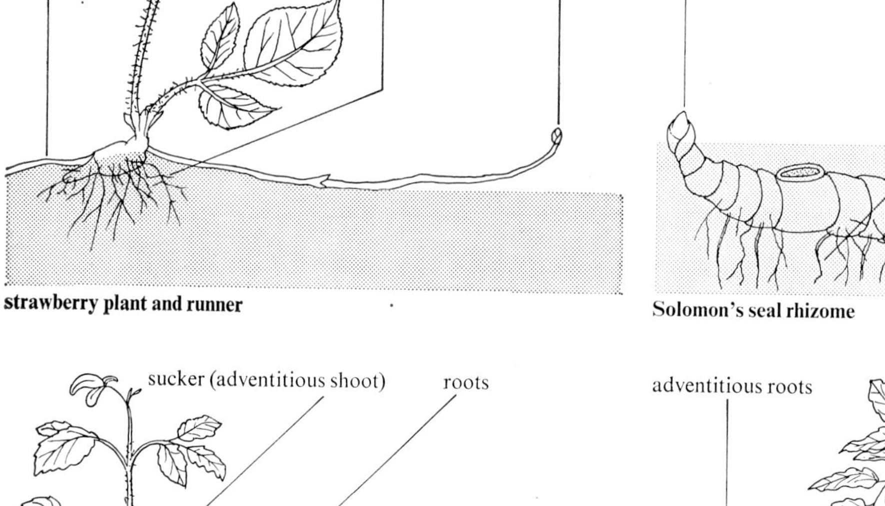

Solomon's seal rhizome

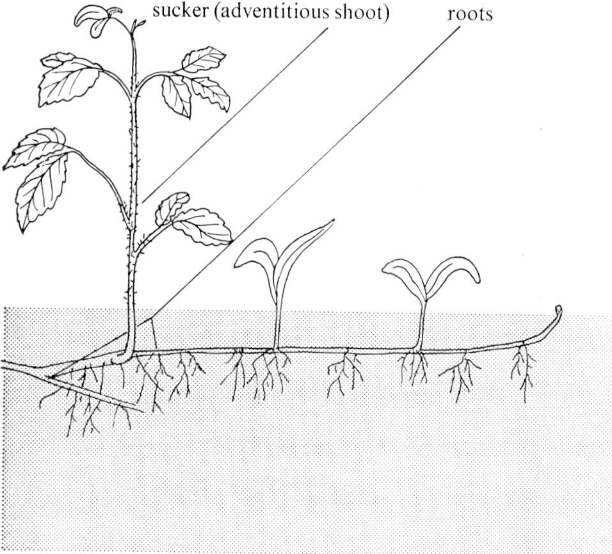

raspberry sucker

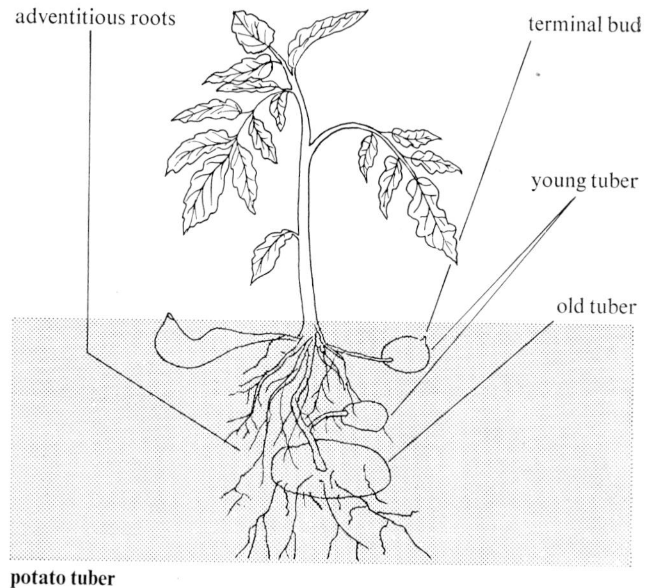

potato tuber

Fig 4.2 **Examples of vegetative reproduction**

An example would be a devastating outbreak of disease (*e.g.* Dutch Elm disease in Great Britain during the 1970's). Clonally produced plants may prove to be susceptible, in which case *all* are susceptible. Seedlings may include some more resistant individuals.

SEXUAL REPRODUCTION In sexual reproduction, two haploid cells (gametes) are formed, the ♂ being called *spermatozoa* or sperms (animals) or *antherozoids* (plants), and the ♀ being called *ova* or eggs (animals) or *oospheres* (plants). Fusion of the ♂ and ♀ gametes forms a diploid organism called a zygote. This then grows by *mitotic division*, but remember that meiosis is involved in the production of haploid gametes from a diploid organism (see p 36).

PLANTS

The life cycles of various plants have been given already (pp 56–63) and you should be familiar with the terms *isogamy*, *anisogamy* and *oogamy*. You should also make yourself familiar with the life histories of a simple plant (such as *Spirogyra*), one showing alternation of generations (such as *Pellia*) and a complex seed plant.

4.2 Germination of seeds

The seed of the broad bean (*Vicia faba*) consists of an embryo protected by a tough leathery skin, the *testa* (seed coat). At one end of the testa, there is a dark scar called the *hilum* marking the place where the seed has broken from the stalk by the *funicle*. There is a small aperture at one end of the hilum, the *micropyle;* it was through this hole that a tube from the pollen grain grew in preparation for fertilisation of the ovule and, during germination, water enters by this route. Inside the testa there is a large embryo bean plant, consisting of the young shoot (*radicle*), the *cotyledons* which are leaf-like but thick and fleshy because of the food (starch and proteins) they contain, and the tip of the young shoot, the *plumule*. During germination in *Vicia*, the *hypocotyl* (part of the radicle immediately under the cotyledons) remains short and the cotyledons remain under the ground, giving up their food to the developing radicle and plumule. The subterranean cotyledons are said to be *hypogeal* (Fig 4.4). The *epicotyl* (axis of *plumule*) stays

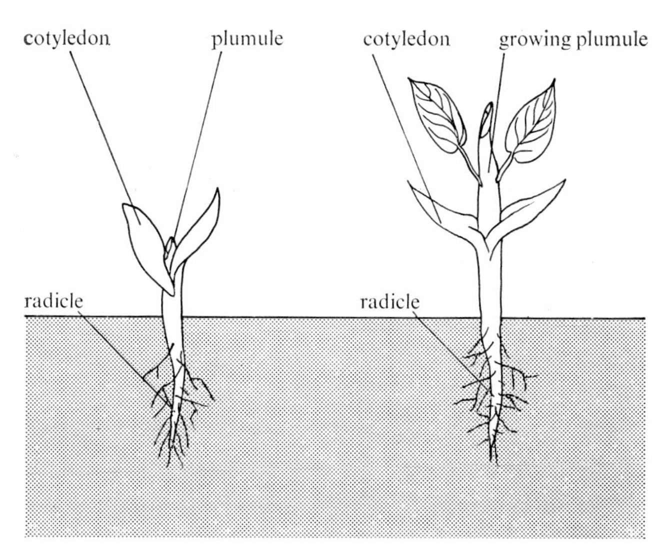

Fig 4.3 Diagrammatic representation of epigeal germination

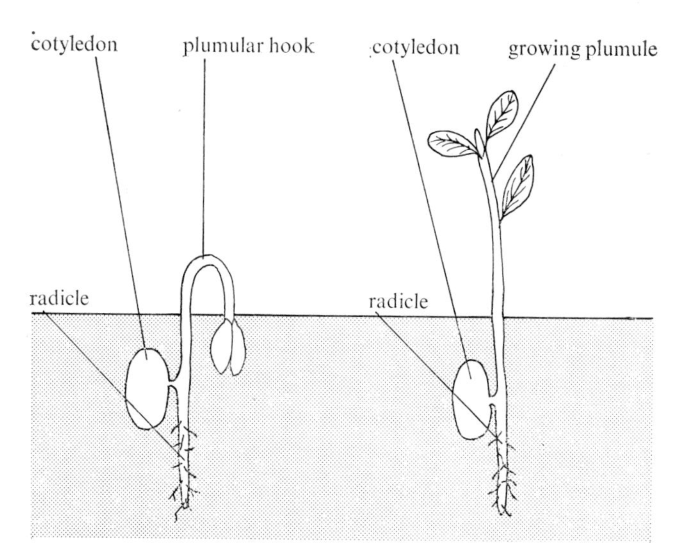

Fig 4.4 Diagrammatic representation of hypogeal germination

bent until it emerges from the soil. In the French bean (*Phaseolus vulgaris*) the cotyledons are *epigeal* (Fig 4.3) – they come above the ground.

In castor-oil plant (*Ricinus communis*) the kernel contains an oily yellow-white *endosperm*, enclosing the embryo. If the endosperm is broken apart, the two cotyledons can be seen as rudimentary leaves. The endosperm stores reserve food material as oil and aleurone grains (dried out dissolved proteins). Seeds of this type, in which the food material is contained in a special tissue embedding the embryo, are called *endospermous*. In the bean, food is stored in cotyledons and not in a special endosperm tissue; such seeds are said to be *non-endospermous*. During germination, the cotyledons absorb food from the endosperm and get larger – the hypocotyl gets longer and the seed is carried out of the ground with the cotyledons, in epigeal germination (Fig 4.3), forming the first green organs.

The maize seed is really a fruit. Pericarp (the wall of the ripened ovary) and testa are thin and are fused together to form one membrane. The embryo can be distinguished as an area of light colour. The remains of the stigma can be seen, demonstrating that the outer covering *is* a pericarp, and that the maize is a fruit. The endosperm is rich in starch and under the testa there is an aleurone layer of protein. The embryo has a large plumule, a radicle and a scutellum; during germination the latter secretes enzymes which hydrolyse the starch and the proteins to provide food for the growth of plumule and radicle. Maize belongs to the family Gramineae (grasses); the 'seeds' of wheat, oats and barley are similar in structure.

Conditions necessary for the early stages of germination are moisture, warmth and oxygen. Later, the young shoot requires light for photosynthesis; without it the shoot becomes *etiolated.*

Practical work

What experiments could be done on germination?

HYPOGEAL Look carefully at a soaked seed of pea or broad bean. Remove the testa and examine the large fleshy cotyledons. Separate these and note the epicotyl and radicle. Line a beaker with blotting paper and place a few soaked seeds between the glass and paper. In the central space pat sand, and moisten at intervals. Sketch as germination proceeds.

EPIGEAL Repeat all of the above for the seeds of castor oil and marrow. (The castor oil seed is endospermous; the marrow seed shows a nucellus remaining after ripening and is called a perisperm). Can you see, for the marrow, the part (a 'heel') on the hypocotyl which aids extraction of the cotyledons from the testa?

Try to germinate seeds of wheat, maize, cress which have been boiled in water for a short while. Try also to germinate seeds in *dry* sand and compare with wet sand. Carry out the experiments illustrated in Fig 4.5. Use a control in both those experiments.

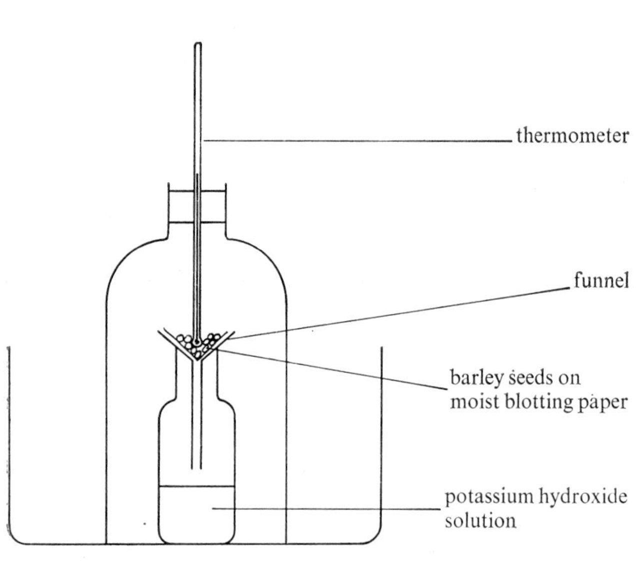

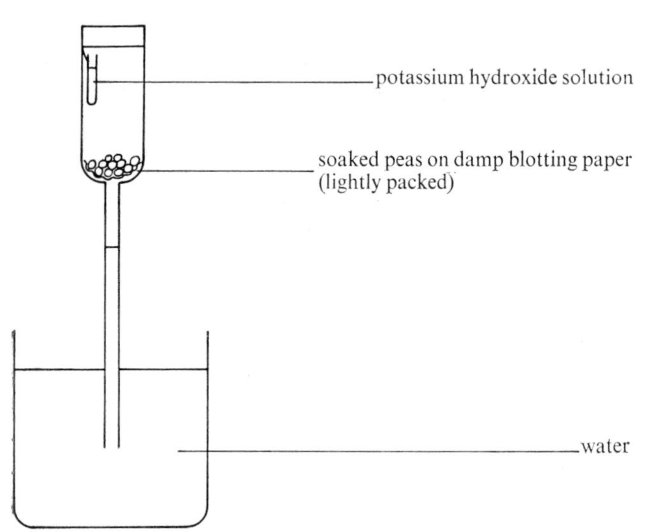

Fig 4.5 **Experimental study of germination**

SEED FORMATION How is the seed actually formed? After pollen grains reach the stigma they absorb water and foodstuffs from the sticky surface, swell and burst, and a tube emerges from each grain and grows down through the style towards the ovary (Fig 4.6), traversing the carpel cavity and entering the micropyle of the ovule; the growth is controlled by secretions from the ovule (chemotropism). The nucleus concerned with formation of gametes (not the tube nucleus) has two male nuclei. When the tube has entered the micropyle, the male nuclei enter the embryo sac, one forming a zygote by union (fertilisation) with the ovum and the other forming a triploid (endosperm) nucleus by union with two secondary nuclei. The nucleus of the zygote divides rapidly by mitosis, forming suspensory cells, and these maintain the embryo which is developing in the middle of the embryo sac. The triploid nucleus divides to form the endosperm (mass of food) which starts to fill the embryo sac. In endospermous plants, *e.g.* cereals and maize, the food store remains where it is – in the endosperm tissue. In non-endospermous plants, which include the majority of dicotyledons (*e.g. Vicia faba*) the food is transferred to the cotyledons when the endosperm tissue fills the embryo sac, and it is then the cotyledons which become very large.

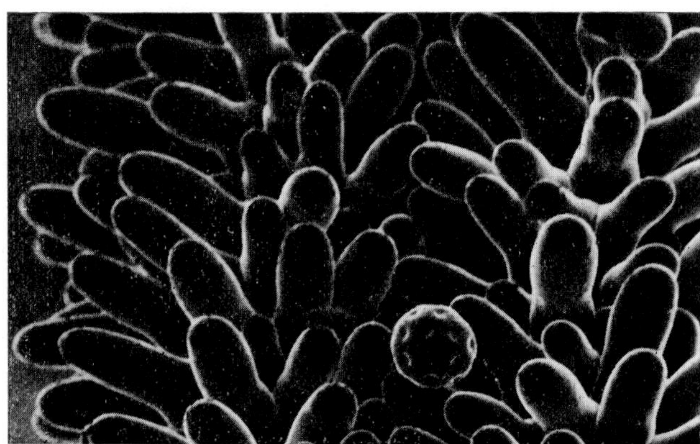

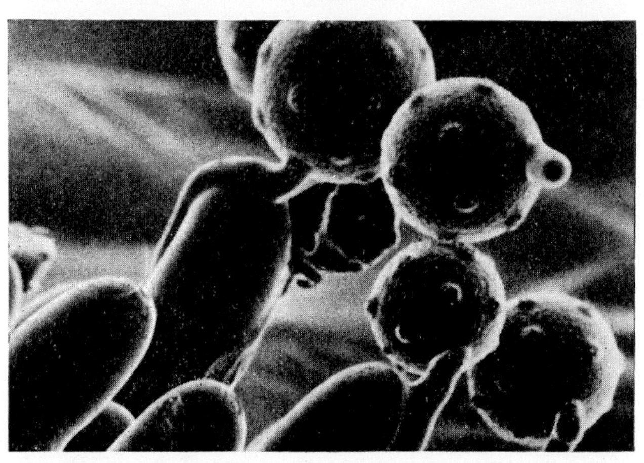

Fig 4.6 **Scanning electron micrographs of stages in pollination. Above: cross pollination occurs as grains are carried to other flowers by insects or the wind. Pollen grains adhere to the stigma after release from the anther. Below: the pollen grains have germinated, pollen tubes are emerging; these grow down to the ovary where fertilisation occurs**

4.3 Seeds and fruit

The material around the ovule in the ovary or carpel of the flower hardens to form the testa, i.e. *the ovule becomes the seed.* The wall of the carpel, the *pericarp* (or seed box), grows under hormonal control to form the *fruit;* when the fruit is ripe, the pericarp can be either tough and leathery, hard and woody or succulent.

Obviously plants must have some means of *dispersal* so that the seeds are moved away from the parent to prevent overcrowding and the resultant competition for resources. Structural adaptations have thus occurred to aid dispersal. In some cases the pericarp has become expanded to aid wind dispersal; some plants that live near water are modified for floating; some fruits are attractive to animals. Methods of dispersal are summarised in Table 4.2.

Table 4.2 **Methods of seed dispersal**

action	*examples*
wind	
1 shaking open seed heads (censer mechanism)	poppy
2 carrying seeds on parachutes	dandelion (Fig 4.7)
3 carrying seeds on wings	lime, sycamore, ash, elm
animal	
1 hooking onto hairs	burdock, goose grass
2 feeding and voiding in faeces	blackberry
3 feeding and rejecting	mistletoe
4 transporting (squirrel)	hazel
plant	
plant itself propels seeds (explosive mechanism)	broom, laburnum
water	
seeds carried by streams etc.	water lily

Fig 4.7 **Dandelion seeds**

Seeds can also be dispersed by chance (think of some examples), but a plant which relies on this as the *only* method of dispersal is unlikely to be successful.

Fruits are classified as *true*, in which part of the carpel is modified, or *false*, in which the receptacle is also modified. Examples of false fruits are *pomes* (*e.g.* apple, pear), the rose hip and the strawberry, all of which are succulent and animal dispersed.

The true fruits can be divided (as in Table 4.3) into those with a *dry* pericarp and those with a *succulent* pericarp (seeds surrounded by a fleshy mass). The former can be subdivided into the *dehiscent* fruits whose pericarp splits open to liberate the seeds, and the *indehiscent* fruits whose pericarp does not split open, at least not spontaneously.

BIONOMICS OF SEED DISPERSAL In both annual and perennial plants available resources can be channelled into many small seeds with little endosperm at one extreme, or few large seeds with ample endosperm at the other; the amount of flesh in fruits also varies widely. The relatively few large seeds contained in a brightly coloured rose-hip may be efficiently dispersed after passage through the gut of a bird, while the many small seeds of a poppy fall a few feet from the parent. However both rely on good conditions for germination and growth – a suitable ecological niche, in fact – if dispersal is to give an effective result. Plants that produce an abundance of small seeds with an efficient inbuilt dispersal mechanism like groundsel are particularly successful at colonising new areas. With large seeded plants new colonies rarely spring up; slow expansion of existing communities is more likely.

Table 4.3 **Classification of true fruits**

classification	*characteristics*	*examples*
dry pericarp:		
a dehiscent		
capsule	several carpels	campion, poppy, willowherb
legume	single carpel splitting down both edges	pea, laburnum
follicle	single carpel splitting down one edge	larkspur
siliqua	two carpels	wallflower
b indehiscent		
caryopsis	testa fused to pericarp	maize, grasses
achene	single carpel, leathery	grass, bullrush
stony	single carpel, hard	hazel
cypsela	single carpel, calyx reduced to a pappus or absent	dandelion
succulent pericarp		
berry	pericarp two layered, several carpels	grape, tomato
drupe	three layered pericarp, single carpel	cherry
	cluster of drupelets	blackberry
	fibrous mesocarp	coconut

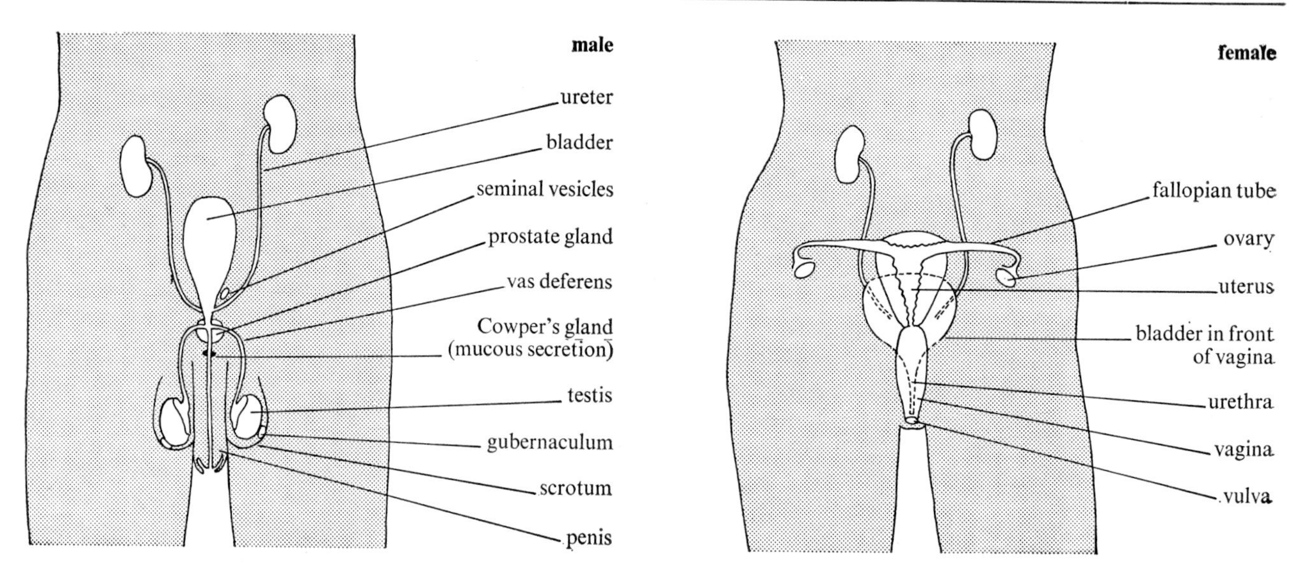

Fig 4.8 Human reproductive organs

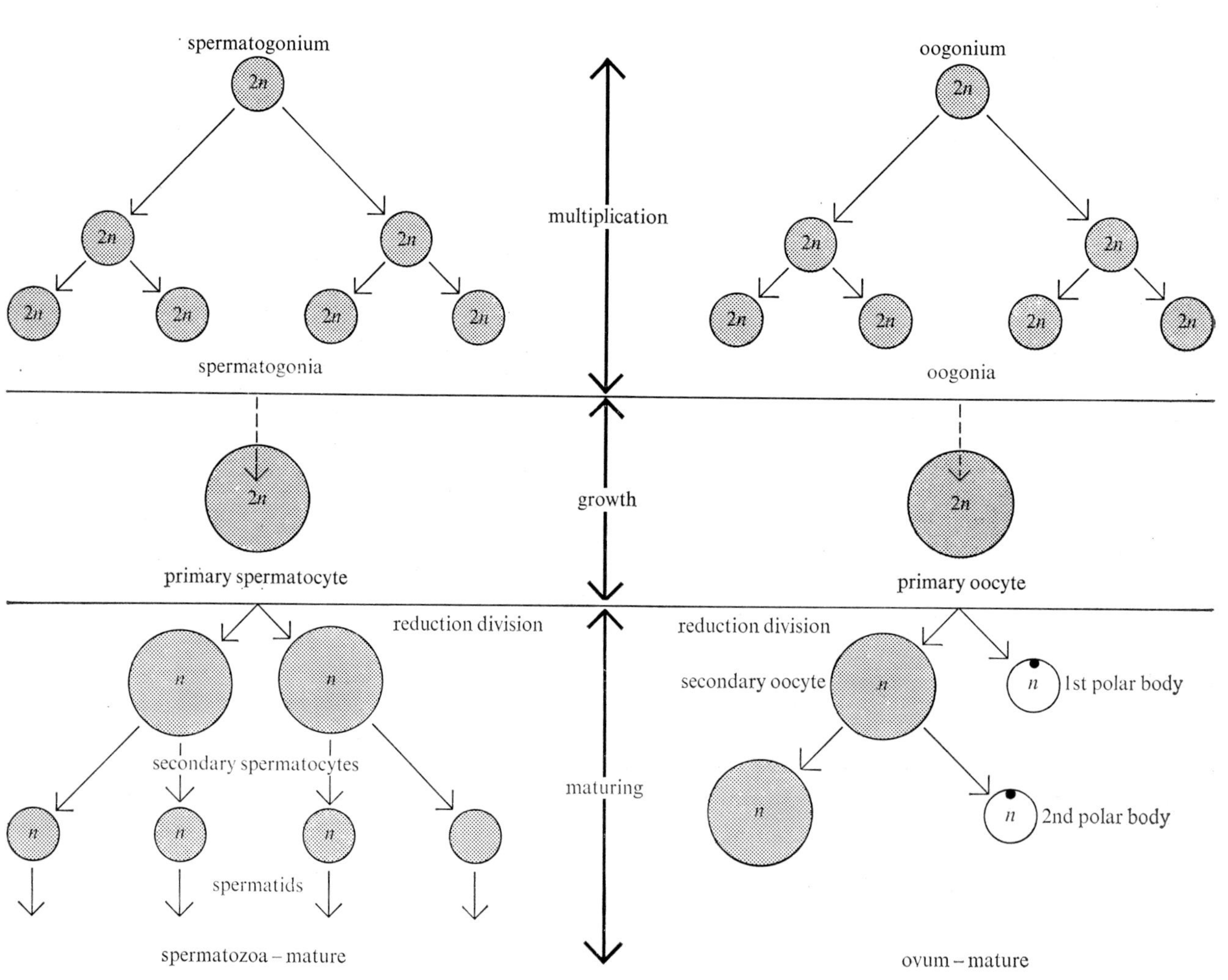

Fig 4.9 Gametogenesis (meiosis)

ANIMALS

4.4 Mammalian reproduction

TESTES The primary sex organs of the male, the *testes*, lie outside the abdominal cavity in the *scrotum*. The testis (Figs 4.8 and 4.10) has several compartments, each containing many coiled *seminiferous tubules*, and it is here that *spermatogenesis* takes place. The seminiferous tubules open into the coiled *epididymis*, which in turn uncoils and forms the *vas deferens*, opening into the *urethra* at the base of the bladder. *Spermatocytes* are formed (Fig 4.9) by *meiosis* of the *spermatogonia* (in outer layers of tubules), and further division gives *spermatids*; these lie on *Sertoli cells* and they mature to give *spermatozoa* or sperm which are formed in seminiferous tubules after which they pass to the vas deferens. In response to the appropriate sexual signals (mating behaviour varying with species), blood flow into the tissue of the penis increases; this organ swells, hardens and becomes erect, enabling it to be inserted into the female vagina. *Copulation* is completed by expulsion of sperms from the penis into the vagina (*ejaculation*). Sperms contain mitochondria which yields ATP to provide energy for the lashing of the tail. This enables sperms to 'swim' on the surface of the uterus, and achieve fertilisation. If *ejaculation* does not take place, sperms are autolysed and become absorbed in the blood.

The male sex hormone, *testosterone* (Fig 4.11), is formed by influence of the *gonadotrophin luteinising hormone* (LH), coming from cells in the tissues surrounding the seminiferous tubules. There are two other male hormones which help to control 'maleness factors' (*e.g.* beards, deep voice), *androsterone* (Fig 4.11) and *dehydroisoandrosterone*. The vas deferens, before entering the urethra, joins the seminal vesicle which secretes a fluid containing globulin. The prostate gland secretes an alkaline solution. The sperm and both these additions make up *semen*.

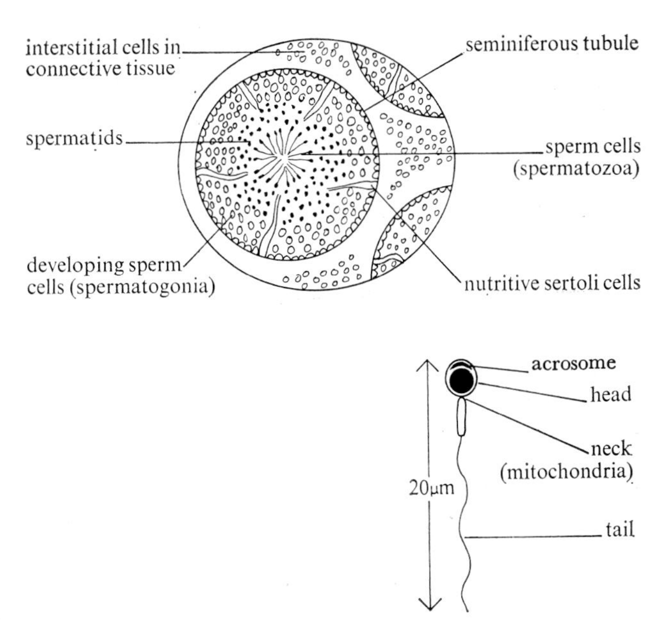

Fig 4.10 **Section of testis and sperm**

OVARIES The *ovaries* are the female sex organs (Fig 4.8). At maturity of the female, the ovaries contain about 7×10^4 potential ova, although only about 400 develop to maturity during the reproductive life of this female. *Ovulation* occurs once every *menstrual cycle*. Mature *ova* are released from the *ovaries* and pass into the *fimbriate funnel* (extension of the Fallopian tube), then along the *Fallopian tubes* by ciliary action. The Fallopian tubes open into the *uterus*, at the centre of the lower abdominal cavity. The uterus has thick muscular walls and its spongy lining (*endometrium*) is supplied with many blood vessels. The cervix (neck of uterus) leads into the *vagina* which opens to the outside at the *vulva*.

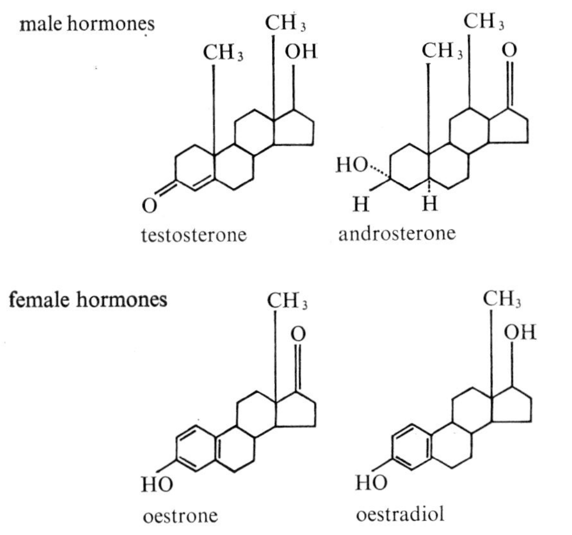

corpus luteum hormones

CH3
CH3
COCH3
O
progesterone

CH3
COCH3
HO
H
pregnenolone

CH3
COCH3
HO
pregnanolone

Fig 4.11 **Some sex hormones (all steroids)**

At sexual maturity, cells from the *germinal epithelium* (at the edge of the ovary) move inwards to form *Graafian follicles.* Development of the follicle and its production of *oestrogens* are governed by *follicle-stimulating hormone* (FSH) from the pituitary gland (p 160). An *oogonium* (Fig 4.9) forms in the middle of the follicle and this grows to form the primary *oocyte;* the first step of meiosis gives a secondary oocyte and the second stage gives the *ovum* (about 200 μm diameter in woman). As the development proceeds, the follicle causes production of oestrogen which causes release of a *luteinising hormone* (LH) by the pituitary. The mature ovum is discharged during the second week of the menstrual cycle, the follicle being ruptured. This is called *ovulation.*

After ovulation, the entire follicular cavity becomes filled with cells from the *stratum granulosum* (in the normal way, this lines the follicle). The LH stimulates this development and the cells are filled with the pigment *lutein;* the follicle is now a gland (*corpus luteum*) and is secreting *progesterone* (Fig 4.11), a hormone which stops any more FSH being liberated and also stimulates activity in the uterine wall, preparing it to receive the ovum if it should be fertilised. Progesterone also prevents maturation of ova and the associated cycle of events during pregnancy. The oestrogens (Fig 4.11) are a series of similar chemical substances (steroids) and function as female sex hormones. They are found in follicular fluid and in other cells in the ovaries. Should there be no fertilisation, the corpus luteum declines and also the flow of progesterone. Another cycle is begun by FSH, and the uterus lining is shed and appears as the menstrual flow.

Progesterone, which causes the development of the endometrium and sensitises it to receive the *zygote,* like the oestrogens is a steroid hormone. It is of interest that steroids such as norethindrone and norethynodrel can suppress ovulation and they are used in the control of excess bleeding during menstruation and also in contraception ('the pill'). The hormone *oxytocin* from the posterior pituitary gland can cause rhythmic contractions to occur in the uterus. Development of the *mammary glands* (in the breasts) is aided by the oestrogens and during pregnancy they develop special ducts for secretion.

After birth of the offspring, milk is secreted by the action of the hormone prolactin from the pituitary gland.

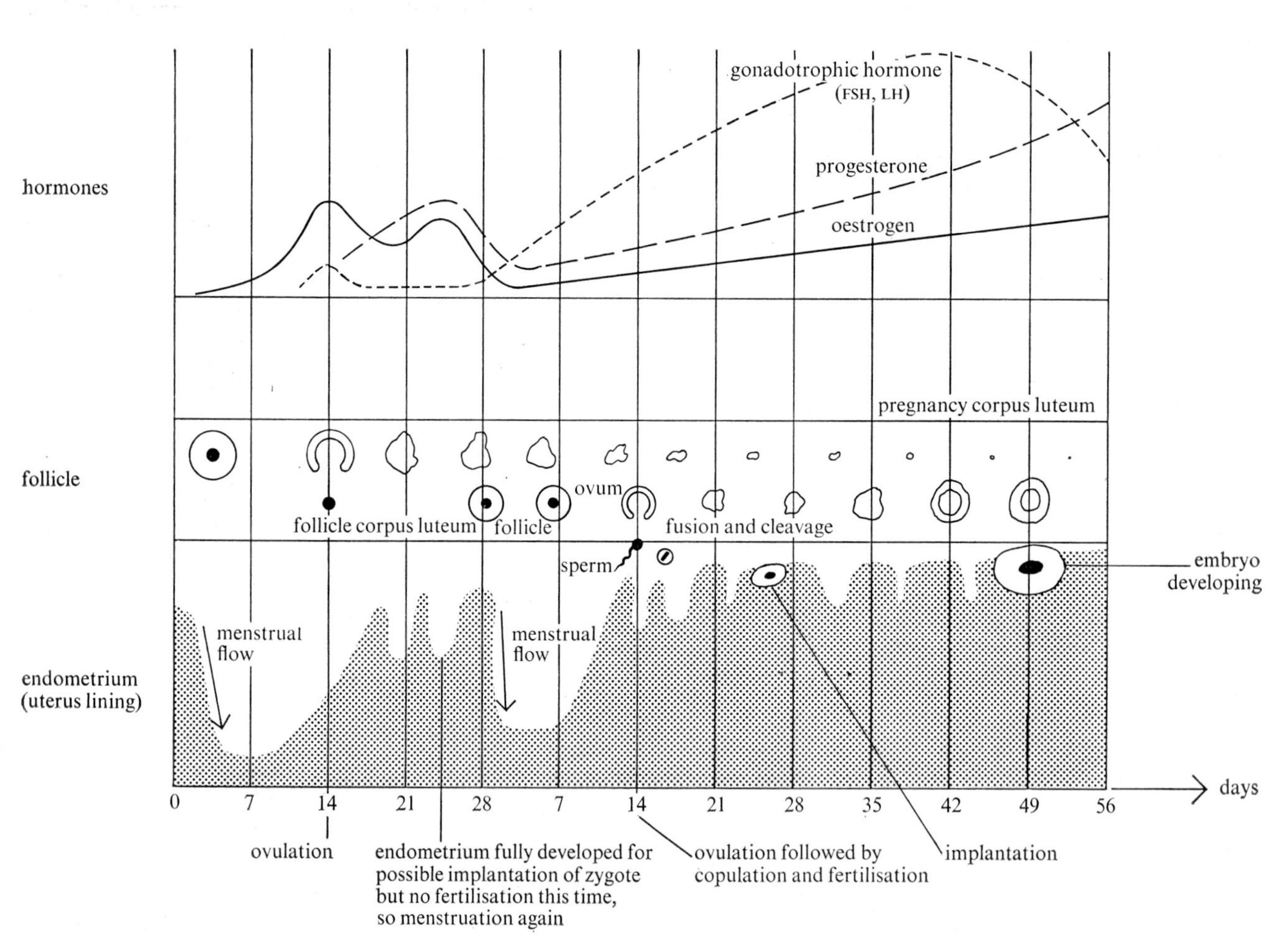

Fig 4.12 Menstrual cycle of human female

FERTILISATION This occurs after sperms have been ejaculated into the vagina during copulation (Fig 4.14). They travel to the uterus and fuse with an ovum in the oviduct. Further sperms are prevented from entry into the fertilised ovum by chemical changes which take place there. The ovum becomes implanted into the enlarged endometrium of the uterus and it acquires food from the uterus through special villi-like projections which have developed on it.

In the human species the female reaches puberty between the ages of 11 and 15 years, and the menstrual cycle begins. A complete cycle, followed by a cycle in which fertilisation takes place, is shown in Fig 4.12. Fig 4.13 shows the hormones involved.

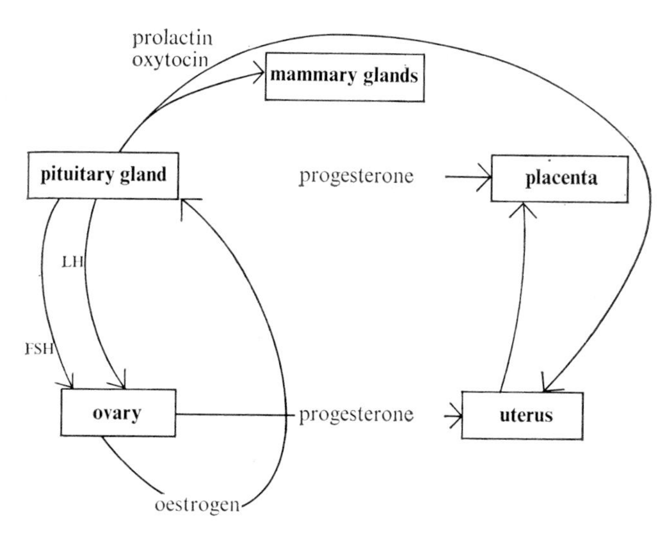

Fig 4.13 **Hormones involved in the menstrual cycle**

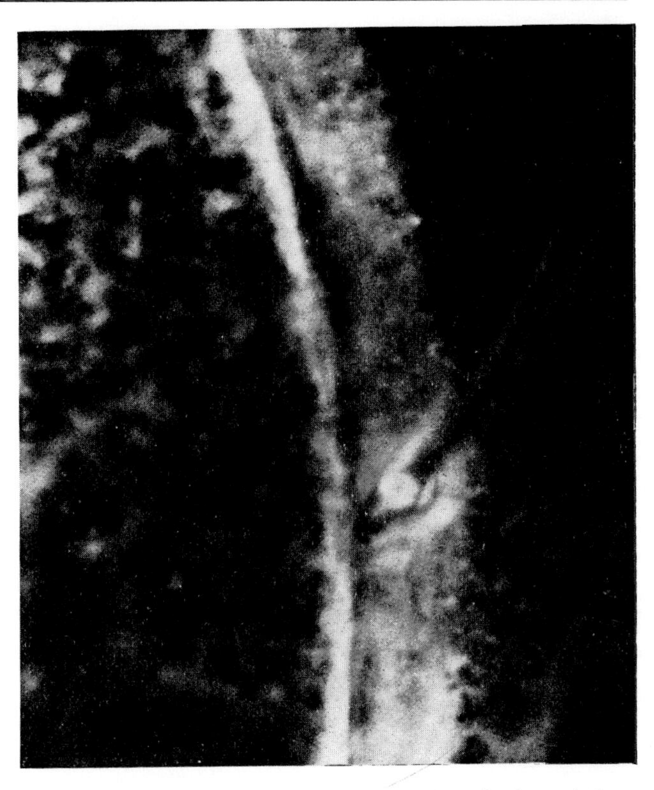

Fig 4.14 **A spermatazoon that has passed partly through the ovary membrane**

GESTATION The pregnancy lasts about 280 days in the human female, starting from the *beginning* of the last menstrual cycle. The implanted embryo develops in the uterus and, after twelve weeks, we can call it a *foetus* (Fig 4.15). It is surrounded by a pair of membranes, the *amnion* and *chorion*. The amniotic cavity (between the amnion and foetus) is filled with a water-like fluid which protects the foetus. As development proceeds, the *umbilical cord* forms a tube that contains a vein and two arteries which have developed from the *allantois* and *yolk sac*. The foetus gets its nutrients via the cord, and the pressure of blood there is high. The chorion surrounds the foetus and its other membranes and, at the end of the umbilical cord, it gets thicker and develops villi-like structures which embed themselves into the uterus wall. The *placenta* is made up from the thickened chorion, the villi-like structures, and the environment of the uterus in which the 'fingers' of the villi are embedded. The umbilical arteries of the foetus branch into many capillaries in the placenta, and these capillaries unite to form a vein through which

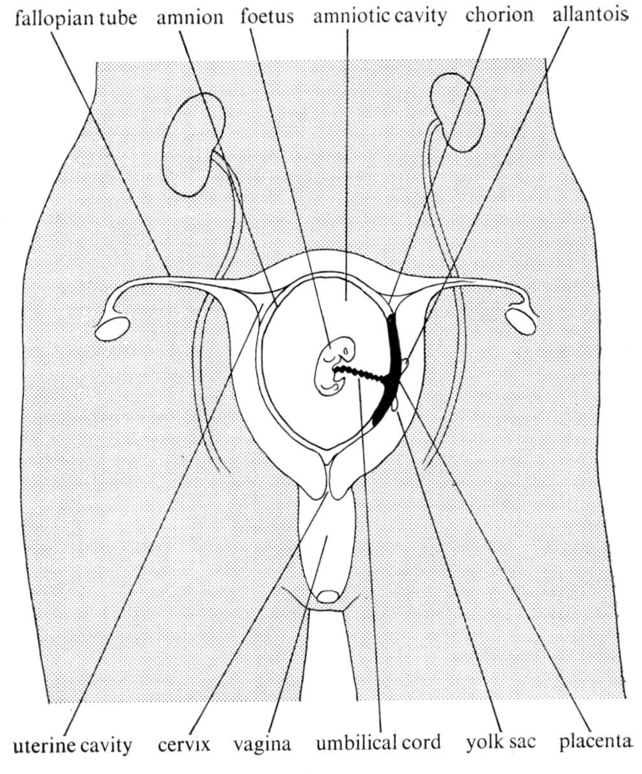

Fig 4.15 **Development of the foetus**

blood returns to the developing foetus. The blood supply of the mother joins the placenta via a capillary pattern coming from the uterine artery; the blood of the mother and her foetus do not mix, however, and the foetus has its own heart. The nutrients and oxygen from the mother diffuse from the capillaries of the mother to those of her foetus and waste of metabolic processes in the foetus diffuse back by a similar route to the blood supply of the mother. Hormones, such as LH are formed in the placenta and can be detected in the urine of the female early in pregnancy.

BIRTH Birth begins when the muscular walls of the uterus begin their rhythmic contractions (*labour*); oxytocin is the hormone that starts the contractions. As contractions get stronger, the amnion ruptures and the embryo is expelled from the uterus. The doctor severs the umbilical cord, that joins the baby to the placenta, and the baby begins to breathe in air after blood has reached its lungs. The placenta leaves the mother a short while later – and is called the afterbirth.

The mammals tend to care for their young for some time after birth, and the baby is fed on milk from the mother's mammary glands (breast).

4.5 Reproduction in other animals

FISHES Most fish release yolky eggs up-current of the feeding ground so that the larvae can drift back to feed. Pituitary hormones control the ripening of gonads, as usual. The fish has an elaborate behaviour pattern so that the gametes are released from both sexes at the same time. Fresh-water fish make nests for their eggs (why?), but marine fish (apart from the herring) have eggs which float on the surface. The dogfish shows internal fertilisation of the ovum, followed by secretion of the well-known 'mermaid's purse', a protective chitinous casing, around it. This purse is attached to objects on the substratum. In general there is no care for the young after hatching, an exception being the stickleback.

AMPHIBIANS Frogs reproduce in a water habitat (spawning ground). After courtship the male grasps the female with *nuptial pads* and fertilises her ova externally, the sperms fusing with the ova before the albumen around the latter swells to form the familiar spawn jelly. Both male and female frogs have a cloaca where eggs or sperms pass out; the male sheds his sperms from the testes, whence they travel down via the kidney and special ducts. Many eggs are laid, sometimes running into

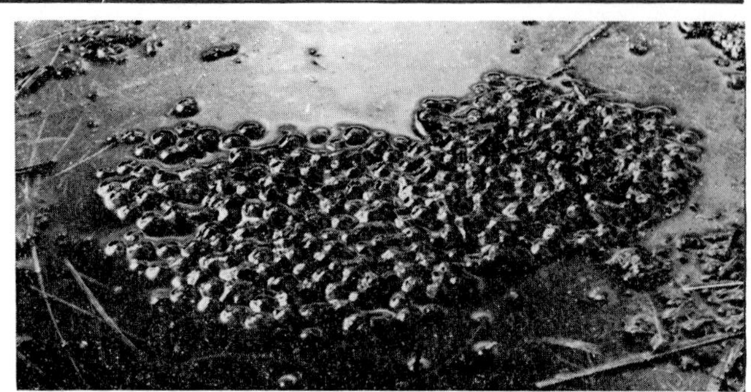

(a) The frogs eggs are invisible in the form of frogs spawn. The jelly round the eggs protects them from drying up or being swept away

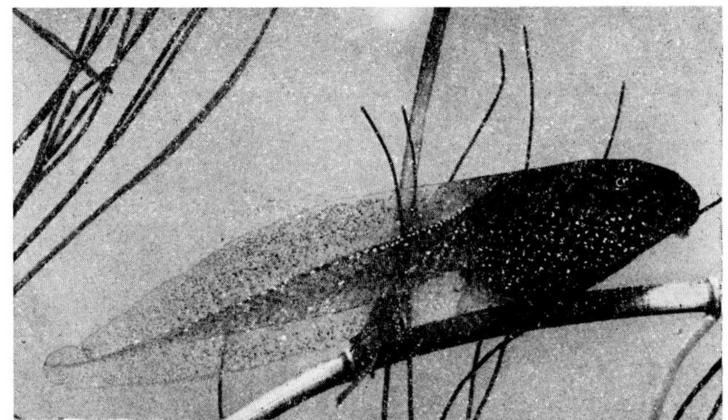

(b) The first stage of tadpole development is the development of the hind legs, at about 8 weeks

(c) At about 11 weeks, all the legs have developed, the eyes become prominent and the tail shortens as it is internally digested

(d) Fully developed young frog, which remains by the side of the pond

Fig 4.16 **Stages in frog development**

batches of thousands. On hatching, about two weeks after laying, the tadpole breathes dissolved oxygen and feeds on the yolk of the egg. After 2–3 days the tail lengthens, the mouth and external gills start to function and it feeds on pondweeds. After two weeks, and up to 3 months, there is a gradual change to a carnivorous diet and to breathing by internal gills. At this time, metamorphosis occurs to give the frog.

The tadpole does not taste pleasant to possible predators and this helps survival. Similarly the eggs, when laid in their jelly, have a black pigment. The development of the frog is shown in Fig 4.16.

BIRDS AND REPTILES The birds evolved from the reptiles and the eggs are similar. The eggs have an amnion membrane and an allantois – with properties of respiratory exchange and excretion. Snakes guard their eggs and turtles and crocodiles bury them. Reptiles show very little parental care. For birds, fertilisation is internal and the eggs can have a wide range of colours and shapes, the eggs are often camouflaged, especially if the nest is open (*e.g.* lapwings have khaki eggs to blend in with the rough ground whereas owls (closed nests) have white eggs). In general, the more advanced the bird the fewer eggs it lays. Birds' nests vary in size and complexity. Parental care and territorial behaviour are well-established.

INVERTEBRATES The details of the reproduction of some selected invertebrates have been given, *e.g. Amoeba* (p 42), *Hydra* (p 44), *Planaria* (p 46), *Lumbricus terrestris* (p 48).

4.6 Embryonic development in animals

It has been seen that the union of a male sperm and a female egg produces a fertilised egg, or *zygote*. The amount of *yolk* (food store) in an egg governs the development of the fertilised egg, including how it *cleaves*. Eggs of hens have large yolks, those of humans have very small yolks, while those of amphibians are intermediate.

AMPHIBIANS (*e.g.* frog) An egg of an amphibian has *polar bodies* (Fig 4.17) formed during *oogenesis*. The heavily pigmented part of the egg adjacent to them is the *animal pole*, whereas the lighter-coloured region at the other end is the *vegetal* (or *vegetative*) *pole;* in the water, the animal pole is uppermost. The yolk-density is greater in the region of the vegetal pole. After fertilisation, mitosis occurs after about 2 ks (30 minutes) and two cells are formed from the one-celled zygote. Further division then takes place to give four cells, and then eight,

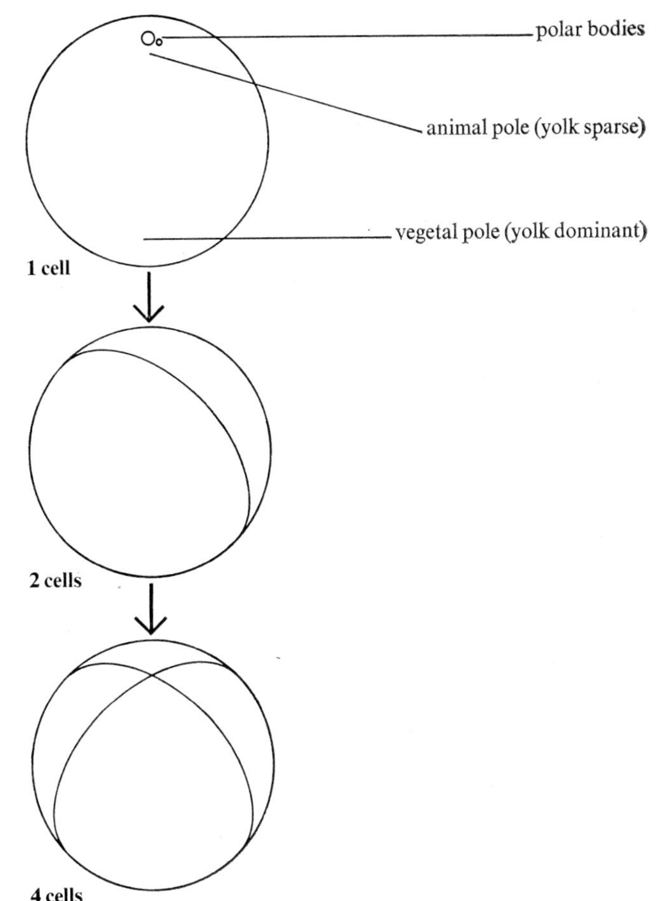

Fig 4.17 **First stages of cell division**

and so on. This period is called the *cleavage period* of development.

Blastulation follows. A hole (*cleavage cavity*) develops in the mass towards the animal-pole zone – the high preponderance of yolk in the other zone slows down cell-division there, with the result that cells in the animal-pole zone are getting smaller more rapidly. The embryo is called a *blastula* (Fig 4.18) when the cavity, or *blastocoele*, is well formed. The cells of the late blastula do not provide any positive clues to the functions that they will

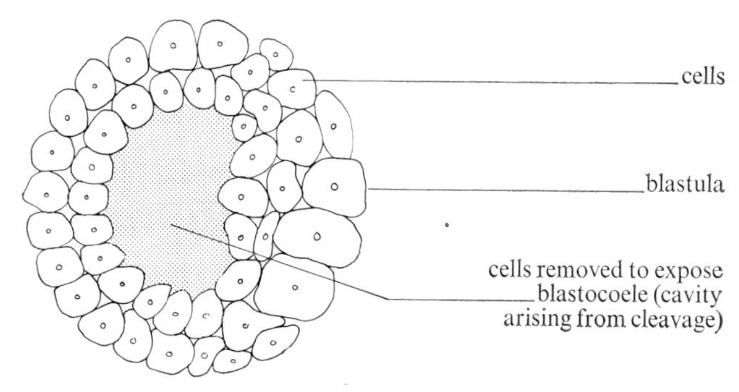

Fig 4.18 **Blastula**

play in future development. This is because there is little distinction between them apart from pigmentation, size and yolk content.

Gastrulation follows, in which there is a rearrangement due to movement of cells. This can be seen by colouring the cells with non-toxic dyes. A groove or *blastopore* (Fig 4.19) develops on the surface of the blastula, and is seen as a line of dye. The blastopore gets longer at each end and forms a semi-circle (Fig 4.19) and a movement of cells takes place inwards, forming a cavity called the *archenteron* (Figs 4.19, 4.20). Eventually, all the lighter coloured yolk cells are covered by pigmented cells (but one small 'plug' remains) and the embryo is called a *gastrula*.

To summarise then, gastrulation is due to three main processes that occur at the same time and is not due to the formation of new embryonic material. The processes are spreading of the cells of the animal half over the vegetal half, intucking of material around the margins of the blastopore and contraction of the lips of the blastopore.

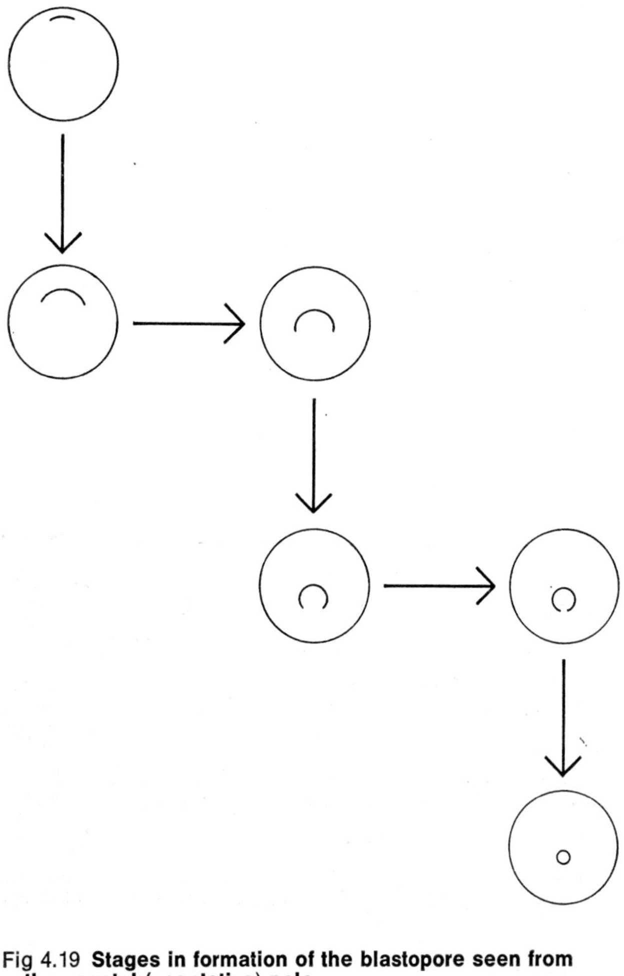

Fig 4.19 **Stages in formation of the blastopore seen from the vegetal (vegetative) pole**

If we were to cut a section through the plug of yolk, it would be found that there were three banks of cells:

ectoderm cells: outer cells of the gastrula.

endoderm cells: partially lining the archenteron cavity.

mesoderm cells: situated in the blastopore region.

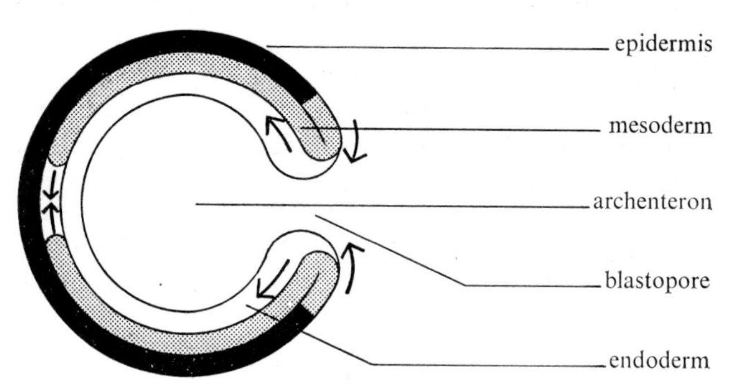

Fig 4.20 **Formation of the archenteron**

Definite parts of the body are developed from each cell bank.

Ectoderm: outer layer of skin, nerves, brain, spinal cord, eye.

Endoderm: lining of alimentary canal, secretory portions of pancreas, liver.

Mesoderm: muscles, reproductive organs, excretary organs, parts of skeleton, heart and circulatory vessels.

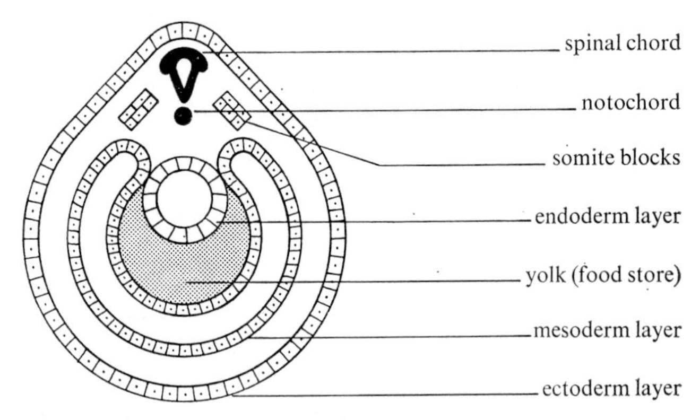

Fig 4.21 **Beginning of tissue differentiation**

Organ development(Fig 4.21) comes after gastrulation. The *notochord* arises from mesoderm cells dorsal to the blastopore. Other such cells migrate towards the centre and others form blocks called *somites*. The archenteron gives rise to the alimentary canal. Some ectoderm cells form what is called the *medullary* (*neural*) *plate* whose edges join together to form the *neural groove*, then (after fusion) the *neural tube*. The larger, anterior region of this tube eventually gives the brain, and the posterior region the

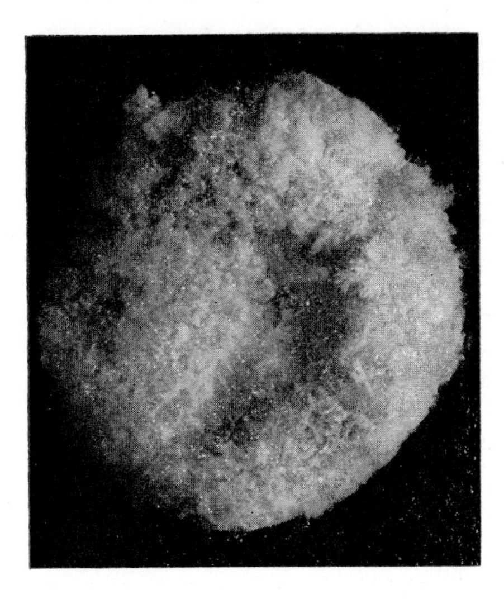

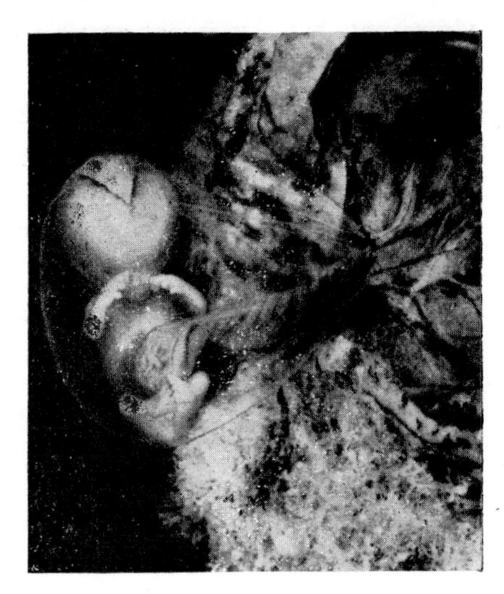

Fig 4.22 **Early stages in human embryonic development**

spinal cord. Cells from the inner edge of each somite block become removed and surround the notochord and neural tube, eventually to give rise to vertebrae. Mesoderm cells lying laterally to the somites give rise to the kidney system, and the formation is extremely complex. In the ventral section of the embryo, some of the mesoderm cells (in contact with the developing alimentary canal) give rise to a pair of thin tubes, which eventually fuse along their length to give a tube which will form the heart-lining. A dense muscular wall is formed around the tube, and thus a primitive heart is produced. Blood enters at the posterior region and leaves the anterior regions. The human heart is derived from this primitive tube-like heart; loops and partitions develop so as to divide the heart into four zones. Venous blood (ex lungs and body) goes in at the anterior region and arterial blood also leaves from this region. You will recall that, in the gastrula, the endoderm encloses the archenteron, and openings soon develop in the endoderm tube (these become the anus and mouth). The layers of cells of the endoderm extending from the mouth to anus develop into the lining of the alimentary canal and the muscular walls and connective tissues arise from the mesoderm. The liver starts off as a projection from the canal near the heart.

REPTILES, BIRDS AND MAMMALS The development here (Figs 4.22, 4.23) follows the pattern outlined for the frog. The amount of yolk governs the nature of blastulation and gastrulation. Since there is a great deal of yolk in the eggs of reptiles and birds, cleavage gives rise to a flat sheet of cells on the yolk (instead of the spherical unit of the amphibian). The flat embryo is called a *blastoderm*. *Extraembryonic membranes* are also formed. The shell and membranes are of vital importance in enabling the embryo to survive on land rather than in an aquatic environment—in the water, excretion, gas-exchange, etc. can be carried out directly between the cells and their en-

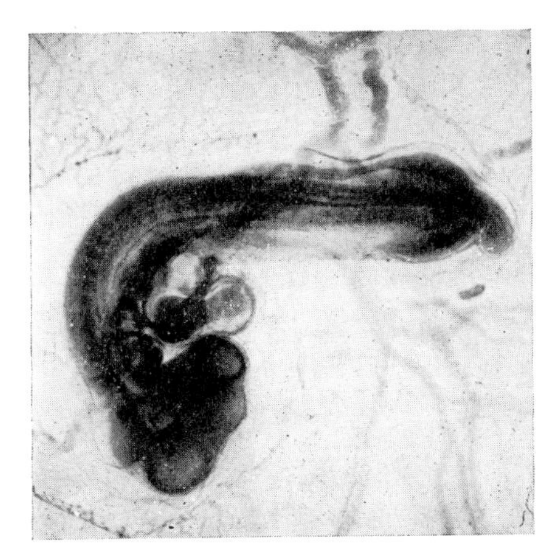
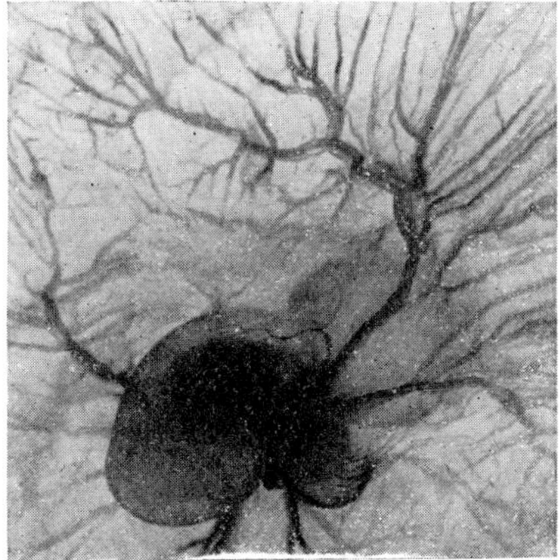

Fig 4.23 **Stages in the development of the chick at three, five and eight days gestation**

vironment, but this is not possible on land. There are four membranes (Fig 4.24):

Amnion: encloses the embryo and contains *amniotic fluid*; protects embryo from injury and stops it from getting too dry.

Chorion: lies just inside shell, surrounds all other membranes (and embryo).

Allantois and Yolk-sac membrane: arise when the intestine is developing. The yolk-sac membrane (with its blood vessels) surrounds the yolk; cells in its walls digest (enzymically) the yolk and the blood vessels carry the nutrients to the embryo. The allantois fuses with the chorion under the shell. Blood vessels, called allantoic vessels, enable a gas-exchange to take place between the environment and the embryo. The fused membrane also serves in excretion, waste materials from metabolism of the embryo being deposited in the hollow part of it.

When the shell of the egg is broken the membranes become attached to the shell rather than accompanying the young creature.

THE HUMAN EGG CELL (Fig 4.22) After fertilisation, this becomes implanted in the wall of the uterus. The embryo there passes through a type of gastrulation, and the connection between the embryo and the wall is reduced to an *umbilical cord;* at the uterus the tissues of the embryo and uterus are merged into a *placenta.* The placenta enables exchanges (fluid, gas) to take place between mother and embryo. The extraembryonic membranes are present; the allantois and yolk sac help to form the umbilical cord; blood vessels corresponding to allantoic vessels link the embryo and mother. The period of development is about forty weeks and, during weeks 1—4, the embryo is similar to those of other mammals. Human-being characteristics appear in weeks 4—8.

4.7 Organizers in embryonic development

In the organism, *organizers* of activity control growth, differentiation and development. Activity of the nucleus, mitochondrion, Golgi-apparatus, etc, may well control differentiation. Much work has been carried out on the relationship between DNA and proteins of chromosomes and the control of development by genes. From a young embryo (*e.g.* newt or frog) it is possible to take a piece of ectoderm from a zone that will, in the normal way, yield epidermis, and transplant it to a new site. It is then possible to carry out specific tests to find out what has happened to the cells present as a result of the change. It has been found that the dorsal 'lip' zone of the blastopore (future notochord and axial mesoderm), when transplanted into

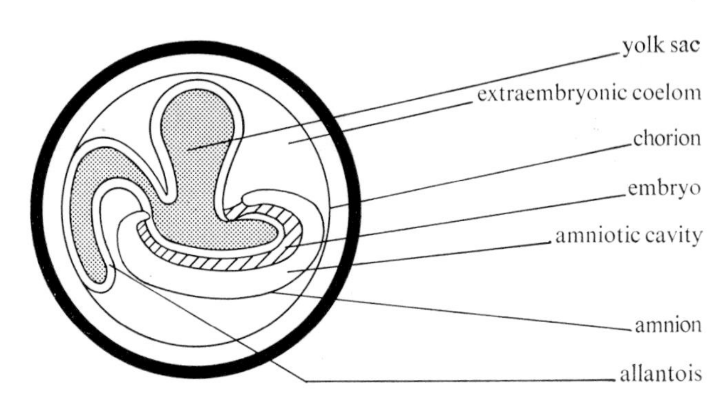

Fig 4.24 **Membrane and embryo**

any other part of a gastrula in an early stage of formation, gives a complete embryo, *i.e.* the dorsal lip of the blastopore is a *primary organizer.* In the development of a frog embryo, it has been shown by this method that, during gastrulation, contact between the notochord and the ectoderm immediately above it gives rise to the medullary plate.

It can be concluded, then, that the tissue invaginated at the dorsal (and to a lesser degree lateral) lip of the blastopore plays a predominant part in the organisation of other cell regions in the embryo. It is known that although no rigid boundary of demarcation exists, the cell area close to the dorsal lip acts as a 'head organiser' while the area close to the lateral lip acts as a 'tail organiser'.

The main embryonic structures, then, are formed under the influence of primary organisers. An organiser coming into operation later is the tissue of the optic cup – this causes the ectoderm lying above it to form a lens. In *Rana temporaria* it has been found that the optic cup will produce the lens normally when grafted into foreign situations (*self-differentiating*). If the optic cup is removed, however, no lens can develop; the lens is a *dependent-differentiating* structure as it needs the optic cup to act as an organiser before it can form.

Organisers act by chemical means; pieces of prospective epidermis, after temporary contact with the dorsal lip of the blastopore, can induce the formation of a neural plate from prospective epidermis. Furthermore, many properties of the organizer can be mimicked by chemical substances, *e.g.* steroids, alkanoic acids, dyes. Only after killing do ordinary tissues acquire the ability to produce a neural plate from prospective epidermis – *evocation.* In chick embryology, the primary organizer is situated at first in the endoderm, which induces a *primitive streak* above it in the mid-line. The primitive streak is an accumulation of cells, seen on the surface of the blastoderm as a dark streak. A swelling of the anterior end of the streak is called the *primitive knot.*

Practical work

As regards practical work on embryology, chick embryos can be prepared as described in any suitable textbook of practical zoology. Slides of chick embryos at various stages of development should be studied and drawn.

4.8 Growth and form

Any change in size in multicellular organisms results from cell division and cell expansion. When cells expand there is no reorganisation of tissues but when they divide, there may be. Thus it is difficult to separate growth (increase in size) from development (differentiation of tissues).

If you plot the height of an organism against time (Fig 4.25) at first the increase in height is small, then it accelerates during the *grand period* of growth, then tapers off as the organism reaches full size. The curve which is sigmoid in shape is often called a *growth curve* but it is actually an *accumulated* growth curve (statistically an ogive or cumulative frequency curve). It is usually best to plot the logarithm of the measured factor (height, volume, mass) against time. This has the effect of 'straightening out' the initial part of the curve; curves obtained for different organisms and sets of conditions can then be more easily compared. This pattern of growth holds for the seasonal growth of plants as well as for animal growth, but the volume and mass of deciduous perennial plants is altered annually by leaf fall.

FORM Huxley worked on the analysis of form in terms of differential growth (allometry). His formula

$$y = bx^k \text{ i.e. } \log y = \log b + k \log x$$

relates to the magnitude (x) of one part of an organism and that (y) of another part; b is a constant ($= y$ when $x = 1$) and k gives the ratio of the growth between the parts. If $k = 1$ then the two parts compared have the same rates of growth. Any departure from 1·0 means that there is a change in form since the rates of growth are different. Huxley showed that, for the mouse, the length of its tail increased by about 1·4 times the whole length of the mouse. Rates of growth of parts of the human body and trees, have been studied in a similar way.

Both at the cell and organism level, there is a typical rate of growth and a maximum size for any species, which is not exceeded. It may be reduced by adverse environmental factors. Animals assume a definite shape and size related to their mobile behaviour and ecological requirements. Plant growth is more plastic, responding to the local environment dictated by their static way of life.

4.9 Plant growth

MERISTEMS Plant tissues originate from regions called meristems, which are aggregates of dividing cells. They initiate growth and differentiation, which is continued by expansion of the new cells. They may be *primary* or *secondary*. Primary meristems are *apical*, i.e. terminal to main or lateral branches in shoots or roots, or intercalary, as in the lower part of the internodes of grasses. Secondary meristems give lateral expansion, increasing the girth of root or shoot tissues that have stopped lengthening.

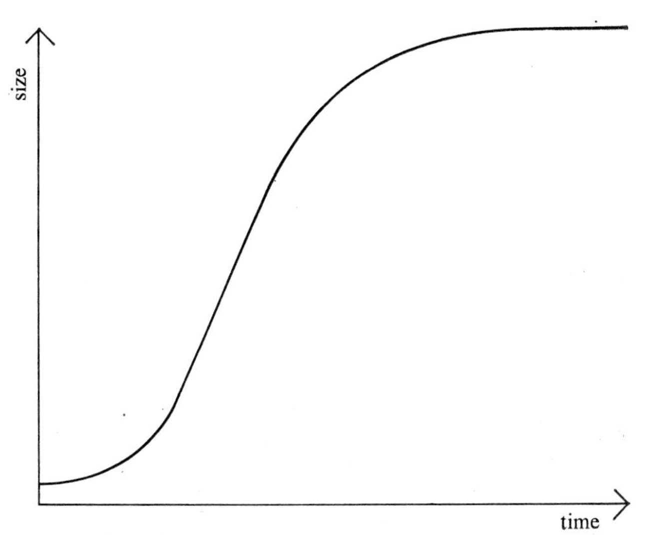

Fig 4.25 **Growth curve**

ROOTS A typical growing root (Fig 4.26) has a *root cap*, which protects the delicate meristem from abrasion by

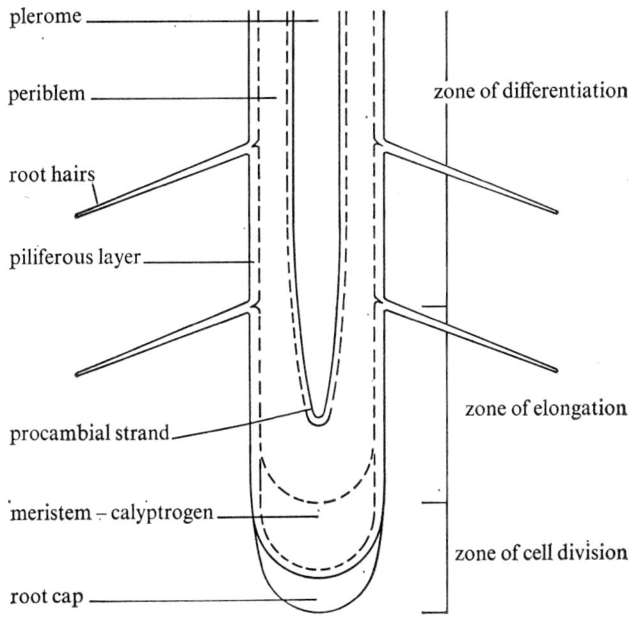

Fig 4.26 **Growing root**

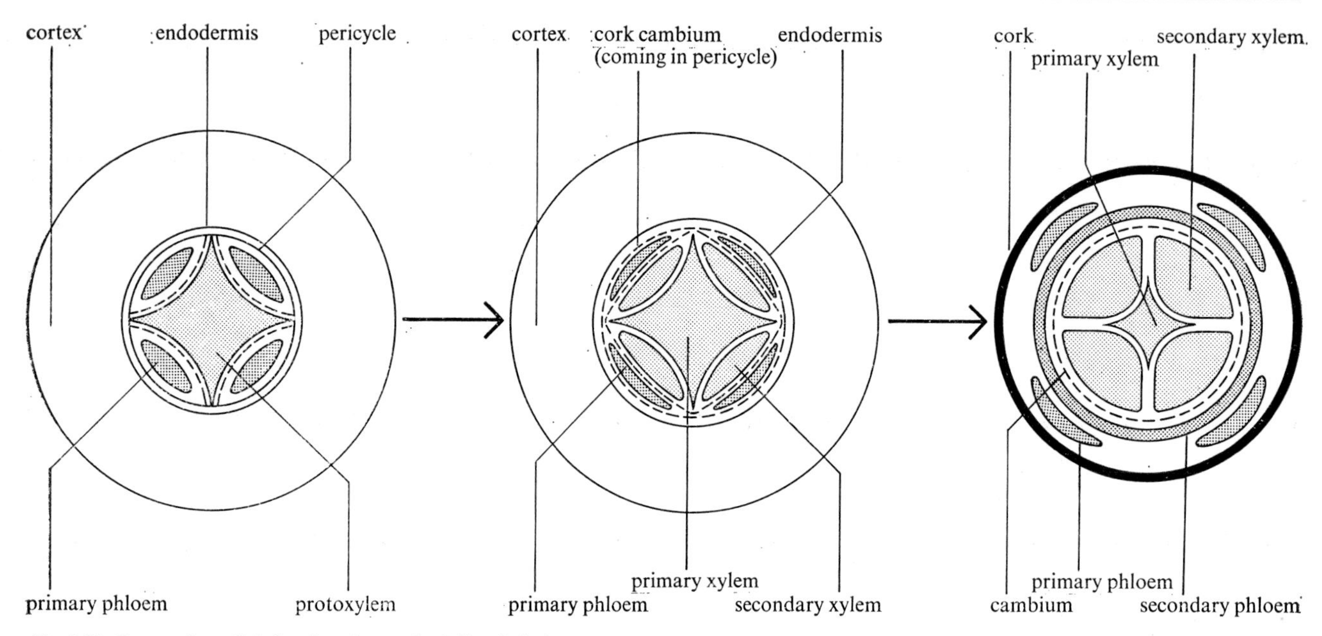

Fig 4.27 Secondary thickening in root of dicotyledon

soil. Behind it the meristem has histogenic areas called calyptrogen, dermatogen, periblem and plerome, which give rise, respectively to root cap, piliferous layer, cortical ground tissue, and the central cylinder or stele which contains vascular tissue (phloem and xylem cells grouped together) with associated tissues.

A procambial strand is also present, from which the cambium will later develop. As the root ages, the hair region dies and secondary thickening (Fig 4.27) proceeds, with new xylem and phloem structures forming, sloughing of the cortex, and formation of a protective cork tissue from the cork cambium.

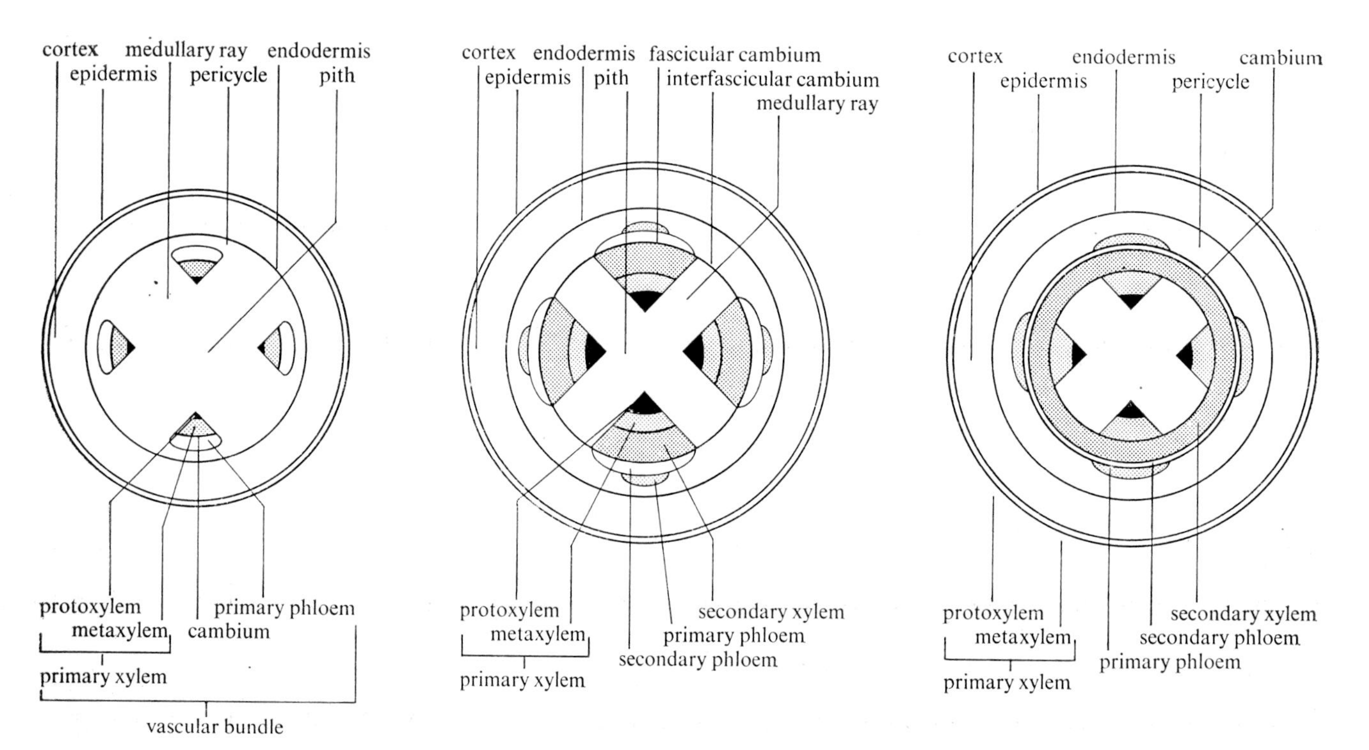

Fig 4.28 Secondary thickening of stem

STEMS There is similar meristematic activity in stems. Histogenic areas give rise initially to epidermis, cortex and stele, within which the vascular tissues form, and the pith. Further differentiaton gives rise to the tissues shown in Fig 4.28; note that the vascular tissues occur in bundles and that strengthening tissues (sclerenchyma, fibres) are present. Further differentiation continues as illustrated, the inter-bundle tissue carrying radial medullary rays. Later stages for the gymnosperm *Pinus sylvestris*, are shown in Fig 4.29.

The *growth in girth* of the stem of a tree or shrub starts with the formation of *cambial cells* across each *medullary ray*. The layer of cambium forms internal and external cell layers, the internal ones being transformed into xylem and the outer to phloem. The thickness of these rings increases as growth continues. Throughout the life of the plant, new xylem and phloem rings are formed but the rate is, of course, uneven throughout the year (different amounts of water are required at different seasons). The different nature of the autumnal and spring xylem forms the *annual rings;* the age of a tree can be ascertained by counting the autumn or spring rings. It should be realised that only the new secondary xylem can conduct water, and we call this *sap wood* (alburnum) to distinguish it from the older *heart wood* (duramen). The latter has ceased to function and has become a storage place for substances such as resins and tannins which give it a characteristic dark colour.

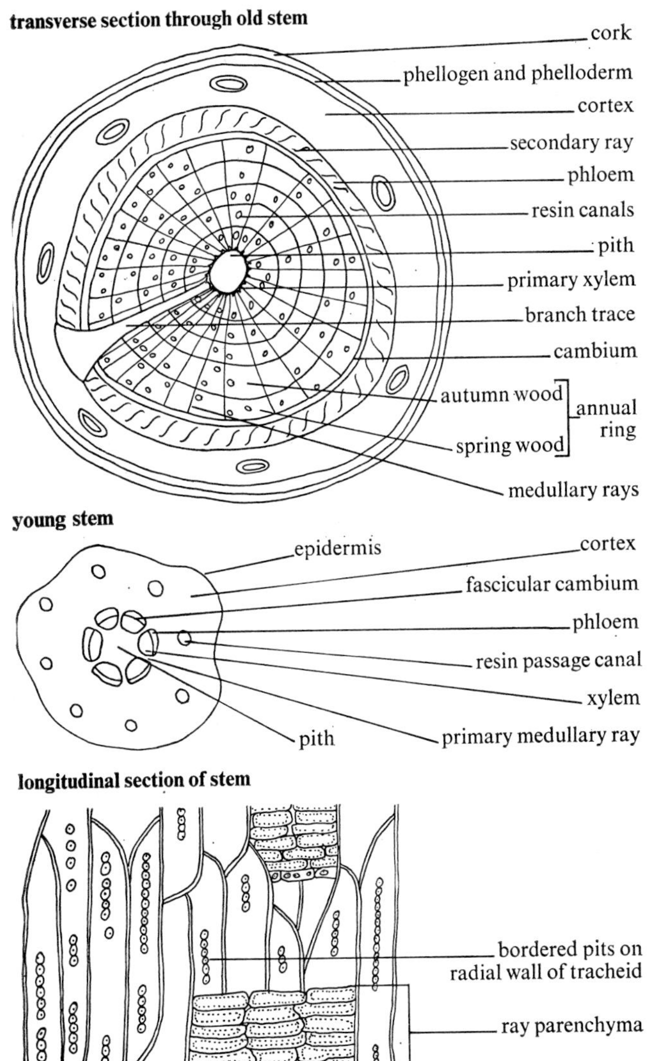

Fig 4.29 **Pinus sylvestris stem structure**

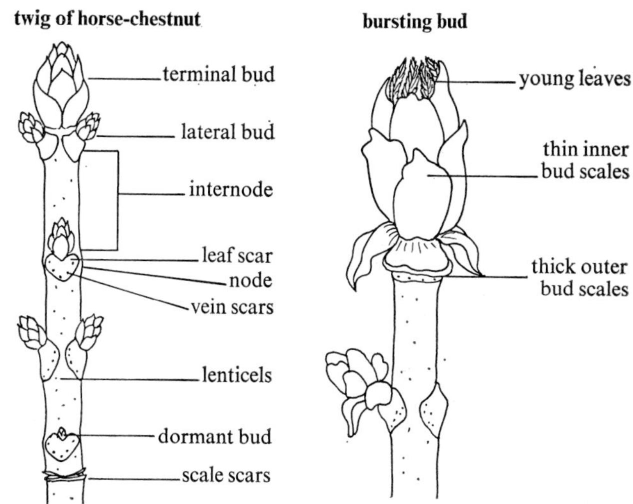

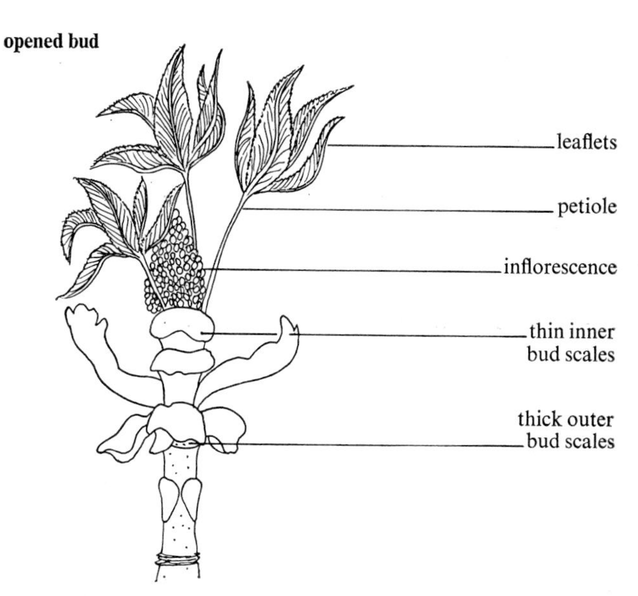

Fig 4.30 **Spring growth of horse-chestnut**

ANNUAL CYCLE After the winter, a woody twig (Fig 4.30) starts to grow again; the foliage leaves swell, the scales are forced apart (and eventually drop off) and a bud ring or girdle of scars is left round the stem – it is possible to discover the age of a tree by counting the girdles. In autumn, food in the leaves enters the stem and roots for storage, chlorophyll is degraded, the leaves assume an autumn tint and fall following the formation of an *abscission layer;* a plate of parenchymatous cells at the base of the petiole (containing dense protoplasm) separates and becomes round, so that the leaf falls off. The exposed surface then develops a protective layer of cork cells from the phellogen, the periderm becoming continuous with the stem. The leaf-scar can be seen.

When growth of animal cells develops without control, and without the correlation of development of different tissues that is present in normal growth, *cancer* may well be involved. In cancers, abnormal numbers of cells are formed and these do not have the usual characteristics of the tissue from which they come. Parallel tissues in plants are *galls* (*e.g.* big buds on blackcurrant, nail galls on sycamore) though these are usually caused by insects or mites.

5 Genetics

5.1 Introduction

In genetics we study *heredity*. The inheritance of characters is described in terms of units called *genes*. The gene is the unit that controls or appears to control a distinct character of the whole organism; the gene is the unit in the chromosome that controls the production of one polypeptide chain, and the characteristics that the geneticist looks at depend upon the activity of that polypeptide.

The genetic constitution of an organism, *i.e.* the total number and kinds of genes encoded into the DNA of a nucleus (see p 30) is described as the *genotype* of the organism (or nucleus). The genotype refers to the information encoded in nuclei, but a description of an organism is usually based on visible, or otherwise detectable, distinct characters; the characters exhibited by an organism are referred to as its *phenotype*.

We have seen that in *mitosis* the parental chromosomes duplicate to form identical copies or *chromatids*, one going to each new cell, the new cells being therefore genetically identical to the cell from which they came. A series of organisms formed mitotically from an individual is called a *clone*, and each member would be potentially identical as regards appearance, growth, etc. Experiments on bacteria using T4 'phage have shown, however, that all individuals of a clone are not necessarily identical, since a mutation can occur, this being a change in the genotype. Anything which changes the sequence of nucleotides along the DNA molecule can cause mutation; experimentally hydrogen peroxide, urethane, manganese (II) chloride, mustard gas, nitric(III) acid have been shown to be mutagens; they act by breaking the DNA molecule or by substituting single bases. Colchicine prevents formation of mitotic spindles. Abnormal temperature may also cause mutations (*e.g.* of *Drosophila*, the vinegar fly). Ionising radiation can also cause them (X-rays or accelerated radiation from a radioactive source, which break the DNA molecule so that substitution may occur).

It is likely that the main cause of natural mutations is chemical. The mutation rate in *Escherischia coli* to T4 'phage resistance is about one per hundred million. In humans, the mutations causing *haemophilia* (impairing of blood clotting) arise about 25 times in every million individuals. Those for giving achrondroplastic dwarfism happens about 80 times per million. When two haemophiliac genes come together in a female, they cause early death of the foetus; such genes are known as *lethal genes*.

Geneticists study the variation in key characters of the species that they are working with. Ideally it is best to study large progenies from parents of known genotype. In man, this is rarely possible but blood groups, fingerprints, eye colour, hair texture and colour, reflex times, ability to roll the tongue, and ability to taste PTU (phenyl thiourea) are all characters that have been studied. Geneticists breeding improved varieties of crop plants may measure such characters as dry weight production, seed number, seed size, fruit quality, and resistance to pests and diseases. You should remember that the variation that is observed is the variation in the phenotype. The genetic material that is handed down from parent to offspring is one factor that is involved, but there also are factors that affect the phenotype from outside; these are called *environmental factors*.

Blood groups are interesting. If blood plasma is allowed to separate from blood by clotting, the liquid formed is the serum. Plasma and serum contain proteins called antibodies, two of which are anti-A and anti-B (A and B are blood proteins). Anti-A causes protein A to agglutinate (form visible red specks) and anti-B does the same to B. Persons with A in red cells could not have anti-A in the plasma; in fact they have anti-B. Those with B red cells have anti-A in the plasma. Those in group O (having neither proteins A or B) have both anti-A and anti-B. Those in group AB have neither anti-A nor anti-B.

Table 5.1 **Summary of blood groups**

blood group	*on wall of red-cell*	*antibody in plasma*
A	A	anti-B
B	B	anti-A
AB	A + B	neither anti-A nor anti-B
O	neither A nor B	anti-A + anti-B

Practical work with blood groups

1 Clean the fleshy part of your second finger with a pad of cotton wool soaked in ethanol.
2 Take one quick jab into it with a sterile lancet.
3 Place a drop of blood at the end of 3 slides which have been labelled as shown in Fig 5.1.

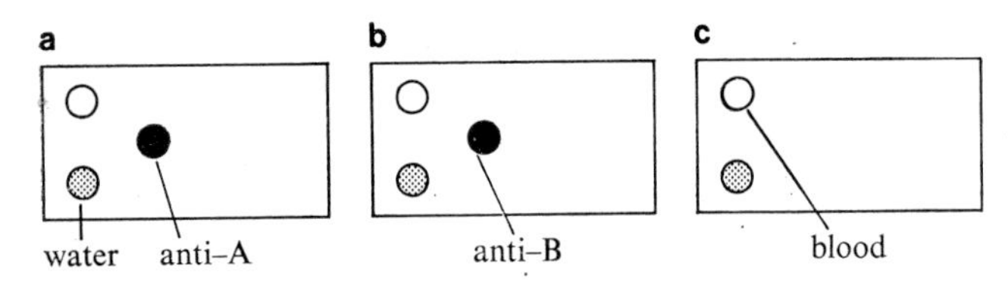

Fig 5.1 **Effect of antibodies a and b; (c is control)**

4 Add a drop of deionised water to each drop.
5 To slide A add one drop of the substance anti-A.
6 Using a *clean* glassrod or dropper, add a drop of anti-B to slide B.
7 Mix the drops on each slide with a separate stick.
8 Observe what has happened to the drops after about 300 s. Rock the solutions back and forth to mix them.

5.2 Mendel's Work

Gregor Mendel (1866) carried out experiments that gave results which were the foundation-stone of the science of genetics. He used the pea plant, *Pisum sativum*, which is self-pollinated. He had the following requirements:

1 The plants must possess constant differentiating characters, *e.g.* some tall, some short and no overlaps. The word 'constant' means the plants could be grown for generations with tall plants always giving rise to tall, etc. A group behaving in this way is called a *pure line* and is said to *breed true*.
2 The experimental plants used for *hybridisation* (making crosses) must, during flowering, be protected from foreign pollen.
3 The *hybrids* and their offspring should suffer no marked disturbance in fertility in any following generations – a good percentage of the seeds should germinate and grow on.

The principal characters he studied were:

size of plant	tall	short
colour of cotyledons	yellow	green
seed coat (*testa*)	round	wrinkled

Mendel made experimental crosses, using artificial hand pollination, between parent plants with these contrasting characters. The seeds produced all gave rise to plants with the characters tall, yellow, round. The firs generation from a cross like this one is known as the *first filial* or F_1 generation. Mendel then allowed the hybrid plants to self-pollinate. The resulting plants of the *second filial* or F_2 generation segregated tall:short, yellow:green, round:wrinkled in the ratio 3:1. These results can be summarised as in Table 5.2.

Table 5.2 **Mendel's results**

parents	tall × short	round × wrinkled	yellow × green
F_1	all tall	all round	all yellow
F_2*	3 tall, 1 short	3 round, 1 wrinkled	3 yellow, 1 green

*from self-pollination of F_1 hybrids

From these results Mendel arrived at his *first law of inheritance* which can be framed:

An organism's characteristics are controlled by genes that occur in pairs. Only one gene of the pair is carried in a single gamete.

We now know that it is the genes that control the expression of characters. A gene normally exists in two forms at a particular *locus* (the site on the chromosome that the gene occupies) and these two forms are called *alleles*. Individuals showing contrasting features (*e.g.* tall, short) are called *allelomorphs*. Thus the gene that affects height in the pea can exist in two forms, designated T and t. We can now represent Mendel's results in terms of the gene content of the gametes and zygotes as in Fig 5.2.

An alternative method of representing the generation is shown in Fig 5.3.

It is now clear why the parents breed true when self-fertilised. Both their tallness alleles are similar in type

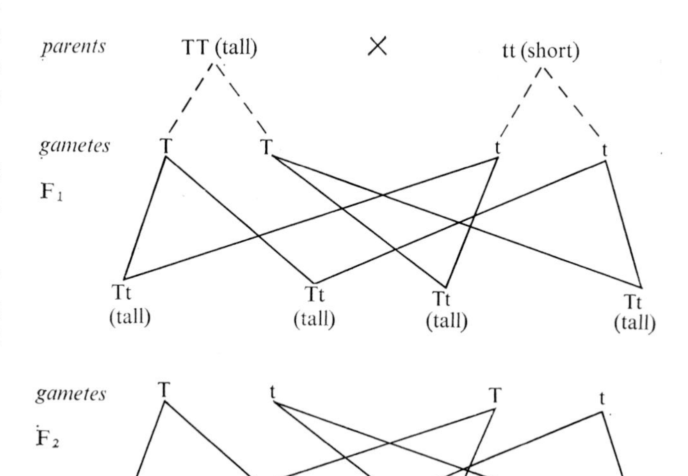

Fig 5.2 **Mendel's experiments**

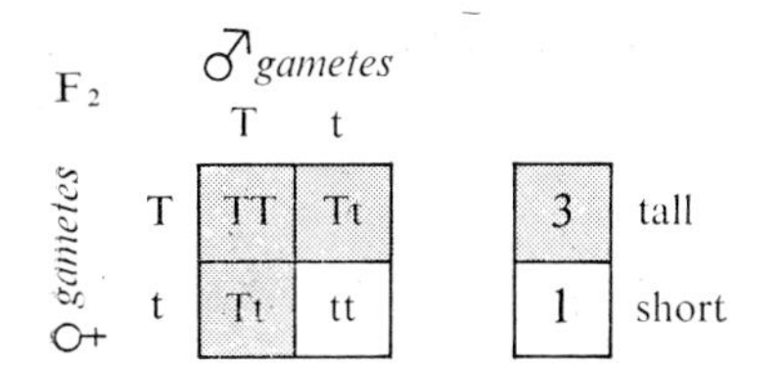

Fig 5.3 **Alternative representation of results**

and the plants are either TT or tt. They are said to be *homozygous* for T or t, and a *homozygote* is a cell or organism having two identical alleles of a gene. A TT plant can never give any short offspring when self-fertilised and a tt plant will never give tall ones. The F_1 plants and two out of the four possibilities in the F_2 contain two different tallness alleles, Tt, and are said to be *heterozygous*. A *heterozygote* is a cell or organism having two different alleles of a gene in its genotype. The heterozygote, Tt, is tall in phenotype. This is because the gene T dominates gene t; *dominant* genes are those which manifest their character when present in the heterozygous condition. *Recessive* genes only manifest their character in the absence of dominant genes, and thus *only* in the homozygous condition. The heterozygote, Tt, can give two different types of gametes, containing T and t in equal numbers. Half the ovules contain T nuclei and half t, and so also for the pollen grains. When self-fertilised, half of each type of pollen grain falls on a stigma containing T and half on a stigma containing t, giving rise to the segregation shown for the F_2 in Fig 5.2.

All the phenotypically tall plants from the F_2 are not identical genetically, one third being homozygous TT and two thirds heterozygous Tt. One way of finding out which is which is to cross them with a short plant. This type of cross, between heterozygote/recessive homozygote, is called a *recessive back-cross*. The TT homozygotes give all tall offspring in this cross (Fig 5.4) whereas the Tt heterozygotes give a 1:1 ratio of tall to short offspring (Fig 5.5).

The various ratios 3:1, etc. are ideal. In practice, the ratios among the offspring from crosses are the results of chance events in fertilisation. The more offspring that are counted the more nearly the experimental results fit these exact ratios.

Mendel's second law of independent segregation can be stated as follows:

Each member of a pair of alleles can combine at random with either allele of another pair.

The experimental results observed by Mendel are set out in Table 5.3, and Fig 5.6 shows the genetic interpretation.

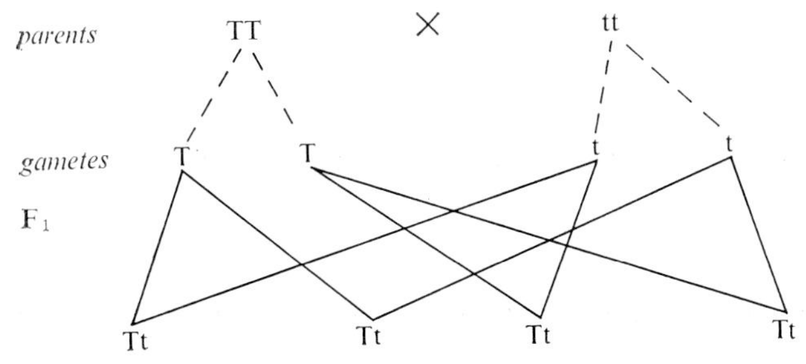

Fig 5.4 **Homozygous tall crossed with homozygous short**

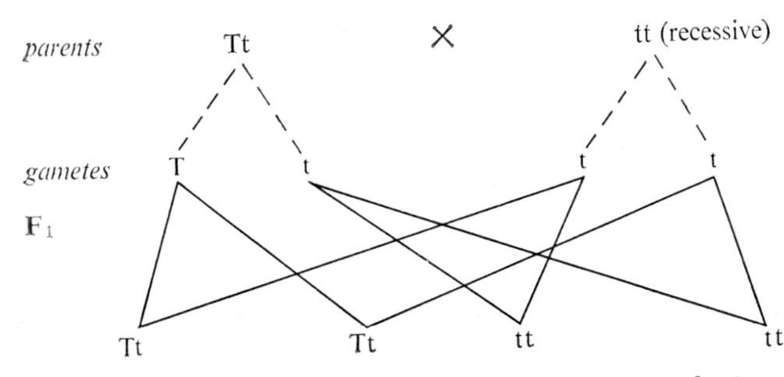

Fig 5.5 **Heterozygous tall crossed with heterozygous short**

Fig 5.6 **Genetic interpretation of results**

♀ gametes \ ♂ gametes	RY	Ry	rY	ry
RY	RRYY	RRYy	RrYY	RrYy
Ry	RRYy	RRyy	RrYy	Rryy
rY	RrYY	RrYy	rrYY	rrYy
ry	RrYy	Rryy	rrYy	rryy

9	round and yellow
3	round and green
3	wrinkled and yellow
1	wrinkled and green

Table 5.3 **Mendel's results**

parents	round testa, yellow cotyledons × wrinkled testa, green cotyledons
F_1	all round and yellow
F_2*	9 round and yellow 3 round and green 3 wrinkled and yellow 1 wrinkled and green

* from self-pollination of F_1 hybrids

Incomplete or partial dominance can give anomalous results as far as Mendel's work is concerned. Neither allele is dominant, giving F_1 offspring phenotypically intermediate between the parents and an F_2 segregating out some intermediates. A familiar example is the pink snapdragon (*Antirrhinum orontium*) produced by a cross of red and white parents (Fig 5.7).

VARIATION CAUSED BY MEIOSIS We know that specialised reproductive cells or gametes fuse in *fertilisation* to form a *zygote*, which has a single nucleus containing the

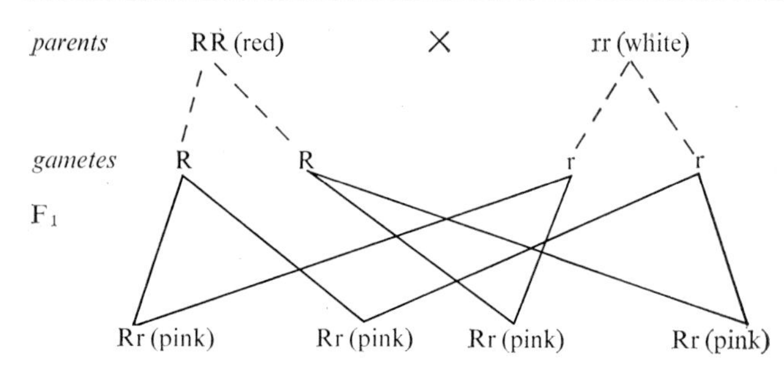

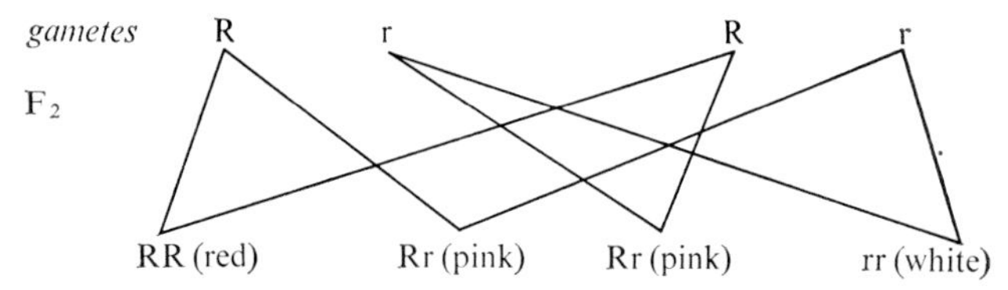

Fig 5.7 **Red and white antirrhinums crossed to produce pink**

genetic information from the two gametes. Production of gametes in the gonads is *meiotic* and the zygote then grows by *mitotic* division. One specialisation the gametes must all have in common is that they shall carry only half the normal (or *diploid*) chromosome number of the adult, and they are said to be *haploid*. Thus, in man the eggs or sperms have 23 chromosomes instead of 46, in *Drosophila* 4 instead of 8, in maize 10 instead of 20, etc. Meiosis can be regarded as a specialised form of mitosis in which the chromosome number is reduced to half in the daughter cell. In a diploid organism the chromosomes are in *homologous pairs*, *i.e.* two similar chromosomes in the nucleus, containing genes which determine the same proteins (and so phenotype), one inherited from each parent.

Drosophila (vinegar fly) is diploid and used widely in experimental genetics. There are four homologous pairs of chromosomes. On each chromosome pair there are many genes, giving rise to characters, *e.g.* grey body versus ebony body, normal antenna versus the mutant aristopedia antenna, normal wings versus vestigial wings, normal eyes versus sepia eyes.

LINKAGE AND CROSSING OVER When two characters are carried on the same pair of chromosomes, *e.g.* the gene for sepia eyes is on the same pair as that for vestigial wings, the phenomenon of *crossing-over* is observed. If genes are close together on the chromosome they may not segregate independently at meiosis. The further apart the two genes are, the more likely they are to become separated by crossing-over. When genes do not segregate independently they are said to be *linked* or to show *linkage*. When they do segregate, crossing-over has occurred and the frequency of this occurrence is the cross-over value.

Fig 5.9 illustrates results like those observed when normal eye, normal wing *Drosophila* are crossed with sepia eye, vestigial wing (Fig 5.8) *Drosophila*. Genes controlling the wing are located at B, those controlling the eye at C. Cross over is more frequent for distant genes (AB) and less frequent for genes which are near each other (BC). Percentage frequencies of cross-over shown are proportional to the relative distance between genes.

Fig 5.8 **Drosophila; left vestigated, right normal wing**

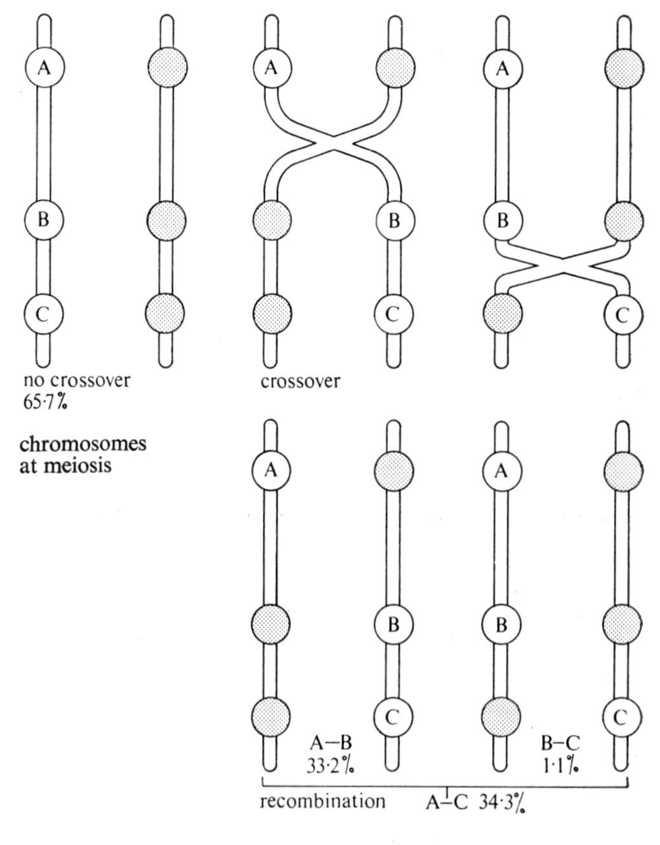

Fig 5.9 **Chromosome diagram showing linkage**

POLYPLOIDY The effect of mutation is to produce entirely new genes. There are other factors giving rise to genetic variations, one being the production of individuals with more than the normal set of chromosomes. These are called *polyploids* and though they seldom survive in animals, many very important crops are polyploid versions of a normal ancestor. Autopolyploids are individuals of a species whose chromosome number has become some multiple of the haploid number other than 2 (*e.g.* tetraploids have $4n$ chromosomes). Colchicine induces polyploidy, being a drug that affects spindle formation during cell division (this stops the chromosomes from separating effectively).

Allopolyploidy is the term used for a hybrid of two species of plant that has doubled its chromosome number. This often enables it to undergo sexual reproduction (Fig 5.10), e.g. *Primula kewensis.*

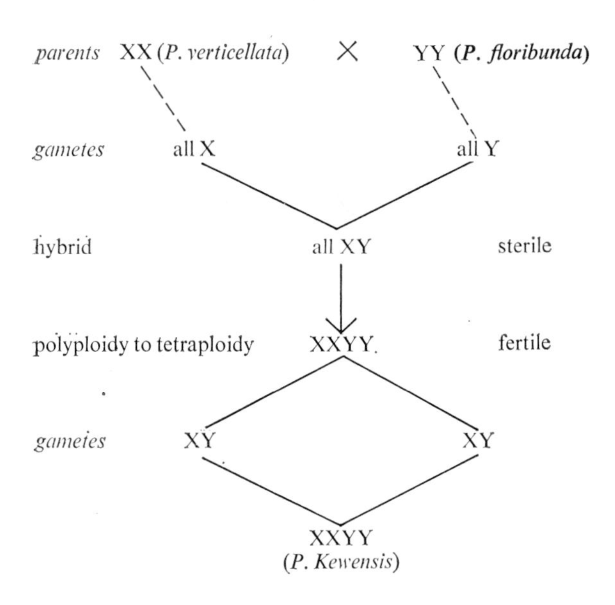

Fig 5.10 **Origin of Primula kewensis**

SEX-LINKAGE Inheritance of unlike X and Y chromosomes determines the sex in many organisms. In mammals (though not in all organisms) the female has XX chromosomes (homogametic) whereas the male has XY (heterogametic). In theory, there should be equal numbers of ♂ and ♀ children in the world (Fig 5.11).

The X and Y chromosomes also carry other genes and these are said to be *sex-linked.* The X and Y chromosomes are thus not solely sex-determining. In *Felis* (domestic cats) the X chromosome bears a gene determining colour (black, ginger) whereas the Y chromosome is inert. Set out in Fig 5.12 is the cross of a ginger male with a black female.

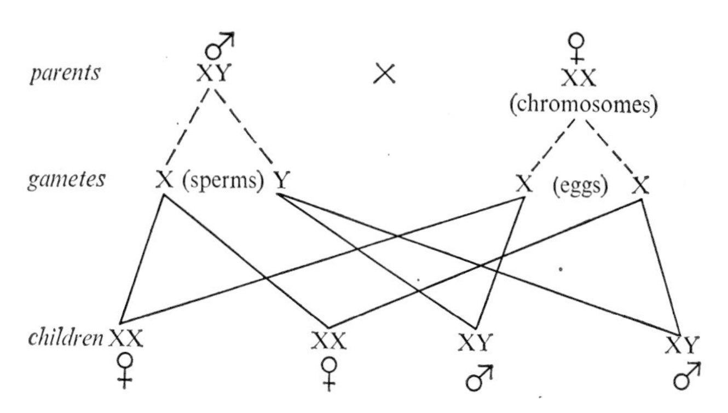

Fig 5.11 **Sex inheritance in mammals**

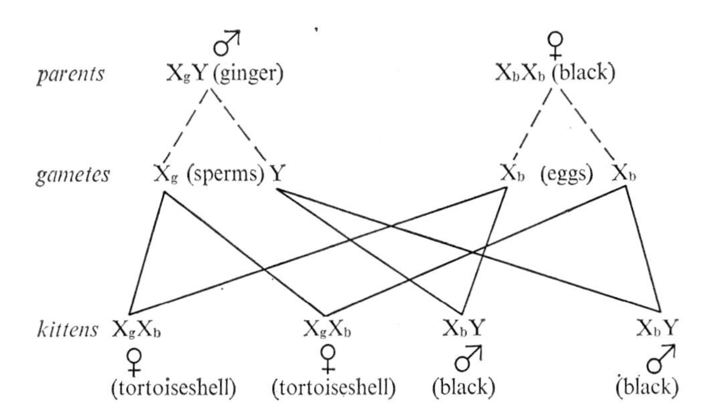

Fig 5.12 **Cross of a ginger male and black female cat**

Haemophilia and colour-blindness genes are sex-linked in man.

5.3 Conclusion

The genetics of populations – *population genetics* – provides a numerical framework within which evolution can be seen to operate.

Population is a loose word meaning a community of individuals which interbreed, *e.g.* races of man. A *species* is potentially an example of a population, but since all races do not interbreed freely (*e.g.* eskimos rarely have the opportunity to interbreed with pygmies) there are smaller populations within each species. Fig 5.13 suggests mechanisms through which the species *Homo sapiens* might have subdivided into new species (speciated) in the past, which no longer apply. When the variation in some character is measured within a population, it usually conforms to a *normal distribution.* An example would be the heights of adult males in the United Kingdom, or males or females of a particular age group in your school. Height is controlled partly by genetic and partly by environmental factors. Where genetic influence is important, as in the F_2 of Mendel's experiment with peas (Fig 5.2) the distribution of height would be discontinuous with two sharp peaks corresponding to TT plants and tt plants. Where there is still evidence of genetic influence but the variation is continuous, control by lots of genes (*polygenes*) each having a small effect on the character observed. Variation not correlated with genetic influence is environmental.

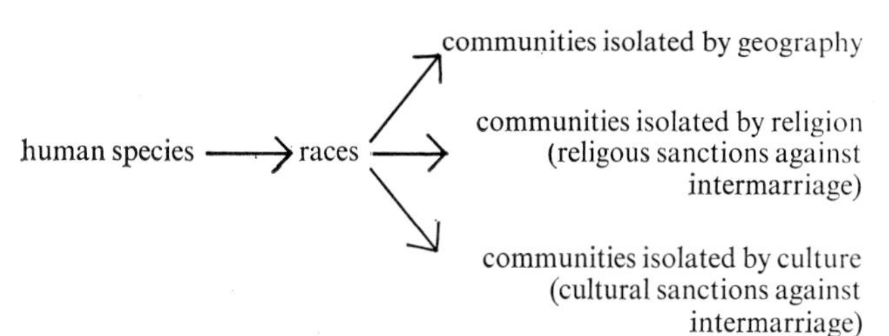

Fig 5.13 **Possible speciation pathways in man**

If we have a gene existing as alleles X and x, each member of the diploid population will have two copies of the gene, XX, Xx, or xx. In a population of size N there will be a total of $2N$ genes of the types X or x. We can calculate the proportions of the genotypes XX Xx and xx if we are given the allele frequencies. If the frequency for X is p and for x is q, where p and q are fractions with a sum of unity, then after one generation of random mating we have XX $= p^2$ of the total offspring, Xx is $2pq$ of the total, and xx is q^2 of it, as shown in Fig 5.14.

Fig 5.14 **Calculation of proportions of genotypes**

	X(p)	x(q)
X(p)	p^2	pq
x(q)	pq	q^2

Let us take a population having three genotypes

AA 64% Aa 32% aa 4%

Let us assume that selection of mates is done randomly; that any mating combination we take is as likely to give offspring as the next; and that A and a experience no mutation. The situation is represented in Table 5.4. Further interbreeding produces the results shown in Fig 5.15.

Table 5.4

parents		AA		Aa		aa	
proportions		64		32		4	
gametes	A	A	A		a	a	a
proportions	64	64	32		32	4	4
		160				40	
ratio		4A		to		1a	

Fig 5.15 **Calculation of the proportion of genotypes in the example of Table 5.4**

	A (4)	a (1)
A (4)	AA (16) (4 × 4)	Aa (4) (4 × 1)
a (1)	Aa (4) (4 × 1)	aa (1) (1 × 1)

The next generation population has a composition (16:8:1 or 64:32:4) exactly the same as that of the original population. This means that change in the make-up of the population cannot take place because there is a state of genetic equilibrium. In fact, the conditions of random mating, equal fertility and no mutation imply a genetic equilibrium. Factors that may disturb this, as well as selective mating, differing reproductive success and mutation, are migration and selection, a process in which certain genotypes transmit their genes to subsequent generations more frequently than others. For instance, if a mixture of a small number of resistant bacteria and a large number of susceptible ones are exposed to the action of a powerful drug such as an antibiotic, the resistant ones will soon predominate. If resistance is under genetic control, the resistant genotypes become more abundant, reversing the previous situation.

NATURAL SELECTION alters the gene frequencies of populations. The individuals best suited to their environment contribute most to the following generation by having the most offspring (Fig 5.16). Environmental factors that tend to change gene frequency exert *selection pressure*.

Fig 5.16 **Man-made selection of apple plants. The plant on the left is resistant to the woolly aphid, the other is not. Selection helps produce better crops.**

6 Evolution

6.1 Introduction

In order to account satisfactorily for the great diversity of species and the complex way in which each is adapted to its surroundings, some form of *organic evolution* is mostly accepted nowadays. Divine creation and spontaneous generation are regarded as unlikely. The ideas of *Lamarck* (1744–1829) that acquired characters can be transmitted have sometimes been revived but there is no convincing experimental data to support them. *Darwin* (1809–1882) and *Wallace* (1823–1913) laid emphasis on *natural selection* as contributing to the evolution of species, maintaining that in the struggle for existence useful variations were preserved at the expense of less useful or useless ones. This leads to the *survival of the fittest*. Mendel's later work provided the key to the problem of how these useful variants were preserved in future generations.

We have seen (§5.3) that a species is a group of organisms which appear similar and cannot interbreed successfully with other species. If species have evolved from earlier ancestors, there must have been heritable variation in the ancestor. One way in which the heritable variation could have been introduced would be if new genotypes had arisen by *mutation*. Mutated genes, in common with all the other genes, undergo constant reassortment through sexual reproduction, because of the fusion of gametes from different parents and because the haploid gametes are produced by *meiosis*. This involves the random sorting of homologous pairs of chromosomes before division, plus crossing-over. This re-shuffles the genetic information in the population at each generation. When there are potentially more individuals than the environment can support, competition intensifies and only the best fitted survive. It has been shown, using *E. coli* bacteria, that a change in environment can lead to a change in the population by favouring some heritable variants within it more than others.

There are many examples of micro-evolution in recent years, *e.g.* selection and multiplication of pests such as housefly strains and 'superlice' (on human head) resistant to DDT, rabbit strains resistant to Myxomatosis virus, rats resistant to the poison Warfarin. The breeding of animals and crops fitted to the environment that the farmer provides for them is another example.

Within a species there are small interbreeding populations within which each genotype contributes to a common *gene pool*. The gene frequency here remains constant until selection acts so as to change it. If some members of a population become isolated and cease to interbreed freely with the remainder, one part of the pool also becomes isolated and selection can then cause a divergence between the main pool and the unit isolated. If, by phenotypic expression of the divergence (*e.g.* increasing tendency to use different feeding grounds or food plants) the exchange of genes is further discouraged, the isolation may become more permanent. The small sub-units of a population among which breeding normally occurs are referred to as *demes*.

The formation of a new species, or merely races or sub-species, may result from this kind of isolation. It is often difficult to distinguish between species and races, but we normally think of a *race* as consisting of a partially isolated unit or sub-unit. A *sub-species* can be regarded as a unit divergent from its species by prolonged isolation but still partially fertile with it. *Speciation* has occurred when, under natural conditions, divergence has proceeded to the point when *fertile* offspring can no longer be produced.

The process is slow, because the demes isolated to begin divergence in a species may reconnect, and diversity is once again restored to the gene pool. Following the enlarged opportunities for travel, interbreeding has stopped any tendency for races of man such as Amerindians and Australoids to diverge further; although differences in physique and physiognomy had become established, interbreeding between races to produce fertile offspring remained possible and the human race clearly remains one species.

The numbers of known species (distinguished on anatomical features) is very approximately as given in Table 6.1.

6.2 Selection and adaptation

NATURAL SELECTION Darwin's theory of natural selection can be summarised as follows:

1 From its bank of variation an organism produces new individuals that differ from one another. Normally the differences are small (known as *continuous variation*), but occasionally a combination of rare recessives or a random mutation produces more extreme differences.

2 The offspring generally have the characters of the parents.

3 In general organisms produce large numbers of offspring, so that each generation is potentially larger than the one before. But in most environments, owing mainly to predation, disease and limited food resources, the population is relatively constant*. This implies that there is a struggle for existence between the members of one species, and between one species and another. Because no two individuals are exactly alike there is differential ability to cope with the environment. Those best equipped survive and pass on their genes to their offspring. In a stable environmental situation extreme variations normally offer no particular advantage and survive, if at all, in small numbers. However, if the environment changes, new selection pressures are set up, which a more extreme variant may be able to exploit.

As an example in action let us consider the case of *Biston betularia*, the peppered moth, commonly found in the British Isles. Collectors in the mid-19th century found that most specimens were white/grey with black dots; black specimens were also found, but these were rare. By 1900 about 98% of the specimens found near Manchester were black. Nowadays the percentages of light and dark *Biston* vary over the country, the light forms dominating in unpolluted rural areas, and the dark forms dominating in industrial areas, where soot has blackened the bark of trees. The colour variation is controlled by a single gene, having two alleles B and b. BB moths are dark; bb moths are light. Natural selection altered the gene frequency in industrial areas, *i.e.* the few black moths existing in the middle of the 19th century survived much more frequently than the light moths. Recently marked moths of both types were released in rural and in industrial regions. Collections at a later date compared the percentages of marked light and dark moths in the collected samples with the percentages of the moths released. In rural areas a higher proportion of dark moths had been lost, whereas in industrial areas it was the light moths which had suffered.

It appeared that the light moths are less well camouflaged against the blackened surfaces of industrial areas, and vice-versa. This is clearly demonstrated in the photographs of Fig 6.1. Less camouflage of course means less protection against predators. Films have been made to show that predatory birds exhibit a preference for black moths in rural areas and light ones in industrial areas. With the expansion of industrial areas this kind of selection probably occurred and intensified between 1850 and 1880, with the consequent increase in the gene frequency of the B allele.

The black moths are called *melanic*, and the phenomenon, investigated by Kettlewell, is called *Industrial Melanism*. It should be noted that non-melanic moths *may not always* necessarily have an advantage in rural areas; this depends on the prevailing winds. A similar relationship between colour varieties of the banded snail, and predation by thrushes, has also been found.

Another example of selection is when an organism *mimics* another. It is interesting to note the peculiar shape and apparently meaningless colour of the Bee Orchid flower. The lip mimics the form and colour of the female of certain bees (different bee orchids mimic different species of bee). The males of the species are tricked and, if one lands on the flower, attempting to mate with it, sticky pollen bags stick to its head. If the same bee then tries to mate with another flower, the pollen becomes transferred from his head to the stigma. Another example is the mimicking by butterflies of others that are distasteful to birds.

The *selective advantage* of any allele is specific for a particular environment, as the above examples show. It was found that when plant breeders incorporated fresh genes for resistance to disease (e.g. *Puccinia* on wheat) practising *artificial* selection, *natural* selection of a race capable of breaking down the resistance followed rapidly. With a further change in the varieties grown, another change occurred in the predominant race of disease organism. Each change in environment led to a change in race; each race is characterised by a single, or at most a few, alleles. Human evolution is an interesting example of natural selection, since the cranial capacity has risen

Table 6.1 **Approximate numbers of known species**

	number of species in thousands
birds	9
mammals	4
reptiles/amphibians	6
fish	20
invertebrates	1050
plants	400
unicellular organisms	75

* The most notable exception being the human world population.

Fig 6.1 **Industrial melanism**

by 300% since *Australopithecus*, an ape-man of 1 million years ago. Of course, bigger cranial capacity does not necessarily mean bigger mental capacity, but there is evidence that this is likely. The expression of the increase in size is the ability of man to make and use implements and to develop sophisticated patterns of speech.

ARTIFICIAL SELECTION Ever since he first domesticated animals and cultivated plants (cows, cereals, apples, etc.) man has been able to reinforce desired qualities by selection and breeding where man chooses which genotype shall be used as parents of the next generation. By so doing he has been able to breed large edible apples from small bitter crab-apples; he has been able to increase the amount of milk a cow will produce in a lactation, 65% in the last 30 years; the sucrose content of sugar beet has been raised by 7% in the last 150 years. After many generations the resultant organism can be quite different from the one with which breeding began. Compare the differences between a modern race-horse and the Arab steed from which it is descended (Fig 6.2 and 6.3).

It was by considering artificial selection that Darwin saw how natural selection might account for the enormous variation in species.

ADAPTATION We have seen that random mutations can cause changes in the form or function of an organism, that either by themselves or cumulatively enable it to grapple more effectively with its environment. This gives it an advantage over its fellows which makes it better fitted to survive. Such changes are called *adaptations*.

Adaptive radiation is a pattern of evolution in which, by means of such adaptations, a common ancestral group has given rise to many new groups which have been able to exploit different ecological niches. The word radiation is used because the divergence has gone in many directions from a point source. Consider a lamp surrounded by a many-faceted polygonal shade, each facet consisting of a different light filter. The light going in different directions is subjected to different influences and the end product, the emitted light, differs. The lamp may be likened to a common ancestor, the facets to mutations which have produced the end products, the different emissions resembling the different adapted variants that have survived. If any of the facets were filled with materials that would not pass light, these could be said to represent mutations that conferred no selective advantage (*blind alley evolutions*), from which no adaptive radiant survives.

Fig 6.2 **Arab steed**

Fig 6.3 **Modern race horse**

6.3 The fossil record

Fossils are studied in the science of palaeontology. A fossil is evidence of the former presence of any living organism, animal or plant. The most common method of fossilisation is the *petrification* (literally turning to stone) of organisms in sedimentary rock. Organisms can also be fossilised if they have been covered in some form of natural preservative. Thus complete mammoths have been found preserved by ice in Siberia; insects bodies have become fossilised in resin from coniferous trees; complete skeletons have been found preserved by the oil in asphalt lakes in California and elsewhere. Footprints of prehistoric animals have been found. (How would these have been preserved?)

Where fossils have been found in rocks (Fig 6.5), we can find the point in time when they were formed, and what the climatic and other environmental conditions were by studying the geological features. Dating by geological means can in some cases be checked using the radioactivity methods outlined below (p 99). From this we can discover when and where the fossilised organisms lived, and sometimes details of their diet and habits as well as their shape and size. We can therefore arrange fossils into chronological order and from this follow the evolution of organisms through time. The main features of the geological succession as revealed by fossils are shown in Table 6.2. In many cases we can find an organism to link together two groups. For example *Seymouria* (Permian) links Amphibia and Reptilia; *Archeopteryx* (Jurassic) links Reptilia and Aves; coelacanth (Holocene, though living examples have been found recently—see §6.5) links Osteichthyes and Amphibia; and cycads (Mesozoic) links Pteridophyta and Spermatophyta. This is important direct evidence to support the theory of evolution.

As an example of how organisms which are derived from one another occur, consider the fossil reptiles of the Mesozoic era (225 to 65 million years ago). The fossil finds are shown in Fig 6.4.

The teleost fishes, the mammals and the birds all evolved during this period. The fauna on land was dominated by reptiles, some of which were very large; most became extinct by the end of the era. Most ichthyosaurs were fish-eaters about the size of a dolphin, although some fossils are more than 10 m long. Plesiosaurs were often larger, ranging up to 17 m; they paddled themselves through the water. The terrestrial reptiles included the pterosaurs (winged lizards, ranging in wingspan from 0·1 m about the size of a sparrow) to 8 m; they ate fish. The largest reptiles were the dinosaurs (ferocious lizards) whose ancestors were probably small, like modern lizards but stood up on their hind legs, balanced by a powerful tail. They were probably ancestors of crocodiles and birds as well. By the standards of modern animals both *Brontosaurus* and *Tyrannosaurus rex* (15 m high, mass 7 tonnes) were enormous.

Several problems arise as regards fossil evidence. Firstly, a *palaeospecies* (based on fossils) differs from a *biospecies* (based on living organisms) since the palaeontologist can hardly test for speciation by the interbreeding method. Secondly, to become a fossil an organism must die under exactly the right conditions and in the right spot, a rare event. Thirdly the fossils must be found by man. It is not surprising therefore, that fossil evidence is patchy and its interpretation difficult.

living creature dies → dead creature covered by sediment → creature decays leaving cavity in hardened sediment (*i.e.* the rock) → more sediment enters cavity via fissures in rock → this hardens to give fossil cast → fossil cast is found millions of years later

Fig 6.3

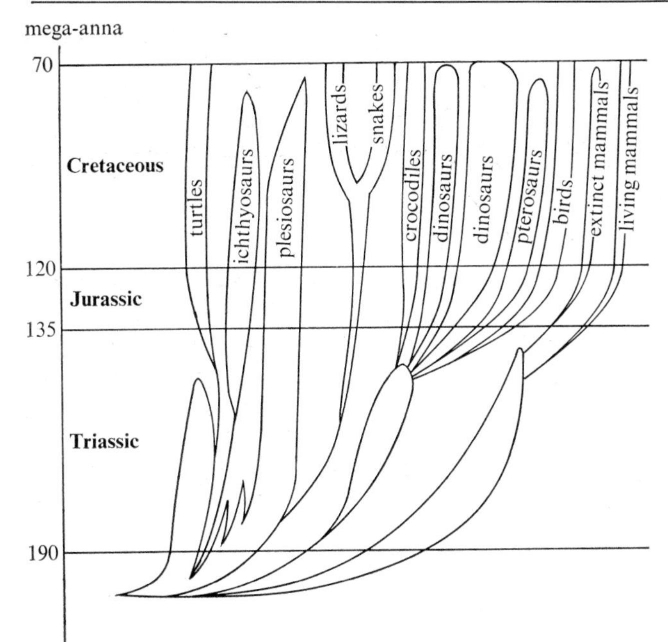

Fig 6.4 **Fossil finds**

RADIOACTIVE DATING The decay of natural radioactive materials provides a means of estimating vast periods of time. The ^{14}C isotope with a half life period of 5 600 years is ideal for dating archaeological specimens containing organic material. The technique was developed by the American scientist Libby. ^{14}C is always present in the atmosphere in constant small amounts in carbon(IV) oxide, and all living organisms take this up – either directly in plants or indirectly in animals. Hence there is an exchange of carbon (via CO_2) between the atmosphere and living organisms; hence the carbon in living organisms contains a constant small amount of ^{14}C. When an organism dies, the exchange of carbon with the atmosphere ceases, and the ^{14}C isotope decays in accordance with the radioactive decay law

$$\log \frac{N_o}{N_t} = \frac{1}{2 \cdot 3} kt$$

where N_o is the number of atoms of the radioactive element at time zero, N_t is the number after time t, k is the disintegration constant. By comparing the disintegrations per second (measured on a *Geiger Counter*) of the specimen with the disintegrations per second of a living organism, the date can be calculated.

The dating of much older rocks can be carried out using a similar method. The naturally occurring isotope of uranium, ^{238}U, decays to form lead with a half-life of $4 \cdot 5 \times 10^9$ years. ^{40}K, a naturally occuring isotope of potassium, decays to argon with a half-life of $1 \cdot 3 \times 10^{10}$ years, these isotopes have half-lifes of the right order of magnitude, and both occur naturally in rocks. How would the measurements be made?

Fig 6.5 **Ogygiocarella; a trilobite of the Ordovician age**

Table 6.2 **Geological succession**

era	*duration (mega-anna)**	*period*	*estimated time since start (mega-anna)*	*life*
Paleozoic	345	Cambrian	570	invertebrates, algae in sea
		Ordovician	500	first vertebrates; lampshells and cephalopods
		Silurian	440	first land plants
		Devonian	395	fish dominant; first amphibia and large land plants
		Carboniferous	345	coal measures, ferns, mosses, amphibia, first reptiles, bony fish (freshwater), sharks (sea), insects
		Permian	280	reptiles replace amphibia; ancestral teleosts
Mesozoic	160	Triassic	225	dinosaurs, ichthyosaurs, plesiosaurs, turtles, conifers, first mammals
		Jurassic	190	reptiles dominant on land, birds, teleosts, fishes, archaic mammals
		Cretaceous	136	flowering plants; large reptiles become extinct
Caenozoic	66	Tertiary	65	placental mammals, birds, insects
		Quaternary	1·5	mammals, dominated by man

* 1 mega-anna = 3×10^{10} ks = 1 million years.

6.4 Distribution of plants and animals

GEOGRAPHICAL ISOLATION If individuals of a species are isolated by natural barriers (*e.g.* on islands) from the main population, a great divergence can often be seen. Darwin studied the fauna and flora of the *Galapagos Islands*, volcanic in origin, which lie on the equator nearly 1 000 km west of Ecuador and 1 600 km south west of Panama. The islands were not inhabited by any living organism when they rose from the sea 2 million years ago. They now have plants and animals living there, including species that are not found anywhere else in the world. The finches, of which there are 13 species, have been extensively studied. The islands were initially mainly colonised by finches from South America, blown there by the wind. As the islands are extremely isolated, colonisation happened only a few times. Owing to the range of foodstuffs available the finches were able to exploit vacant ecological niches; from their 'bank' of variation, characters that helped them exploit the new environment were selected. The divergence of the finches is shown in Fig 6.7.

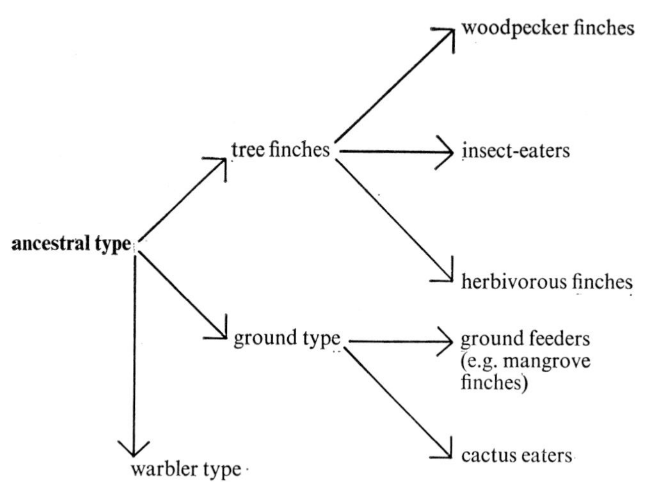

Fig 6.7 **Divergence of the Galapagos finches**

CONTINENTAL DISTRIBUTION Wallace studied the distribution of the earth's animals, particularly birds and mammals, and divided them into six groups, Australasian, Oriental, Palearctic, Nearctic, Neotropical and Ethiopian, each associated with a geographical area (Fig 6.6). The areas are separated by natural geographical barriers, such as mountains and oceans. The organisms in these areas have overall similarities, but show tremendous differences in detail (*e.g.* the llama and the camel – see p 101). The differences in detail are most pronounced in the areas which are the most isolated. Australasia, South America (Neotropical) and Africa (Ethiopian) have a similar range of habitats, but have quite different fauna and flora. It is possible to explain the differences by postulating a common ancestor, which has migrated, been cut off, and then evolved differently to fill the available ecological niches, as was the case with the finches in the Galapagos Islands. This theory is supported by both the geological and the fossil evidence. Consider the camel family; this is represented in Asia and Africa by

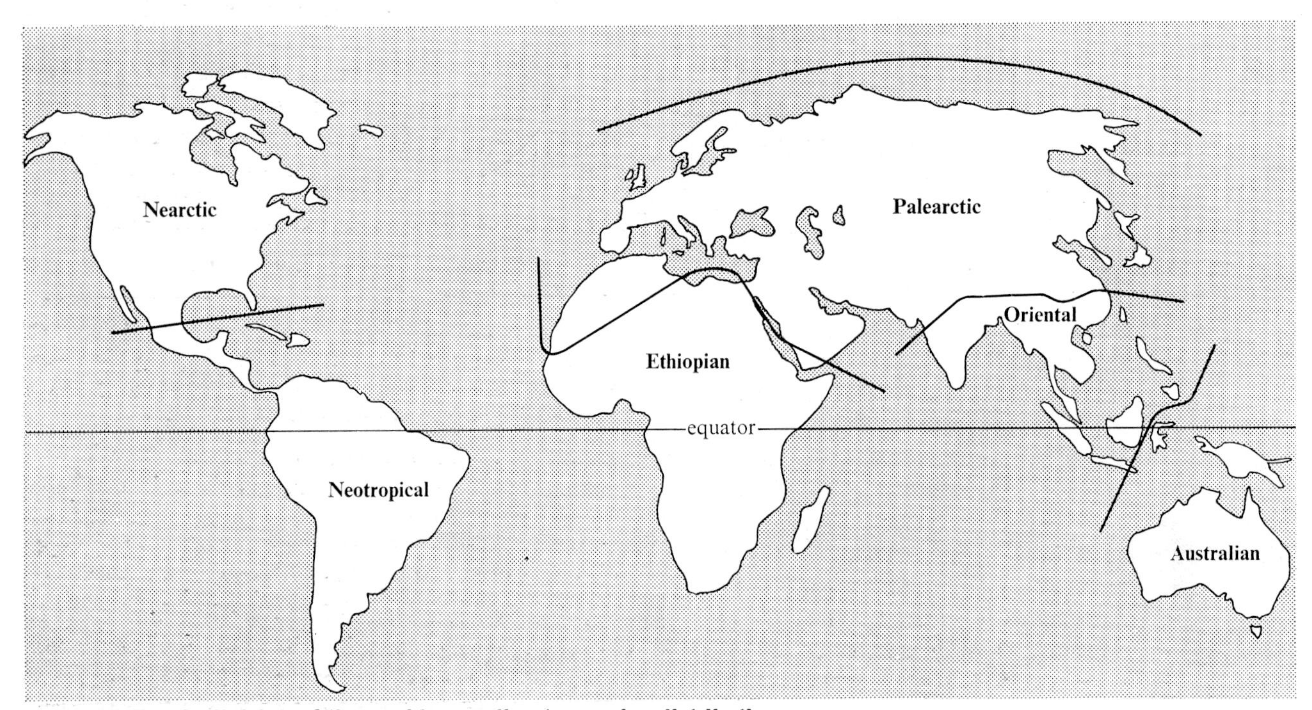

Fig 6.6 **Wallace's division of the world according to species distribution**

the Bactrian and Arabian camels (Fig 6.8 and 6.9) and in South America by the llama (Fig 6.10). However, the oldest camel fossils have been found in North America. By plotting the finds of more recent fossils it is possible to follow the migration of the camel south into South America and west into Asia (across the Bering straights which the geological records show at that time formed a bridge between America and Asia) and hence through Central Asia, Arabia and into North Africa. Because there are no living links between the camel in North Africa and the llama in South America, this is called a *discontinuous distribution*.

Fig 6.10 **Divergence in the camel family; the llama**

OTHER FORMS OF ISOLATION Apart from natural barriers there are other forms of isolation which prevent individuals of the same species breeding, and so cause divergence. Griffith-Smith (1967) studied the North-Polar gulls, *Larus*, and found that some species live together and look alike, yet do not interbreed. Experiments have shown that the populations remain separate by failing to recognise subtle visual signals during mating rituals. In

Fig 6.8 **Below Bactrian Camel**
Fig 6.9 **Above Arabian Camel**

animals with complex behavioural patterns, the wrong responses during courtship may prevent copulation. This is an example of *behavioural isolation*; it is most likely to occur if some natural barrier is removed (such as a river drying up) and two previously isolated populations are reunited. In some cases fertilisation cannot occur even when copulation has taken place. This is known as *genetic isolation*. Amongst flowering plants, where pollination is often entirely at random (as with wind pollinated plants), this is particularly important. The genes from different populations may give the wrong biochemical signals to each other. Sometimes even if fertilisation does occur, the zygote may fail to develop properly.

6.5 Adaptive radiation in higher groups

ANATOMICAL AND BEHAVIOURAL DIVERGENCE In the vertebrates there are structures that show similarities, suggesting that they have diverged from a basic common pattern. These include a supporting skeleton in the body, a pump-like heart forcing blood round the body, and (early in development) slit-like openings between the gut in the region of the throat and outside the body; also, in many vertebrates, a tail. Fish, frogs, snakes, mammals and birds all conform to this pattern, and we may conclude that they have all evolved by divergence from a single ancestral group. Major developments in this branch of evolution were as follows.

The ancestor was marine; the major step in evolution must have been the development of a *notochord* (support rod) with blocks of muscle arranged along the body on each side of the rod. Contractual movements of these muscles, causing a wave movement along the animal, made swimming possible. Development of a vertebral column, fins, and a streamlined body shape accompanied the development of swimming. Elasmobranchii, of which sharks and rays are modern examples, were marine, breathed oxygen through *gills*, had flexible vertebral columns and skeletons made of cartilage. Other fish that lived in primaeval swamps breathed air via *lungs* and water via gills; the modern lung-fishes are derived from them. It is thought that when these fish moved into the sea, lungs were no longer needed and were developed into the *air-bladder* that provides buoyancy control. The elasmobranch dogfish has no air bladder whereas the *teleosts* (trout, eel) whose ancestors were common about 225 million years ago possess one. Special mention should be made of the coelacanth, the 'living fossil' fish. The species found recently belong to a vertebrate group thought to have been extinct for 70 million years. It was discovered in 1938 in a South African trawler's catch. Since then other coelacanths have been found. It cannot survive for long at sea level as it normally lives at great depths.

Terrestrial vertebrates developed from the ancestral fishes. Paired *limbs* evolved from fins. The earliest limbs had a short central axis, and hands and feet (instead of the web of the fin toes and fingers developed). The earliest to evolve were the Amphibia and the oldest fossils of these are about 350 million years old. Respiration was like that of the modern lung-fish. They ate water food and laid *eggs* in water. The frogs egg provides an example from a modern amphibian. These are laid in water surrounded by jelly (frog-spawn). The embryo develops using the yolk and the hatched tadpole feeds in water and respires by gills, using its tail to swim. The tadpole grows lungs and legs, breathes air, and after losing its tail, lives on land and feeds on terrestrial animals. Reptiles were the first vertebrates that laid eggs able to develop without water, about 340 million years ago. Unlike amphibians, they breathed exclusively by lungs. Evolution of laying eggs on land probably avoided predators, most of which were then in the water. Remember that even now crocodiles and turtles live in water but lay eggs on land. Further development is evident in the hen's egg, in which the embryo, floating on the yolk, uses it and the white as food, using oxygen from the air space that has diffused through the membrane. The hatched chick resembles the adult hen in structure.

CONTROL OF BODY TEMPERATURE All modern reptiles, fish and amphibians are *poikilothermic* (body temperature varying with environment). Two groups of primitive reptiles evolved adaptations which produced *homoiothermy* (maintenance of constant body temperature). This is a selective advantage where activity in a wide range of temperatures is necessary. Further adaptations in mammals are capillaries that bring warm blood near to the skin surface where heat is lost through radiation, and sweat glands that secrete a fluid that provides cooling as it evaporates. Muscular contractions (*e.g.* shivering) provide heat energy and most bodily processes produce heat. Layers of fat and hair conserve heat. In hibernating animals (hedgehog) the winter is spent in a lethargic state and the blood temperature falls. Birds lack sweat glands and pant to lose heat. Feathers conserve heat.

WINGS, CLAWS AND BEAKS The earliest fossil birds are in rocks 160 million years old. Their ancestors leapt from branch to branch of trees, holding their fore-limbs out in front. With the development of feathers and modification of the limb, first gliding and then flapping flight evolved (Fig 6.11). Modern birds are of many types, *e.g.* soaring (gulls), manoeuvrable (swifts, swallows), swooping (hawk, cormorant). Feet have become adapted for different purposes and so have beaks (Fig 6.12).

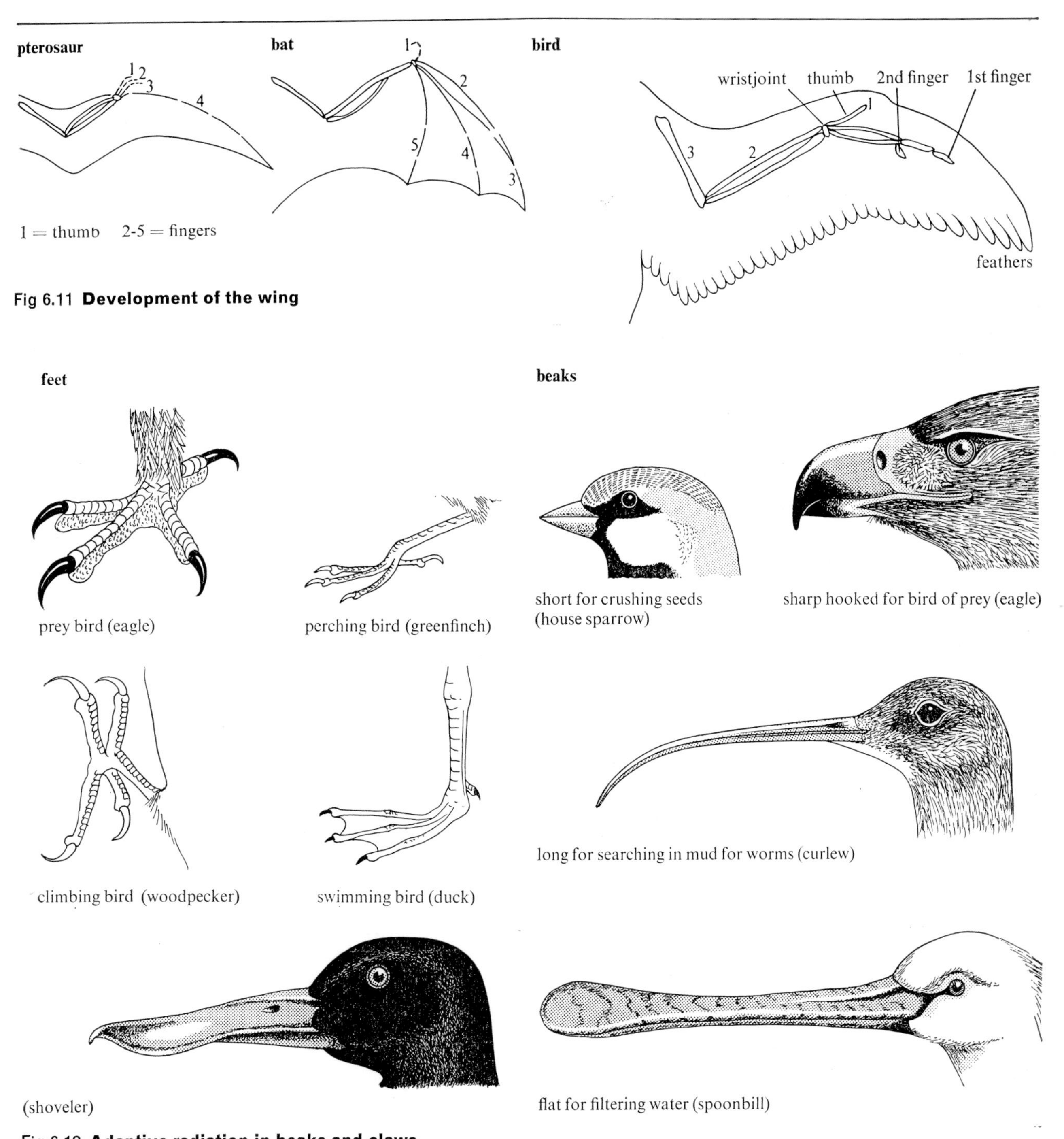

Fig 6.11 **Development of the wing**

Fig 6.12 **Adaptive radiation in beaks and claws**

VIVIPARITY Here, the embryo is retained in the body of the mother and nourished via the placenta, the food being carried there by the mother's blood and withdrawn in the blood of the embryo. The blood of the mother and embryo does not mix – the nutrients diffuse. Oxygen and waste travel the same path. Viviparity is also valuable in affording a protected environment, at constant temperature, for the young. Viviparity is not confined to mammals, *e.g.* some sharks, insects and reptiles show it.

MAMMARY GLANDS Only mammals, birds, and a few other species feed their young after birth. The duck-billed platypus is called a mammal because it has fur and produces milk, but it lays its eggs, incubates them, and then feeds its young on milk after hatching. It represents a parallel evolution to other mammals coming from similar reptilian ancestry. Most fossil mammals come in rocks about 900 million years old, and are recognised by features of jaws, teeth and ears.

EMBRYOLOGY AND THE RECAPITULATION THEORY Strong supporting evidence for organic evolution comes from *embryology*. Embryos of fish, mammals and reptiles bear a very close resemblance to one another (Fig 6.13). in early stages of development. In many cases, the development of individual organs is similar too. Consider the embryonic development of pharyngeal pouches and pharyngeal slits. Each gets larger until only a thin tissue separates the structures. Then in *fish*, the tissue breaks forming gill slits. But in *mammals*, *birds*, etc. only the first pair remain, the rest disappearing soon after formation; the first pair give rise to *1.* eustachian tube of ear, *2.* opening of the ear, *3.* adult tympanic membrane (ear drum).

In animals, differentation and development of the main structures has been completed when the embryo is fully developed, whereas in plants the process goes on all through life (stems and roots grow from apices). In angiosperms, the organisation in the embryos is very similar and these similarities remain in the young seedlings, differences only appearing much later in development. The *recapitulation theory* (biogenetic law) suggests that an organism, in the development of its embryo, always passed through certain stages characteristic of its ancestors. Although the theory is not now accepted completely, it is interesting that often seemingly useless embryonic structures persist in organisms, which may be derived from ancestral forms.

TEETH Another example of adaptive radiation is the specialisation of mammalian teeth to suit diet.

CEREBRAL CORTEX AND BEHAVIOUR Contact with parents allows the young to learn types of behaviour and this development has been paralleled by evolution of *centres* (p 154) in the brain, most in the cerebral cortex. Neither reptiles nor birds have a very extensively developed cerebral cortex; birds, however, do have some behaviour which can be modified. The first primates were arboreal animals, feeding on a diet of insects and fruit. Monkeys descended from these 'tree-shrews'. More advanced still are the apes, which have no tails. Chimpanzees, gorillas, etc. can walk on hind legs, run and swing from trees using long arms. Man differs in that his legs are longer than his arms and he always walks upright (associated with evolution of the skeleton). The brains of the later primates become increasingly complex, and some monkeys and apes have fairly elaborate social behaviour.

HEART A study of the hearts and arterial arches shows a clear evolutionary series, the organs of the bird and mammal being derived from the common ancestral form of the fish, via the amphibian.

PENTADACTYL LIMB This is characteristic of tetrapod vertebrates. There is a single proximal bone (humerus or

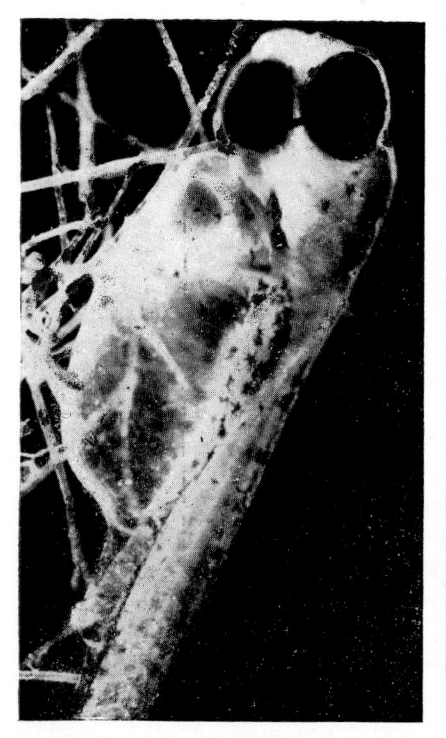
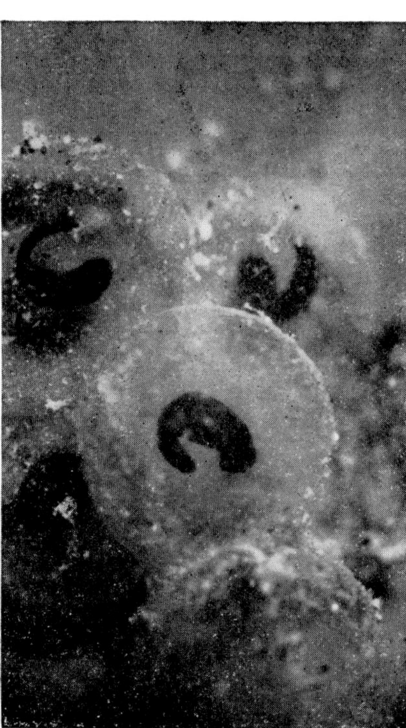
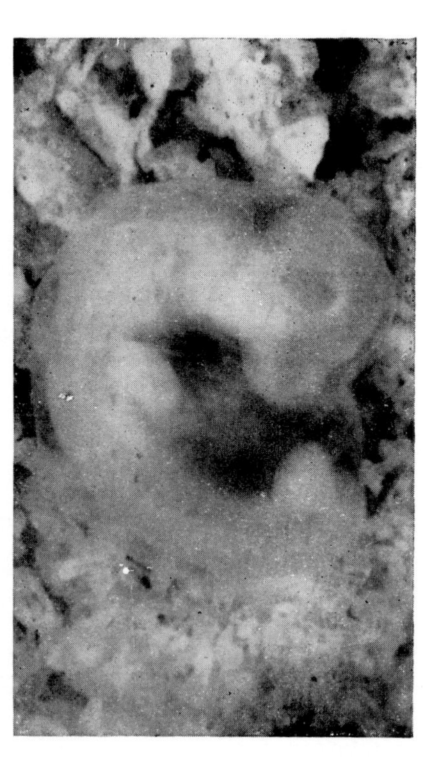

Fig 6.13 Embryos of fish frog, and human

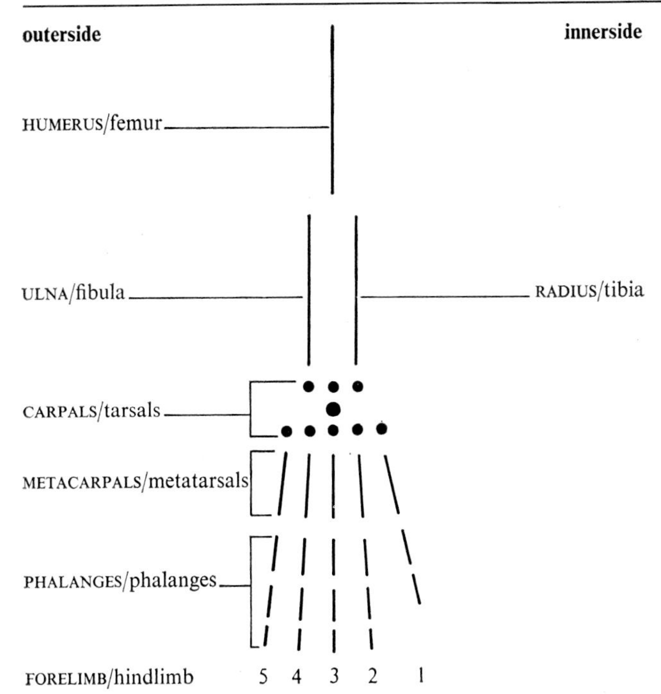

Fig 6.14 **The pentadactyl limb**

femur), two distal bones (radius and ulna or tibia and fibula) and carpals and tarsals, followed by five digits or phalanges (Fig 6.14). It is modified according to habitat, *e.g.* running (deer), flying (bat), jumping (kangaroo), climbing (monkey); the basic form is for walking.

Modifications in the limbs of the ancestors of the modern horse, well represented in fossils, provide a good series illustrating adaption and specialisation. Fig 6.15 shows the limb extremities of the palaeospecies *Eohippus*, the less primitive *Mesohippus*, and the type of the modern *Equus*. There is a progressive reduction in the number of digits.

VESTIGIAL ORGANS The loss of the tail and the retention of hair on the head appear to be examples of complete and incomplete adaptations. Flightless birds like the kiwi have a fully structured but non-functional wing.

6.6 Adaptive radiation in other groups

AMPHIOXUS This primitive chordate closely resembles the lamprey larva, which shows how fishes may be related to other chordates. Thus, the brain of the fishes has probably evolved from the very primitive brain of *Amphioxus*.

TROCHOSPHERE The parallel evolution of the annelida and mollusca is evident from the fact that both show a trochosphere larva.

PLANT LIFE CYCLES The development of the life cycles through bryophytes, pteridophytes to spermatophytes is evidence for evolution.

PARALLEL RADIATION OF INSECTS AND PLANTS Early insects had biting mouth parts like present day beetles and locusts. Sucking mouth parts as in bees, butterflies and moths, aphids and mosquitos have evolved to suit the needs of the insect. Many flowers are cross-pollinated by insects, and this helps maintain genetic diversity to meet the needs of environmental changes. Such flowers often have *nectaries* that secrete a sugary fluid

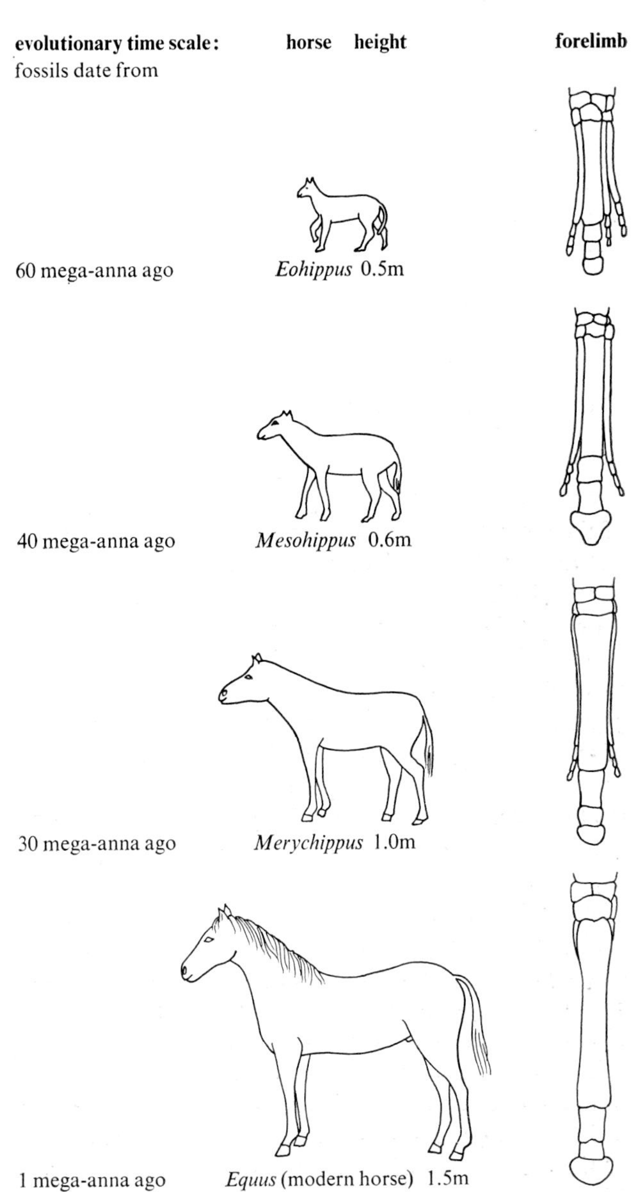

Fig 6.15 **Adaptation and specialisation in the ancestors of the modern horse**

which provides food for visiting insects. In some flowers the corolla forms a long tube and only long tongued insects can reach the nectar. Other flowers are less specialised but in all, the floral structure is such that the visiting insect rubs against stamens, collects pollen, and rubs against the stigmas of subsequent flowers visited to effect pollination.

The colour sense of insects has evolved in parallel with flower colours. Insect colour vision extends further into the ultraviolet but less far into the red than human. Wind pollinated flowers are often green, but insect-pollinated flowers are highly coloured. Winged aphids show a preference for yellow colours, which is probably an adaptation for selecting the young growing parts of the plants on which they feed.

There has also been parallel evolution of aphids and their food. Related aphid species, subspecies and races have each been found to feed only on certain plants which, although they differ for each aphid type, are all closely related. Examples are so many that it seems an inescapable conclusion that there has been parallel divergence from a common aphid ancestor on a common plant ancestor.

6.7 Evolution in cells and chemical constituents

BIOCHEMICAL EVOLUTION There is a similarity in biochemical pathways (*e.g.* respiration) in most cells, indicating common ancestry. When blood is transferred from one animal to another, antibodies build up in the plasma of the second animal. Sera can be prepared and used to test against the blood of other animals; the degree of clotting indicates the closeness of the relationship. Results show that the gorilla is closely related to man. Anti-sera prepared by injecting rabbits with viruses are widely used to classify viruses. Cytochrome occurs in animal and plant mitochondria. It has been shown that the sequences of amino acids in the proteins of cytochrome c are more similar in closely related species than in more distantly related ones.

Evolution also operates at molecular level. *Euglena* is a plant and can perform photosynthesis (autotrophic) whereas *Paramecium* is an animal (heterotrophic). Yet if *Euglena* is kept in the dark and provided with ethanoic (acetic) acid it can grow and survive. The organelles of simple living things are alike, which is not surprising. If a protein of one species were composed of *d*-amino acids and another of *l*-amino acids, or *d*-carbohydrates in one species and *l*-carbohydrates in another, each one lacks the enzymic apparatus to deal with the chemicals of the other, and each would be poisonous to the other. Modern theories favour the idea that organelles are symbionts and that the origin of the process of evolution

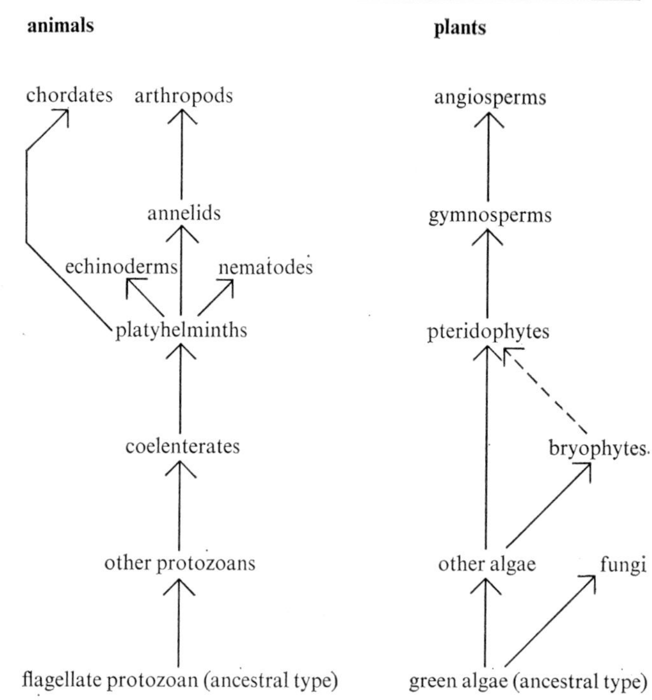

Fig 6.16 **Ancestral relationships**

(the base of the family 'tree') is not a unicellular organism but biocnemical materials.

6.8 Classification

Systems of classification based on homologous features show evolutionary trends* (Fig 6.16).

At generic and species level, confirmation of the validity of close relationships established by observing homologous features is often obtained from cross-breeding results, serum diagnosis and other biochemical affinities.

* Some classification systems are based on evolutionary relationships. These clearly cannot be used to support the theory of evolution.

7 Ecology

7.1 Introduction

Ecology is the study of living things in relation to their *environment*. Older text books refer to the 'natural' environment but man's activities have affected so much of the Earth's surface, sea and land, that the term is outdated. When the environment of single species is studied it is called an *autecological* study; when the whole environment is studied, the study is *synecological*. It has been found that certain areas (*e.g.* sand dunes) have similar animal and plant species composition and interactions and this has led to the term *ecosystem*. Ecologists also refer to *closed communities* when they mean areas in which there is minimal interaction between the organisms present with those in the neighbouring areas. One of the Galapagos Islands is a good example, but smaller units such as a pond, a rock pool, a section of hedgerow or stream, a ball of moss, a garden compost heap, a patch of soil, pasture or lawn are also ecosystems. Real ecology is long term and often related to economic factors. But almost any ecological study can make an original contribution to knowledge and several A-level Boards have scope in their courses for project work.

Any ecosystem contains many, usually very many, different species, from viruses, bacteria and protozoa up to flowering plants and large vertebrates. Each will be affected by *climatic factors* (temperature, light, humidity, windspeed) which for the smaller organisms affect the *microclimate* immediately about them (*e.g.* between leaf-hairs for a fungal spore). Those associated with soil will experience *edaphic factors* (pH value, mineral salts and trace elements). Interactions with other organisms (*e.g.* as food or in competition for living space) are *biotic* factors. Man will almost always be one important biotic factor. An ideal ecological study will reveal the species present; their numbers, and the changes in the numbers within each season, and from one season to the next; and what interactions there are between species that cause the numbers to change.

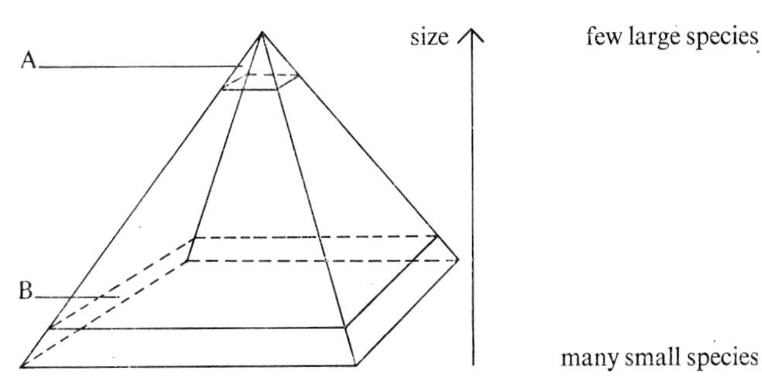

Fig 7.1 **Number pyramid of species**

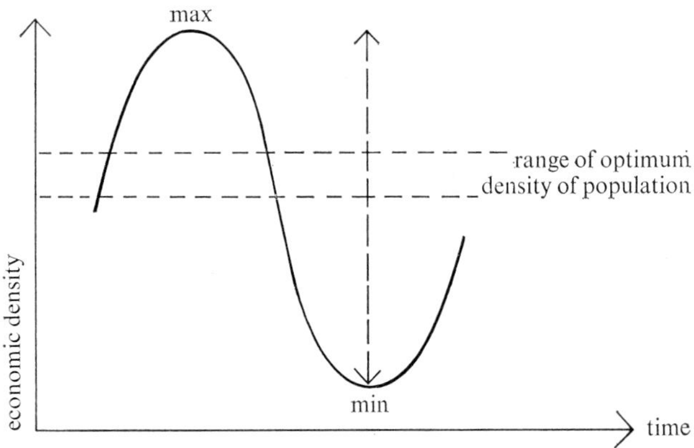

Fig 7.2 **Population fluctuations in a species**

In an ecosystem there will be a *community* of animals and plants. The numbers of each species will form a *number pyramid* (Fig 7.1). At the top of the pyramid (A) are a few large species, each with few individuals; at the base (B) many small species, each abundant. The numbers of individuals of each species undergo changes called *population fluctuations*. The size of a population can be expressed in several ways, the *density* being most commonly used. The density means the number of individuals per unit area of habitat. The *economic density* (Fig 7.2) refers to the number of individuals per unit of inhabited area. Sometimes the numbers in a population increases greatly, as when a pest attacks a crop. There are maximum, minimum and optimum densities; the optimum density is the number the environment can support without over *or* under exploitation.

Each species occupies an *ecological niche* which satisfies its needs for nutrition, space and protection from

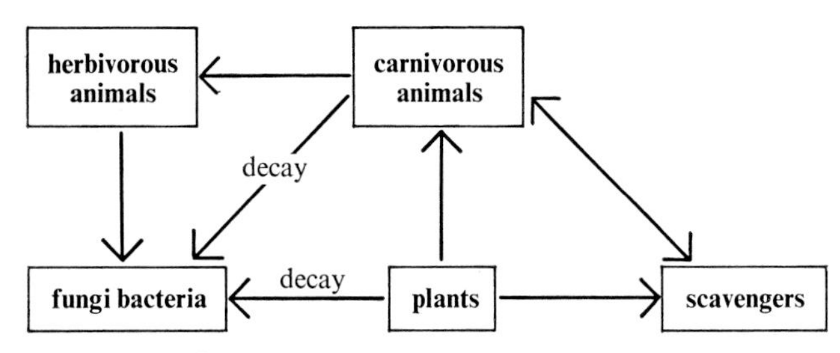

Fig 7.3 **Interactions in an ecosystem**

POND
tiny water plants → tadpoles → sticklebacks → kingfishers
(fish) (birds)

MEADOW
clover → bumble bees → field mice → cats

SEA
diatoms in plankton → *Calanus* → herrings → men
(crustacean)

FIELD
wheat → field mice → barn owls

Fig 7.4 Food chains

enemies. Of these, nutrition is most important. Table 7.1 shows the varied niches, in relation to feeding, that animals in woodland might occupy (after Neal *Woodland Ecology*).

All animal communities ultimately depend on plants since these are the organisms capable of autotrophic nutrition (see p 131). Fig 7.3 shows the general relation between plant and animals, producers and consumers in any ecosystem. A complex *food chain* is usually present within any community – large carnivores might prey on smaller ones, and you can trace the chain back to the herbivore stage, then back to the green plant.

Some very simple food chains are shown in Fig 7.4.

Fig 7.5 shows, from examples in Table 7.1, a *food web;* food webs can be even more complex than this.

7.2 Energy in the ecosystem

We have seen that nitrogen, carbon and other elements including minerals taken up from the soil circulate round an ecosystem. Energy, by contrast, flows in one direction only. The energy flow starts with the sun. Radiation from the sun falls on autotrophic organisms (mainly green plants, but some photo- and chemo-synthetic bacteria as well – p 138). These convert some of the energy into organic compounds that can be used by heterotrophs. Among the heterotrophs the herbivores eat green plants, absorbing some of the available energy and dissipating the rest. Carnivores obtain their energy from herbivores; larger carnivores from small carnivores etc. (not literally according to size of course!). Decomposers absorb energy from the remains of other organisms. At each stage there is a considerable loss in useful energy – photosynthesis is less than 1% efficient, and between the other stages useful energy transfer is of the order of 10% efficient. Eventually all the useful energy is lost, either in the form

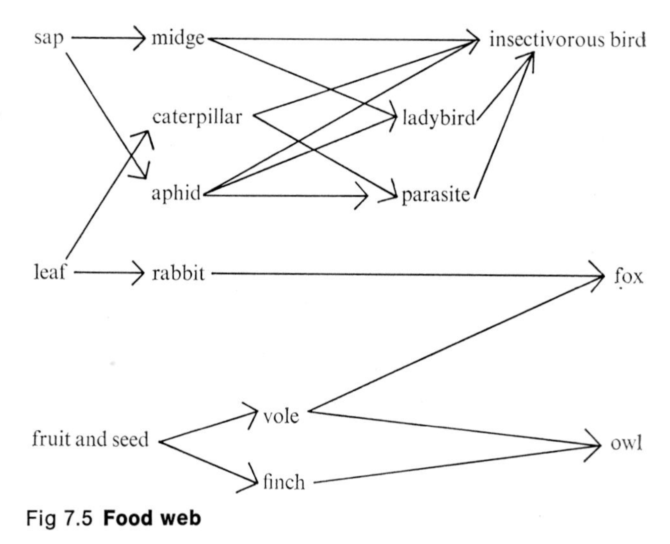

Fig 7.5 **Food web**

Table 7.1 **Niches of woodland animals**

	type		*examples*
herbivores	leaf, shoot, bud feeders	(miners)	caterpillar
		(defoliators)	caterpillar, snail, rabbit, deer
	plant juice suckers	(nectar)	butterfly, bee
		(sap)	aphid
	gall formers		gall midge
	bark feeders		wood louse, grey squirrel, rabbit
	fruit and/or seed feeders		caterpillar, vole, badger, chaffinch
	root feeders		larva (moth, beetle)
	wood borers		wood wasp, larva (moth, beetle)
	fungus feeders		beetles
carnivores	parasites		fleas, ichneumon flies, ticks
	blood suckers		mosquitoes, spiders, flies
	predators		ladybird beetles, centipedes, fox, badger, stoat
scavengers	omnivores		badger
	animal remain feeders		carrion beetle
	vegetable remain feeders		earthworms

of heat, or in materials (such as bone) that are of no use to other organisms*.

Because of the loss in energy at each stage, there must be a reduction in *biomass* (mass of living matter) at each level. We can construct a pyramid of biomass in the same way as the pyramid of numbers (Fig 7.1). Table 7.2 shows the results† of an experiment on biomass carried out in a river in the USA.

Table 7.2 **Biomass in a river**

level	*biomass in* g m^{-2} (*dry mass*)
primary producers (microscopic algae and *Sagittaria*, an aquatic Angiosperm)	809
herbivores (snails, shrimps, midges, caddis flies)	37
carnivores (mainly fish)	12
decomposers (mainly bacteria)	5

7.3 **Identification**

Correct identification of animals and plants is essential. Standard works of identification should be used or more specialist publications if necessary. '*Keys*' are provided by these which by presenting a succession of choices (such as evergreen/deciduous, woody/herbaceous, flowers yellow/flowers red, etc.) lead the reader to the correct name. For instance, a habitat containing fresh water snails was studied and the classification of them provides a key for their identification.

FRESHWATER SNAILS

1 Shell flat	*Planorbis*
Shell spiral	go to question 2
2 Shell sinistral (mouth on left when held vertical)	*Physa fontinalis*
Shell dextral (mouth on right)	go to question 3
3 Operculum (hard mouth-covering) absent	go to question 4
Operculum present	go to question 5
4 Under 15 mm high	*Lymnaea pereger*
Over 30 mm high	*L. stagnalis*
5 6–12 mm high, unbanded	*Bithynia tentaculata*
Over 20 mm high, banded	*Viviparus spp.*

This key provides positive identification among these six species except for the *Viviparus* species. This would need a further key to positively identify this species.

7.4 **Studying climatic and other physical factors**

Standard meteorological measurements are useful if a meteorological station is close at hand and figures are made available.

TEMPERATURE In ecology, you might want to record continuously the changes in temperature in the bark of a tree or in the soil. Electrical thermometers, both *thermocouple* (measuring current variations) and *thermistor* (variations in resistance) are available. The *thermistor bridge circuit* (Fig 7.6) incorporates a pair of matched thermistors (resistors of high negative temperature coefficient of resistivity) with a balancing potentiometer which is adjusted until a zero recording is obtained on the galvanometer. Any temperature difference results in an unbalancing and the galvanometer registers a flow of current.

LIGHT For comparisons of light illumination, a *photo voltaic light meter* is normally used, and this can be calibrated against a commercial light meter or using a standard electric lamp. You might, for example, want to measure the illumination at the surface of a pond. The SI unit of luminous intensity is the candela and for

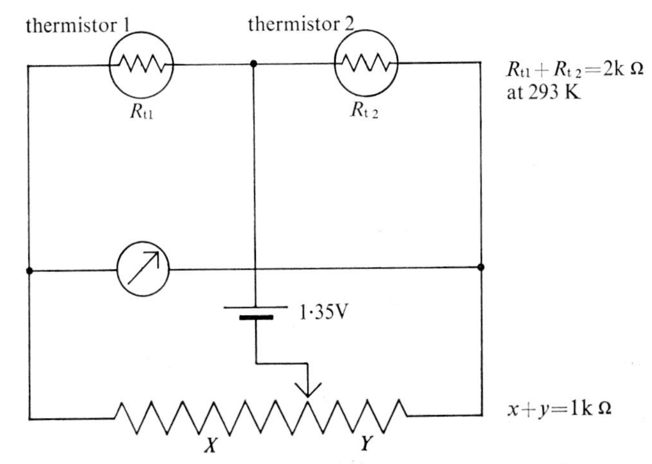

Fig 7.6 **Measure of temperature using a thermistor bridge circuit**

* Man has of course been able to use organic compounds such as coal and oil to provide energy. But this is not energy that is used directly to maintain his bodily functions.

† Figures from Simon, Dormer and Hartshorne *Lowson's Botany* (UTP, 1973).

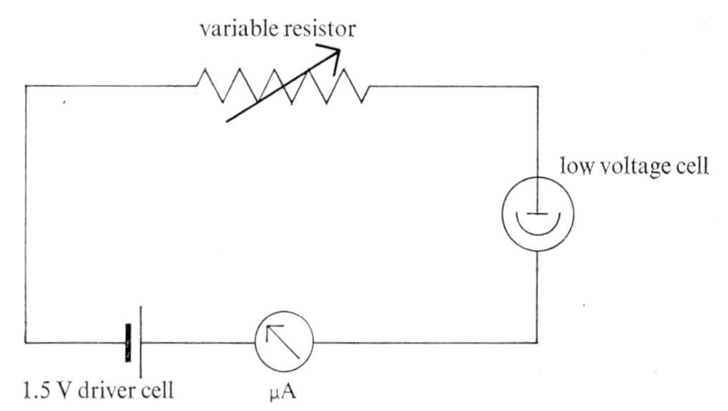

Fig 7.7 **Photostatic light meter**

illumination it is the lux. You may have a grasp of this from A-level physics. The circuit for a cadmium sulphide photoelectric cell (Fig 7.7) has a resistor which allows readings to be made from the ammeter over a wide range of illumination.

HUMIDITY This can be measured by a *hygrometer*. One type depends on the change in the length of a hair (or paper strip) according to how much water vapour is in the air. The wet and dry bulb is often convenient, consisting of two thermometers mounted side by side with the bulb of one covered with an absorbent material and dipping in a water reservoir; the specific latent heat of evaporation results in the wet bulb being at a slightly lower temperature than the dry bulb and the difference can be used to tabulate relative humidities of the air.

A whirling hygrometer is very convenient for field studies. In microhabitats paper impregnated with cobalt(II) ions (as the chloride) is often used. The amount of moisture in the air can be determined from the colour of the paper – if much moisture is there the hexaquocobalt(II) ion $[Co(H_2O)_6]^{2+}$ is present and is pink; if the moisture content is low the cobalt is present as the hexachlorocobaltate(II), $[CoCl_6]^{4-}$ which is blue. There are a series of lilac colours in between.

EVAPORATION In order to measure the total evaporation of the air in a given ecosystem over a long period of time, an *atmometer* is used. A porous tube is attached to a capillary tube and reservoir, and a second capillary tube (fitted with scale) is joined to the base of the reservoir. The apparatus is filled with distilled water; evaporation from the porous surface results in water coming up from below and the movement can be read off on the scale.

WATER FLOW The velocity of flow of water in a stream can be measured by floating light objects downstream and timing their motion. For more accurate work, *current meters* are available; a simple one consists of an L-shaped tube immersed in the water, the horizontal limb upstream; the height of water in the vertical limb is measured. The velocity of the current (v) can be calculated from the formula $v^2 = 2gh$, where h is the height of water and g is the acceleration due to gravity.

OXYGEN IN WATER Various sampling apparatus is available for taking water from different localities in a water habitat. Oxygen content can be determined by *Winkler's method.* A stoppered bottle is filled with the water sample and some concentrated alkali released near the bottom with a long jet burette, followed by some concentrated manganese(II) chloride. The stopper is allowed to fall into place and the bottle tumbled about to swirl the contents which are then allowed to settle. Some potassium iodide crystals are inserted, then concentrated hydrochloric acid, and again the contents are swirled. Iodine is released and this can be determined by titration with standard sodium thiosulphate(VI) solution. The volume of the stoppered bottle should be determined and the result quoted in grams per cubic decimetre of oxygen.

$Mn^{2+} + 2OH^- \rightarrow Mn(OH)_2 \downarrow$

$Mn(OH)_2 + \frac{1}{2}O_2(aq) \rightarrow \underset{\text{brown}}{MnO_2 \downarrow} + H_2O$

$MnO_2 + 2I^- + 4H^+ \rightarrow Mn^{2+} + I_2 + 2H_2O$

$I_2 + 2S_2O_3^{2-} \rightarrow S_4O_6^{2-} + 2I^-$

i.e. $\frac{1}{2}O_2 \equiv MnO_2 \equiv I_2 \equiv 2S_2O_3^{2-}$

An alternative method involves the use of *phenosafranine;* the principle is that dissolved oxygen oxidises iron(II) to iron(III); the phenosafranine dye (red) is reduced to a colourless form by the iron(II) solution (reducing agent).

CARBON(IV) OXIDE IN WATER This can be determined in acid waters by titration with sodium carbonate solution to give hydrogen carbonate ions, phenolphthalein being used as indicator (it is colourless in the presence of HCO_3^- and changes to pink when free CO_3^{2-} appears).

$CO_3^{2-}(aq) + CO_2 + H_2O \rightarrow 2HCO_3^-(aq)$

In neutral alkaline waters the carbon(IV) oxide is present as HCO_3^- ions, (plus Ca^{2+} or Mg^{2+}) and it can be determined by titration with standard acid in the presence of a suitable indicator (*e.g.* methyl red and bromocresol green in ethanol, where the colour change, pH 4·5, is from blue to greyish pink).

The pH scale is the scale of acidity/alkalinity and runs approximately from 1 to 14.

1 ——————— 7 ——————— 14

ACID NEUTRAL ALKALINE

The pH is the potential of the hydrogen ions in the solution and is defined as the negative logarithm to the base ten of the hydrogen ion concentration.

$pH = -\log_{10}[H^+]$

Colorimetric determination of pH is very convenient and many indicators are available, *e.g.* BDH soil indicator (covering pH 4 to 8). Indicators can be used as solutions or booklets of test papers. For more accurate work the *pH meter* (involving the use of glass electrodes) is employed. The glass electrode adsorbs H^+ ions, a potential difference is set up and this is amplified and read off on a pre-calibrated scale.

DETERMINATION OF DISSOLVED IONS This is often carried out in coastal and estuarine regions. Chloride ions can be determined by titration with standard silver(I) nitrate(V) using chromate(VI) ions as indicator. Calcium and magnesium ions (which cause hardness of water) can be determined by titration with EDTA (ethylenediamine-tetra-acetic acid or bis[di(carboxymethyl)amino]ethane), which forms stable complexes with the Ca^{2+} and Mg^{2+} ions. If the dye erichrome black T is added to a solution containing Ca^{2+} or Mg^{2+} ions, a pink colour results by complex formation. Titration to an end-point with EDTA results in a change back to black again since the Ca^{2+} or Mg^{2+} ions form more stable complexes with the reagent. The dye ammonium purpurate reacts similarly but can be used to determine Ca^{2+} ions on their own since it forms a pink complex with these ions only. The end point with EDTA is from pink to purple.

7.5 Studying edaphic factors

Soils contain mineral materials derived from the rocks of the earth's crust, and have arisen from *weathering* (the agents of denudation are rain, wind, snow, ice, etc.). Soil particles can be classified into the following types, the figures referring to the particle diameters in micrometres: gravel (above 2 000), coarse sand (200 to 2 000), fine sand (20 to 200), silt (2 to 20), clay (less than 2). Humus, the rotting remains of plants and animals, is also present, and it provides material for the activity of nitrifying bacteria (p 138). It has great water-holding capacity. A rough *mechanical analysis* of a soil can be made by shaking up with water and allowing it to sediment (Fig 7.8).

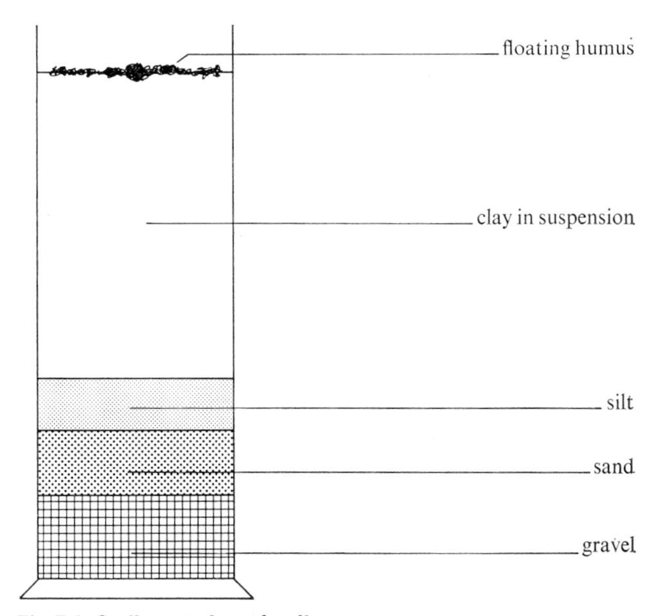

Fig 7.8 **Sedimentation of soil**

Particles of different size are separated by shaking or washing through sieves of different mesh sizes.

The main *natural* soils of Great Britain are shown in Table 7.3. A classification of soil that is useful for comparisons is given in Table 7.4.

WATER When rain water sinks through a soil, the soil particles become covered with a layer of water (capillary water). When a soil can hold no more water it is at *field*

Table 7.3 **Main natural soils of Great Britain**

type	*appearance*	*pH*	*depth*	*drainage*	*subsoil*	*occurrence*	*vegetation*
brown earth	dark, much humus	5–6	0·5–1 m	rarely really free	often lighter	widespread	deciduous trees
podsol	surface humus grey layer red brown layer light brown layer	2–3	variable about 0·5 m	excessive	light brown to rock	sandstone areas	conifers heather heathland
rendzina	dark layer with dark grades to lighten colour	7–9	shallow 0·1 to 0·2 m	free	often chalk	chalk areas Carboniferous limestone or Great Oolite	beech whitebeam old man's beard
fen peat	dark and uniform to great depth	8–9	great	poor, sometimes waterlogged	same	fens	fen plants
acid peat	dark and uniform to great depth	2–6	great	poor, waterlogged	same	often uplands	bog plants
gley	mottled brown layer green blue lower layer	2–6	average	impeded over waterlogged	waterlogged	near rivers and streams	marsh plants

Table 7.4 **Percentage composition of soils**

type of soil	*water*	*humus*	*coarse sand*	*fine sand*	*clay*	*calcium carbonate*
sandy loam	2	5	10	70	13	0
clay loam	4	6	4	30	56	0
normal loam	2·5	15	0·5	50	32	0
chalk	2	7	20	20	11	40
peat	9	31	25	15	20	0

capacity. The field capacity of soils rich in humus and clay is greater than that of sandy soils. If *drainage* is unimpeded further water will drain away, but if the drainage is impeded by impervious layers the soil will become waterlogged. For agriculture, soils with too much clay or humus have to be artificially drained to get rid of excessive water; soils with too little humus require the addition of organic matter to improve their field capacity, since plant roots cannot grow in soil that is too wet or too dry. To estimate the water content of the soil, the soil (a definite mass) is heated in an evaporating basin on a water-bath or in an oven (378 K). The sample is then cooled and reweighed.

HUMUS Under natural conditions humus is supplied by the decay of plant and animal remains in or on the soil. Worms bury quantities of humus by dragging decaying leaves down their holes, and other organisms, mainly bacteria, continue the process.

One gram of soil contains about 5×10^9 bacteria! Remember also that soils are *made* and maintained by the occupants. Farmers supplement the humus in the soil by adding farm yard manure or ploughing in green manure such as a mustard crop. Organic matter can be estimated by taking oven-dried soil (weighed amount) and heating to red heat in a crucible for about 2 ks (30 min) cooling and then reweighing.

MINERALS The soil water contains many minerals in solution, of immediate availability to plants. Other ions are 'bound' on the surface of soil particles. One method of analysis is by the *ring-oven technique* (see Abbott *Education in Chemistry* Vol 2 number 4, July 1965). Standard inorganic chemical techniques can also be used and standard kits are available for determining what elements are present. The main elements required by plants, obtained from the soil, are hydrogen, oxygen (in water), nitrogen, phosphorus, potassium, calcium, magnesium and iron, with traces of other elements. Grasses also contain silicon.

AIR A degree of aeration is necessary for root growth of most plants. The volume of air in soil is easily found by taking a known volume of water in a measuring cylinder, a known volume of soil and mixing the two. If the volume of the soil = 250 cm³, the volume of the water = 250 cm³, and the volume of the mixture = 400 cm³, the volume of the air is 100 cm³ (40%).

Table 7.6 **Testing soil acidity**

pH	*colour*
4·0	red/brown
4·5	brown
5·0	orange/brown
5·5	pale orange
6·0	yellow
6·5	pistaccio
7·0	pale green
7·5	leaf green
8·0	dark green

testing procedure

ACIDITY Plants may be *calcifuge* (lime-hating) or *calcicole*. Acid and alkaline soils have a typical flora and fauna. Acid soils contain no lime whereas alkaline soils are rich in lime. One way of testing acidity (Table 7.6) is to place a little barium sulphate(VI) plus soil in a test-tube, add de-ionised water to a definite mark, then pour 'soil indicator' up to a second mark. The contents of the tube should be well mixed, allowed to stand, and the colour

Table 7.5 **Soil comparisons**

soil	*particle size*	*structure*	*pore space*	*aeration*	*drainage*	*water holding*	*pH*	*nutrients*	*temperature*	*tillage*
sandy	coarse	no clods few crumbs	large	good	good	bad	acid to neutral	low, rapidly leached	warm in spring	easy, usually possible
loam	medium	crumby	medium	good	good	good	neutral	intermediate	intermediate	intermediate
heavy clay	fine	large clods	little	poor in winter*	poor in winter*	excessive in winter*	acid except over chalk	high colloid retention, low leaching	cold in spring	difficult, often wet

* or rainy weather at other seasons.

of the indicator recorded. An indicator chart is provided with the indicator sample and matching of colours will give the pH.

TEMPERATURE Root growth only takes place between maximum and minimum temperatures. The drier, sandier soils tend to warm up earlier in the spring. Soil temperatures are generally found using an electrical thermometer, such as a resistance thermometer.

SOIL SURVEYING As well as analysing the soil, you should study its physical characteristics. Remember when examining samples that soil type may change quite considerably over a few metres. The first examination should be made *in situ* by digging a trench and examining a profile. A podsol heathland soil should reveal several layers; a surface layer of unrotted humus, a sand/humus layer, a layer of coarser sand or gravel, an organic pan (hard layer) formed by components of the humus washed down, an iron pan rich in oxides of iron, and a layer of sand or gravel below. Measurements and analysis can be done on samples from the different layers. The results will help you form ideas on why the soil supports the flora and fauna you observe in it. The comparison in Table 7.5 will help you to see where your soil 'fits'.

Loams are the best soils for farmers and gardeners. Light soils tend to dry out during the summer, and heavy soils are difficult to work. Both are improved by addition of organic material, to improve the field capacity in the one case, and improve aeration in the other.

Soil conditions affect the animals as well as the plants living there, both in type and numbers. If you are going to investigate such a community you should first survey the area and draw a map, putting in the areas occupied by plants (this gives an indication of the condition of the soil). A complete breakdown of the soil should then be done, *e.g.* mechanical analysis, mineral ions, pH, temperature, air content, water content, humus content, and so on. Root zones, root densities and earthworm holes should be noted. The animals in the soil will vary; litter layers may contain millipedes, centipedes, ground beetles, and fly larvae. Root zones will contain root feeders such as moth and beetle larvae and root-feeding aphids. Burrows of moles and earthworms may be present. Mites, springtails and beetle larvae occupy these burrows and other air spaces. The soil water will contain flagellate protozoa, turbellarians, nematodes and rotifers. Fungal mycelium may be present on roots or in organic matter.

7.6 Studying biotic factors

In both autecology and synecology studies of *life cycle* and *nutrition* are important. These often give the clue to the preferred habitat of the organism. Individual stages in animal life cycles should be recognised, *e.g.* for an insect, egg, larva, pupa, and adult. The season(s) of appearance of each stage, and its duration, should be observed. The food required, any special mode of nutrition, (see p 131) the mating habits and any other interesting features of behaviour, such as time of activity, should be noted. With plants, the season of growth and flowering are important. *Reproductive success* should be assessed for plants and animals. These factors should be related to short term and seasonal climatic changes and edaphic factors.

Most components of ecosystems will show changes in numbers during the year or from year to year. The smaller animals may breed through several generations each year. If reproductive success is greater than mortality factors, such as predation, parasitism, disease or climatic disasters, the population will increase; or the reverse may happen. What is good for one species may be bad for another. The proportion of each species in the ecosystem is constantly changing. There will also be *immigration* and *emigration* of the motile animals, and the seeds and spores of plants, and these will cause changes in population density. Those organisms best fitted to the environment will be more successful in *competition* with others. Overcrowding may lead to increased *aggression* in some species (*e.g.* predators) and to mass *dispersal* in others (*e.g.* aphids). Parasites and their *hosts* (the animals on which they live) may show cyclic *population fluctuations* (Fig 7.2). The parasite reaches maximum density shortly after the host, and then as the numbers of the host decline, the parasite population also falls. With fewer parasites the host population then starts to increase. However, most organisms show a wide range of *immunity* to potential enemies, so that pests, parasites, predators and diseases are either extremely specialised, being adapted to attack a single host often with lethal consequences, or generalised feeders that can adapt to a variety of foods. Animals may become immune to diseases after an attack (antibody reactions). *Resistance mechanisms* have also evolved; after a defoliating attack by caterpillars, oak trees produce a fresh set of leaves (Lammas growth). Some animals have also developed *social organization* for mutual assistance with feeding, protection against enemies, care of the young, etc. Hive bees, herds, and breeding colonies of birds are examples.

Factors such as these emphasise the *dynamic* nature of the ecosystem, both in the autecological context of each member and the synecological context of the whole community. Phases can be recognised; virgin territory, or vacant ecological niches, are colonised (e.g. *Buddleia* on building sites in cities). As better fitted organisms arrive or increase in size the first colonists may be displaced; these progressive changes in flora and fauna are called *successions* and each successive community is called a *sere*. Finally a community of species that is permanent and stable may become established, and this is called a *climax* or *completion* community. Although there will be seasonal changes in the plants and animals present, population fluctuations in such communities are less violent and the numbers of each component of the ecosystem are near optimum density.

Man as a biotic factor is so important that this is considered separately in §7.13.

7.7 Sampling methods

SMALL ANIMALS Flying insects can be collected with *nets*, and those resting in herbage with a *sweep-net*. Insects on trees and shrubs can be jarred into *beating-trays* or inverted umbrellas. Small insects in flight can be trapped in *suction traps*, night flying insects in mercury-vapour *light traps* and crawling animals in *pitfall traps*. Small animals in litter can be collected at the base of various *funnels* when gentle heat is applied over a layer of the litter held in the funnel. *Flotation* is used for small creatures in soil, which mostly float in solutions 1·1 to 1·2 as dense as water (25 % sodium chloride has a density of $1{\cdot}19 \times 10^3$ kg m^{-3}). These animals can also be separated from soil by washing through nests of *sieves* of decreasing mesh size. *Chemical extraction* (chemicals poured on the surface of the soil) is used, *e.g.* aqueous 'formalin' (*i.e.* methanal dissolved in water) for earthworms. Zooplankton can be sampled with a plankton pump in which suction is generated by revolving plastic blades in a chamber fitted with inlet and outlet tubes.

ANIMALS IN WATER *Drag nets* are other nets used to sample animals in water. Soft mud can be sampled using a cylindrical tin container, fitted with a ferrule (for penetration) and closed by a swivel-lid operated by a piece of wire running down the handle attached to the tin.

PLANTS These are easier to sample. *Quadrats* (areas of known size) situated at random in the study area can be pegged out, and the plants present counted at intervals (*sequential sampling*). Alternatively, sample areas or volumes can be removed and sorted in the laboratory (*destructive sampling*) and the animals and soil can then be included. *Transects* (lines across the area along which the distribution of plants and animals is noted) can be used.

7.8 Mark – release – recapture

This method is used to assess population size. It is only suitable where large-scale immigration or emigration is unlikely, or can be accounted for. Suppose we are studying squirrels in a wood. We catch 50 specimens, mark them, and release them at random positions in the wood. Some time later we again catch 50 squirrels of which 10 are marked. The total population (T) is related to the number caught (C) as the total marked and released (M) is to recaptured (R).

$$\frac{T}{C} = \frac{M}{R} \text{ or}$$

$$T = \frac{CM}{R} = \frac{50 \times 50}{10} = 250$$

Animals can be remarked, released, and recapture estimates become more accurate as the process is continued. Randomisation of release and sampling points is essential.

Marking can be done with *rings* of light metal (aluminium) or plastic (birds and mammals). New Forest ponies are painlessly *branded*. Fishes can be tagged with a metal label. Marking *paints*, *e.g.* quick drying cellulose-in-ester are used for land mammals, arthropods and molluscs. They are very effective for the wings of butterflies and moths. Radio-active tracers can be used for land animals.

Marked specimens can be detected in the ecosystem by use of a portable Geiger-counter. Isotopes must be selected with *great care* and must be readily absorbed (*e.g.* sulphur ones) by the species. They should only be used under the guidance of an instructor who has the Department of Education and Science certificate for radioactive tracer studies.

7.9 Numerical methods

PLANNING EXPERIMENTS Only very rarely will it be possible to count a whole population of animals or plants. We have to resort to sampling. There are no hard and fast rules for sampling but results are likely to be more accurate if we first ask ourselves a number of questions.

1 *Is the distribution of what we sample likely to be fairly uniform*? If it is, properly randomised sampling points are indicated. If not (zones on the sea shore) transects might provide more information, or sampling points arranged on a grid.

2 *Is our species very abundant or scarce*? *Is it big or small*? By thinking about this we can decide whether a large number of small samples or a smaller number of large samples would be most helpful.

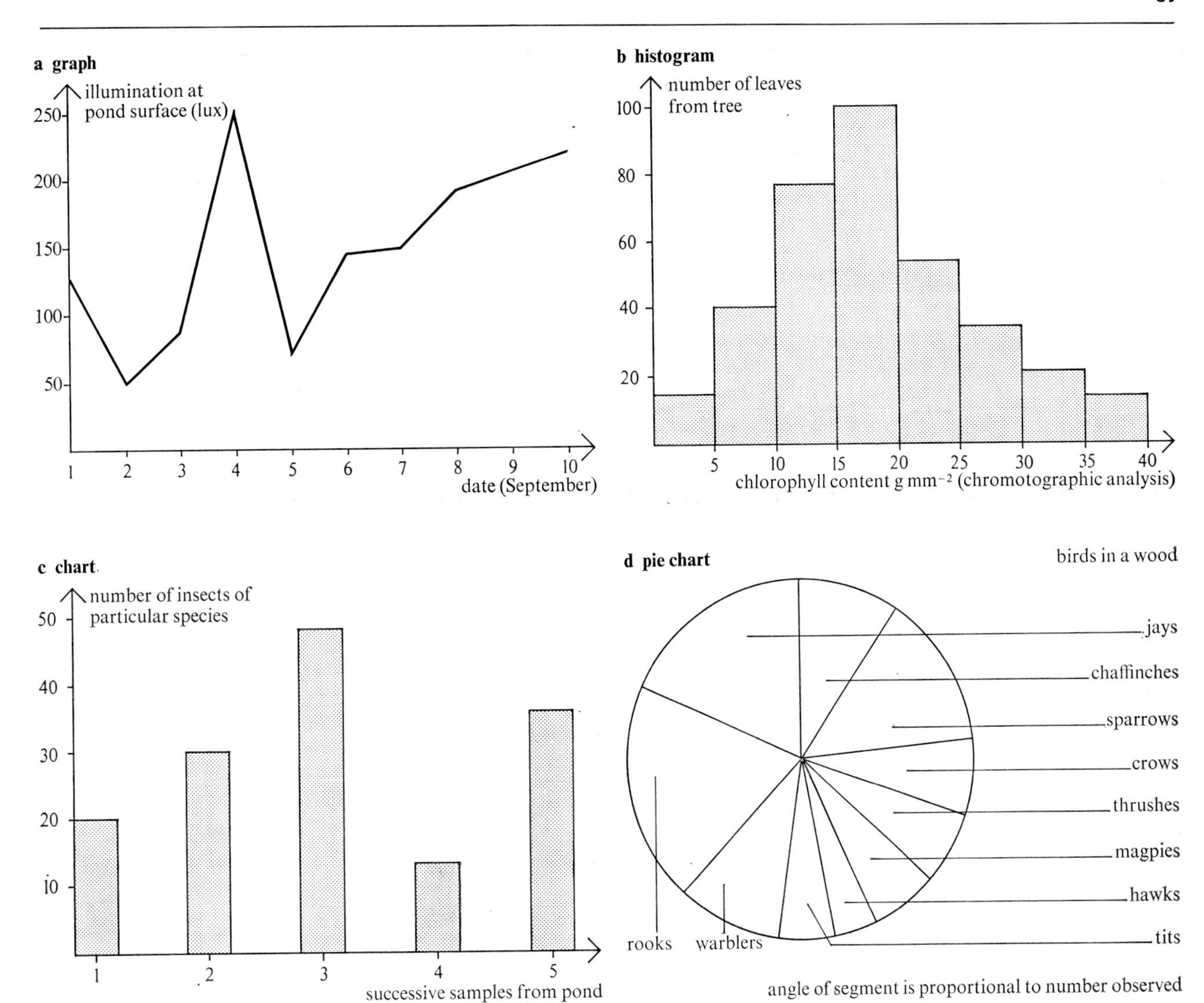

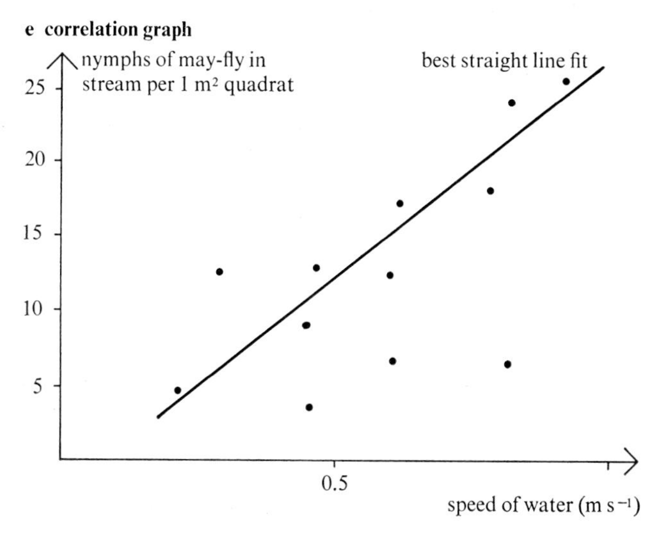

Fig 7.9 **Presentation of results**

3 *Is there likely to be clumping*? Think of estimating the number and sorts of weeds in a lawn. Dandelions mostly grow from seed and so occur at random. Creeping weeds grow new plants near the original coloniser. If a quadrat covered a group of creeping plants like this, you would tend to overestimate the frequency of this weed. Well-separated point samples would give a better result.

4 *How many samples should we take*? This depends on the time available, the problem, and the size of your study area. Too few samples will lead to a high variance (see below) and too many may cause the environment to be changed by trampling.

Having decided how to sample, we should map the area and superimpose a plan of the experimental design onto it, showing the points at which samples are to be taken, on each sampling occasion.

PRESENTING RESULTS The results of experiments should always be *tabulated*, and if possible presented *graphically*, so that the reader can take in immediately what you are trying to convey. Some types of graphical representations are shown in Fig 7.9.

a to **d** explain themselves. **e** is an example of a *direct* relatonship – faster water, more nymphs. We say the relationship is a *positive correlation*. If the situation had been faster water, fewer nymphs, the relationship would have been *inverse* and the correlation *negative*. If the scatter diagram showed a confused cloud of points with no trend, there would have been an absence of correlation.

ANALYSIS OF RESULTS Biological experiments are unlike the majority of those in the physical sciences in that the answer is seldom 'yes' or 'no' but is a *probability* that the result has been obtained by chance rather than an actual repeating occurrence. Biologists must know enough *statistics* to find whether their results are merely chance. It will almost always be a comparison we are considering, between populations in different places, or at different times.

In § 7.10 we introduce a few useful statistical tools with which some readers will already be familiar. No attempt is made to go into the mathematical theory on which the methods are based. For this consult your Mathematics Department. Two further books which non-mathematicians in particular may find helpful are Bishop *Statistics for Biology* Longmans 1966, and Moroney *Facts from Figures* Pelican 1956.

7.10 Some statistical tools

ARITHMETIC MEAN For a series of numbers (X), of which there are N, the arithmetic mean is defined as the sum of the X values divided by N. In algebraic notation

$$\overline{X} = \frac{1}{N}(X_1 + X_2 + \dots X_N) = \frac{1}{N}\sum X$$

As an example, a survey of buttercups was carried out in two different localities A and B. Five fields in each locality were sampled by marking off a plot 5 m × 5 m in each field and counting the numbers of buttercups.

	A	B
Field 1	74	22
2	120	100
3	60	140
4	79	16
5	95	80
TOTAL	428	358

The mean number of buttercups per plot in locality A is given by $\overline{X}_A$ and in locality B by $\overline{X}_B$, where

$$\overline{X}_A = 428/5 \simeq 86 \quad \overline{X}_B = 358/5 \simeq 72$$

To make calculations simpler it is often useful to take an assumed mean $\overline{X}_o$. The mean is then calculated as follows.

$$\overline{X} = \overline{X}_o + \frac{1}{N}\sum(X - \overline{X}_o)$$

What does this mean in practice? Consider the number of buttercups obtained in the samples from locality A above. Take an assumed mean of 80 buttercups per plot, *i.e.* $\overline{X}_o = 80$.

	X	$X - \overline{X}_o$
Field 1	74	− 6
2	120	+40
3	60	−20
4	79	− 1
5	95	+15
	TOTAL	28

$$\overline{X} = 80 + \frac{28}{5} \simeq 86$$

Be careful when using computational short cuts that you do not lose sight of the data itself. Look for extreme results, and see if there is any special reason that might account for them.

STANDARD DEVIATION The standard deviation (s) is commonly used as a measure of spread in observations. It is often called the *root mean square deviation from the mean*. It is defined in algebraic terms as follows.

$$s = \sqrt{\frac{\Sigma(X - \overline{X})^2}{N}}$$

or using an assumed mean $\overline{X}_o$

$$s = \sqrt{\frac{\Sigma(X - \overline{X}_o)^2}{N} - \left(\frac{\Sigma(X - \overline{X}_o)}{N}\right)^2}$$

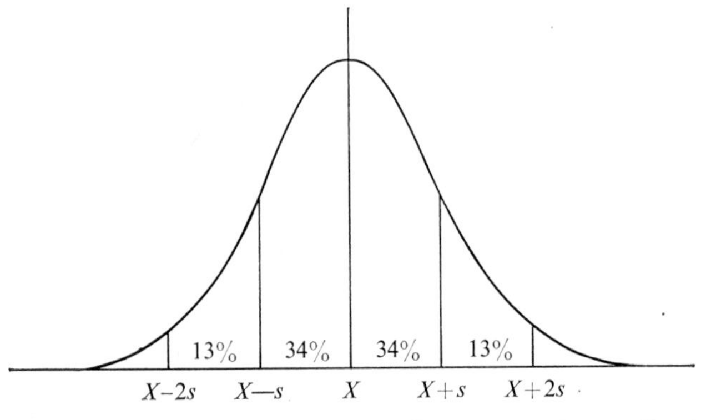

Fig 7.10 **Graphical representation of spread in observations**

As an example consider the numbers of buttercups obtained in the samples from locality A above. We have shown that $\overline{X} = 86$.

	X	$X - \overline{X}$	$(X - \overline{X})^2$
Field 1	74	−12	144
2	120	+34	1156
3	60	−26	676
4	79	− 7	49
5	95	+ 9	81
			2106

$$s = \sqrt{\frac{2106}{5}} = 20{\cdot}5$$

The variance is the square of the standard deviation.

The usefulness of standard deviation lies in the fact that for any symmetrical distribution such as that shown in Fig 7.10, 66% of the population lies within one standard deviation from the mean, and 95% within two standard deviations. Thus if we took a number of samples 66% would lie in the range (86 ± 20·5) and 95% in the range (86 ± 41·0). Estimates of population from these samples would have the same limits.

When an assumed mean is taken the second formula for s given above is useful. Going back to the buttercups again, let us use an assumed mean of 80 again.

	X	$X - \overline{X}_o$	$(X - \overline{X}_o)^2$
Field 1	74	− 6	36
2	120	+40	1600
3	60	−20	400
4	79	− 1	1
5	95	+15	225
		+28	2262

$$s = \sqrt{\frac{\Sigma(X - \overline{X}_o)^2}{N} - \left(\frac{\Sigma(X - \overline{X}_o)}{N}\right)^2}$$

$$s = \sqrt{\frac{2262}{5} - \left(\frac{28}{5}\right)^2}$$

$$s = \sqrt{452 - 31} = 20{\cdot}5$$

FREQUENCY DISTRIBUTION In ecology, statistical results are often obtained in the form of a frequency distribution. For instance suppose we are sampling the number of tomato flowers per truss in a greenhouse. We decide to take a sample of 1 000 trusses, and we can see that the number of flowers per truss varies from 0 up to 100 or so. Rather than record the number of flowers on each truss separately, from which the mean and standard deviation could be calculated only with considerable computational effort, we can record the number of trusses within groups of 0–9, 10–19, 20–29 etc, flowers as shown above. The calculation of the mean and standard deviation is then relatively simple.

number of flowers per truss	*mid-interval value* (X)	*frequency* f	$X - \overline{X}_o$*	$f(X - \overline{X}_o)$	$f(X - \overline{X}_o)^2$
0–9	4·5	1	−40	− 40	1600
10–19	14·5	3	−30	− 90	2700
20–29	24·5	19	−20	− 380	7600
30–39	34·5	131	−10	−1310	13100
40–49	44·5	346	0	0	0
50–59	54·5	340	10	3400	34000
60–69	64·5	134	20	2680	53600
70–79	74·5	20	30	600	18000
80–89	84·5	4	40	160	6400
90–99	94·5	2	50	100	5000
	TOTALS	1000		5120	142000

* Note that we have taken an assumed mean of 44·5.

$$\overline{X} = \overline{X}_o + \frac{\Sigma f(X - \overline{X}_o)}{N}$$

$$\overline{X} = 44{\cdot}5 + \frac{5120}{1000} = 49{\cdot}62$$

$$s = \sqrt{\frac{\Sigma f(X - \overline{X}_o)^2}{N} - \left(\frac{\Sigma f(X - \overline{X}_o)}{N}\right)^2}$$

$$s = \sqrt{142 - (5{\cdot}12)^2} = 10{\cdot}76$$

How could a scale factor be used in the $(X - \overline{X}_o)$ column to make the numbers even more manageable?

STANDARD ERROR One way of deciding whether there is a significant difference between the means of samples, or whether the differences are due to chance is to use the standard error (SE). The standard error between two means is calculated as follows.

$$\text{SE} = \sqrt{\frac{s_1^2}{N_1} + \frac{s_2^2}{N_2}}$$

where s_1, s_2, N_1 and N_2 are the standard deviations and sizes of the two samples.

In the example of the buttercups from localities A and B

$$\text{SE} = \sqrt{\frac{20{\cdot}5^2}{5} + \frac{47{\cdot}1^2}{5}}$$

$$= \sqrt{527} \simeq 23$$

It is known that on average, chance differences in sample means greater than 2 × SE occur only once in every 22 trials. When difference exceeds 2 × SE they are said to be *significant*, and when they exceed 3 × SE they are highly significant.

In the case of the buttercups the observed difference between the means is (86 − 72) = 14. This is less than one standard error so the differences could well be due to chance. There is no evidence to suggest a difference in buttercup density between the two localities.

COEFFICIENT OF CORRELATION We have seen that the standard deviation gives a measure of *variance* between samples. When thinking about whether two variables are related (*e.g.* numbers of cigarettes smoked and incidence of lung cancer) we are considering *covariance*. A method of deciding whether two variables are functionally related is to calculate a *coefficient of correlation.* The coefficient normally used is the *product moment correlation coefficient* (r). If the two variables whose relationship we want to test are X and Y, then the correlation coefficient is defined as

$$r = \frac{1}{N}\sum\left(\frac{X-\bar{X}}{s_X}\right)\left(\frac{Y-\bar{Y}}{s_Y}\right)$$

where s_X and s_Y are the standard deviations of X and Y respectively. Mathematically it can be shown that r can only have a value in the range $-1 \leqslant r \leqslant +1$. A value of $r = +1$ indicates a perfect functional relationship, with X increasing as Y increases; a positive correlation. $r = 0$ indicates that X and Y are not functionally related at all. A value of $r = -1$ indicates a perfect functional relationship with X increasing as Y decreases; a negative correlation.

As an example let us consider the result of an experiment in which mayfly nymphs were counted along a stream. Is there a relation between the number of nymphs and the speed of the water? Table 7.7 shows results of the experiment.

Table 7.7 **Mayfly nymphs and speed of water**

number of nymphs per m² quadrat (X)	*speed of water in* m s⁻¹ (Y)
4.5	0.21
12.5	0.29
3.5	0.45
13.5	0.46
6.5	0.6
12.5	0.6
17.5	0.61
6.5	0.78
24.0	0.81
25.0	0.91

To make the calculations simpler it can be shown that if we take an assumed mean of X and Y, transforming them into x and y, then

$$r = \frac{\frac{1}{N}\Sigma xy - \bar{x}\,\bar{y}}{s_x\, s_y}$$

where $\bar{x} = \frac{\Sigma x}{N}$ $\quad \bar{y} = \frac{\Sigma y}{N}$ $\quad s_x{}^2 = \frac{1}{N}\Sigma x^2 - (\bar{x})^2$

$$s_y{}^2 = \frac{1}{N}\Sigma y^2 - (\bar{y}^2)$$

Let us take an assumed mean for X of 12.5 and Y of 0.6.

x	y	xy	x^2	y^2
−8.0	−0.39	3.12	64.0	0.1521
0	−0.31	0	0	0.0961
−9.0	−0.15	1.35	81.0	0.0225
1.0	−0.14	−0.14	1.0	0.0196
−6.0	0	0	36.0	0
0	0	0	0	0
5.0	0.01	0.05	25.0	0.0001
−6.0	0.18	−1.08	36.0	0.0324
11.5	0.21	2.42	132.25	0.0441
12.5	0.31	3.87	156.25	0.0961
1.0	−0.28	9.59	531.5	0.463
Σx	Σy	Σxy	Σx^2	Σy^2

$N = 10$

$\bar{x} = 1.0/10 = 0.1$ $\qquad \bar{x}\,\bar{y} = 0.0028$

$\bar{y} = -0.28/10 = 0.028$

$\frac{1}{N}\Sigma xy = 9.59/10 = 0.959$

$$s_x{}^2 = \frac{531.5}{10} - (0.1)^2 = 53.14$$

$s_x = 7.29$

$$s_y{}^2 = \frac{0.463}{10} - (0.028)^2 = 0.0455$$

$s_y = 0.214$

$$r = \frac{0.959 - 0.0028}{7.29 \times 0.214} = 0.61$$

A graph of the results is plotted in Fig 7.9.

By consulting statistical tables we find that the probability of such a result occuring by chance is less than one in one thousand ($p < 0.001$). We may confidently conclude that there is a direct relationship between the mayfly nymph population and the speed of the water *in this instance*. We should beware of assuming that the correlation would necessarily hold good under other conditions.

CHI-SQUARED (χ^2) TEST This test is very useful in biology. It often happens that we know nothing about the distribution of our organisms; perhaps all we know is the number of individuals coming into several categories. χ^2 tests whether the observed frequencies differ significantly from what we would expect according to some assumed hypothesis.

χ^2 is calculated as follows. If O is the observed frequency of any variable quantity and E is the frequency expected from our hypothesis, then

$$\chi^2 = \sum \frac{(O - E)^2}{E}$$

Suppose we count the number of insects in a sample from a pond habitat, and find that, of 100 insects of a particular species, 40 have long wings (LW) and 60 have short wings (SW). Is there any significant departure from a 1:1 ratio of short:long wings? Our hypothesis is that we would expect to find equal numbers of long and short winged insects; *i.e.* 50 long wings and 50 short wings in a sample of 100.

	LW	SW	Total
O	40	60	100
E	50	50	100
$O - E$	-10	$+10$	
$(O - E)^2$	100	100	
$\left[\frac{(O - E)^2}{E}\right]$	2	2	4

$$\chi^2 = \sum \frac{(O - E)^2}{E} = 2 + 2 = 4{\cdot}0$$

Tables of χ^2 are readily available (*e.g.* Lindley and Miller *Cambridge Elementary Statistical Tables* CUP 1953). Before using the tables we have to deduce the *degree of freedom* (DF). The calculation of the degrees of freedom is largely a matter of practice, but in general the DF is the number of classes minus the number of restrictions imposed in calculating the expected frequencies. In our example, once the number of long-winged insects has been calculated the number with short wings is fixed. Therefore DF $= 2 - 1 = 1$. From the tables we find that with DF $= 1$ and $\chi^2 = 4{\cdot}0$ the probability level $p \simeq 0{\cdot}05$. This means that this result would occur by chance in only five trials in a hundred if wing length were in a 1:1 ratio, or that 'odds' are 95:5 that there is a departure from a 1:1 expectation. When the probability p is less than 0·05 (1 in 20) we infer that there is evidence that the data are significantly different from the hypothesis being tested.

Normally this hypothesis assumes that there is no association between the variable quantities, and it is then called a *null hypothesis*.

χ^2 can also be used to test *association* between contingencies. Set out below are the numbers of observed matings between ♂ insects with long or short wings and ♀ insects with long or short legs.

♀	♂ short wings	♂ long wings	TOTAL
long legs	52	37	89
short legs	28	40	68
	80	77	157

Is there an association of mating partner and wing/leg length? The data above form a 2×2 contingency table. In general such a table can set out as follows.

	X_1	X_2	Total
Y_1	a	b	N_1
Y_2	c	d	N_2
	N_3	N_4	N

It can be shown that

$$\chi^2 = \frac{(ad - bc)^2 N}{N_1 N_2 N_3 N_4}$$

In the wing/leg length example

$$\chi^2 = \frac{(52 \times 40 - 37 \times 28)^2 \times 157}{89 \times 68 \times 80 \times 77} = 4{\cdot}55$$

There is only one DF in this example. $\chi^2 = 4{\cdot}55$ indicates $p \simeq 0{\cdot}03$. The association, an apparent preference of short winged ♂ for short legged ♀ and vice versa, is therefore significant.

7.11 The study of different habitats

Whatever the habitat, wherever possible the area should be mapped, surveyed, and an experimental approach should be thoroughly thought out, including the design of sampling and the method of recording. Arrangements should be made to record relevant climatic factors, including microclimate, and to do necessary soil/freshwater/seawater analysis for biotic and chemical components. Biological observations should be concerned with nutrition, life cycle, mating, behaviour and population dynamics, including reproductive success, mortality factors, and immigration/emigration.

TERRESTRIAL HABITATS These are very diverse but the following may be important – food webs, identification of primary food producers, changes in biomass, plant succession, and human influence.

FRESHWATER HABITATS Microhabitats may include: stones associated with filamentous algae fed on by herbivores: sand/silt/mud in slower water, with burrowing animals, and rooted plants with a fauna of insects and planarians: the water, with fish, plankton and insects such as water boatmen (*Corixa*) and *Dytiscus*, the ferocious, carnivorous great diving beetle: the water surface, with pond skaters (*Gerria*) and water gnat (*Hydrometra*). Adaptations that should be looked for include ways of attachment to prevent being swept along by currents and present gills to the flow, *e.g.* hooks on legs of caddis flies and secretion of adhesive mucus by planarians. Once again human influence (water management, fishing, pollution) should be remembered.

SHORE HABITATS Intertidal habitats include rock pools, rock faces, sand, graded pebbles, or in estuaries mud, brackish pools and salt-tolerant plants. Zonation of plants and animals according to salt-tolerance, and avoidance of desiccation when the tide is out, are two important features. Seaweeds (*Fucus serratus* – *Ascophyllum nodosum* – *Fucus spiralis*) and periwinkles (common – flat – rough – small) show zonation between low tide and splash areas, and lugworms and cockles bury themselves to keep moist and maintain respiration.

The animals can be classified according to diet. Among the carnivores there is the rock-pool dwelling blenny fish, with its powerful teeth, the sand-goby fish with its crustacean diet, the common starfish with a bivalve mollusc diet, and the sessile sea anemone (e.g. *Actinia equina*) eating worms and crustaceans. Among the herbivores are the seaweed eating gastropod molluscs, periwinkles etc., the weed eating shanny fish and the plant eating bristle worms (*Nereis*). Among the detritus feeders are the filter feeding bivalve molluscs and the mud swallowing lugworms.

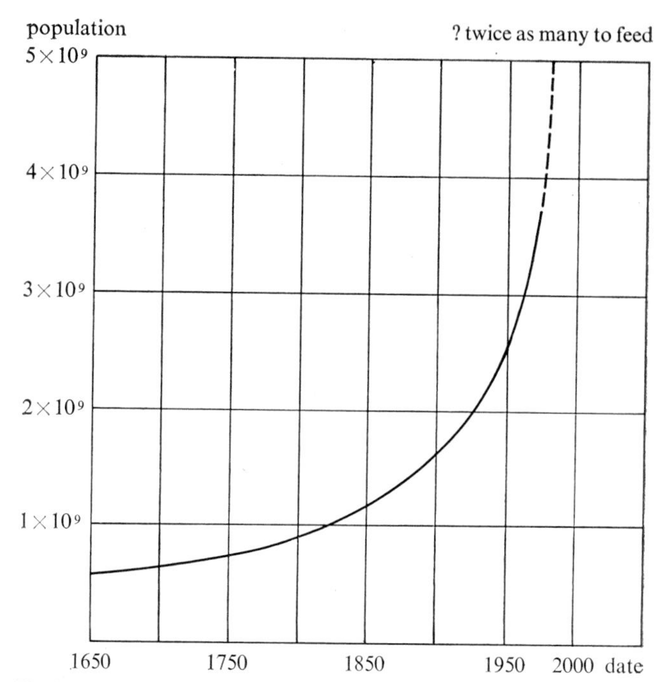

Fig 7.11 **Growth of human world population**

7.12 Autecological studies

These may be very suitable for the rather limited time available for VI form work. A garden or playing field insect, bird, or plant should be available to most students, and animals or plants in hedgerows, woodland, bog, stream, fen, heath or moor can be studied without too much fear of interference. The local Natural History and Conservation societies may be able to help as well as your tutor or teacher. You should be able to answer the following questions, with a well illustrated set of results that stand up to statistical examination.

What niche does my species occupy?
Why?
What does it live on? How much does it eat?
What lives on it and why? How much gets eaten?
How does it reproduce? How much? How often?
How is it adapted to its environment?
Is the population rising/stable/falling?
Why?

7.13 Man as an ecological factor

As we have said before man is likely to be an important biotic factor in almost every environment. Man is different from the other species in that he is not adapted to any particular environment; he appears at first sight to be particularly ill-adapted to any environment. Yet he can survive (after a fashion) at the North Pole, in the Sahara desert, even on the Moon. Since he first discovered how to use tools, he has been able to change the environment around him.

POPULATION Man is also special in that he is one of the few species whose numbers are increasing rapidly. It has taken tens of thousands of years for the world's population to achieve its present level (3 500 million). But the increase is exponential (Fig 7.11) and so this figure could be doubled in less than 50 years. The main reason for the rate of increase is the decline in the mortality rate, particularly infant mortality, and a consequent increased expectation of life at birth. This means that more people are reaching reproductive age, and so increasing the birth rate. The major cause of the decrease in mortality has

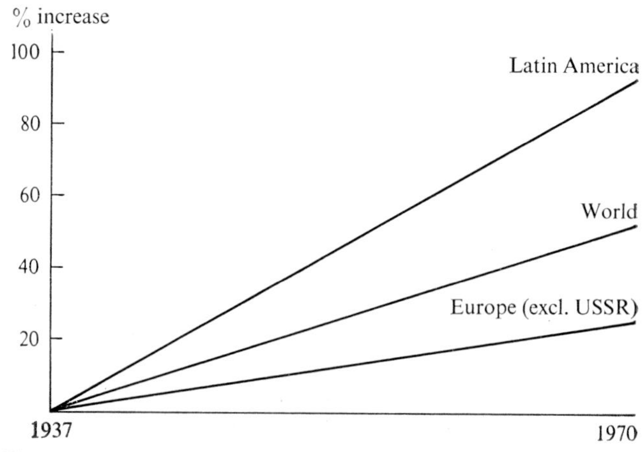

Fig 7.12 **Comparison of population increase by area**

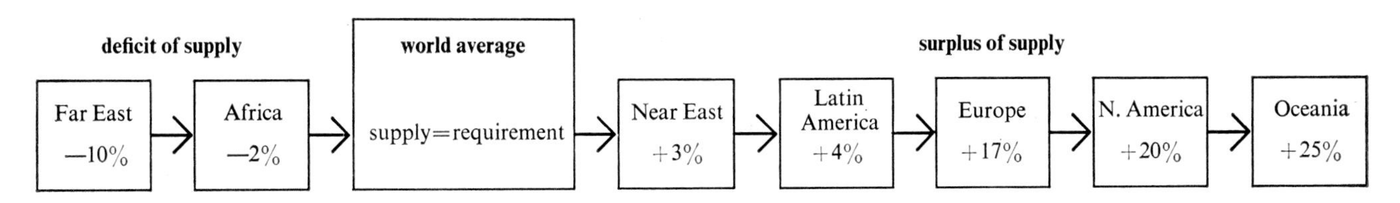

Fig 7.13 **Deficit and surplus of world food supplies**

been the advances made in medical science, and the improvements in public health services.

The greatest increase in the population has been in the poor countries of the Third World (Fig 7.12), increasing their problems still further.

World food supplies have risen to match the increase in population, but the distribution is unequal (Fig 7.13). But it is not at all clear that food production can continue to keep pace with population if it continues to increase at the present rate although there are still large areas of the world's land surface that might be used for farming. However, present day thinking is more towards increasing productivity through mechanisation, more efficient conversion of leaf protein into human food, fertilisers, new varieties and stocks, particularly in underdeveloped countries. However the problems of raising farm output in the developing world are as often social as technological. Many societies do not see any value in producing more than they themselves need; extra time spent in the fields lessens the time that can be spent in social activities which are customary. In other societies there are actual social sanctions against over-production.

To avoid the problems of lack of food, measures are having to be taken to cut the birth rate, such as education in methods of birth control and the supply of contraceptives.

MAN AS PREDATOR Early man was a hunter and gatherer relating to other species as other omnivores do today. Today he is still a predator on many species, but only in a few parts of the world (*e.g.* bushmen in the Kalahari desert) is this the principle means of subsistence.

MAN AS HUSBANDER In the Near East between about 10 000 and 7 000 BC man first began to domesticate animals (*e.g.* sheep and goats) and plants (*e.g.* wild wheat). From these early beginnings man has come to rely almost exclusively on husbandry to provide his food, and in so doing has changed the ecological face of the earth. By means of artificial selection and breeding he has improved favoured species and protected them from their natural enemies. Now, for instance, 75% of man's plant food requirements come from only 20 species.

But the consequences for other species have been far reaching. Many have become extinct. The cutting down of forests, the increasing use of chemicals in agriculture, the pulling out of hedgerows have all affected the delicate interactions in nature.

As the population of the world increases the problems of providing enough food increase the pressure on the land. If more is taken from an ecosystem (in the way of crops) than is put back, then the system becomes degraded. There is the temptation to plough up marginal grazing land, which if conditions are wrong can cause the top soil to be blown off, as happened in the Oklahoma 'dustbowl' during the early part of this century. Even the practice of monoculture, the growing of areas of a single crop, has produced special problems. Insects to which the crop is a host can multiply rapidly when the host is present in such vast quantities. The insect becomes a *pest*, and may drastically reduce yields if not controlled. A particularly effective method of control is the spraying of chemical insecticides. Unfortunately it is difficult to produce a selective insecticide, *i.e.* one that kills the pest only. Harmless insects, which may be the food of other animals and birds, perish as well.

Another problem associated with chemical insecticides is the problem of stability. One of the most effective groups of insecticides are the polyhalogenohydrocarbons of which DDT is an example. They have extremely stable molecules, and so have a *cumulative* effect. Animals at the end of food chains are particularly vulnerable. For instance the chemical may be present in very small concentrations in insects; but insectivorous birds eat many insects, so the concentration in their bodies is higher; the concentration in the body of the predator that feeds on these birds could be fatal. Recently deaths and low fertility in predacious birds such as sparrowhawks has been shown to be due to the accumulation of polyhalogenohydrocarbons. Very low but measureable quantities of these chemicals have been found in rain water, in the sea and in the polar ice. Remembering that man himself is on the end of a food chain many countries, including Britain, have severely restricted the use of these compounds. But the benefits of DDT should not be underestimated. It has been extremely successful in controlling malaria in many parts of the world, and so saved many thousands of lives, and made previously uninhabitable areas habitable.

MAN AS POLLUTER There is always a risk that a side effect of attempting to produce more food, more roads and buildings, and more consumer goods will damage the environment. This is what we mean by the term pollution.

Pollution takes many forms, one of which is the *eutrophication*, or nutrient enrichment of water. When excess nitrogenous compounds such as sewage are discharged into a lake or river, they are quickly broken down by micro-organisms (by a process that is *aerobic* or oxygen consuming) into nutrients that can readily be used by aquatic plants such as algae. As a result of extra nutrients the biomass of the algae increases dramatically, making the water more opaque and reducing the amount of light that reaches deeper water; this then becomes less biologically active. The dense, unhealthy algae are themselves attacked by bacteria, also an aerobic process, so that even less oxygen is available for other organisms such as fish.

Eutrophication is also caused by the use of artificial fertilisers which man applies to the land to increase food production. Nitrates and phosphates are applied to supplement levels or make good deficiencies of these nutrients. However some is drained from the land in solution and flows into rivers.

The point to note is that sewage, nitrates, phosphates etc. are not in themselves harmful. In small quantities they are beneficial to all forms of life. But it is when they are dumped indiscriminately in such quantities that the normal cycles are broken that enrichment becomes pollution.

Apart from the polyhalogenohydrocarbons mentioned above many poisons such as lead and mercury, which are also cumulative, are released into the air as an undesirable by-product of industry. The lead content of the air, contributed to by many centuries of lead smelting, is causing particular concern. Many governments have banned its use as an anti-knock device in petrol.

There are also many pollution problems associated with the energy requirements of industrial society. As well as the production of poisonous gases by the internal combustion engine and the damage to marine and bird life caused by accidental or deliberate discharge of oil at sea (Fig 7.14), there are the problems of waste from nuclear power stations. As oil and coal reserves run out and the political problems associated with them become more acute nuclear power becomes increasingly attractive to governments. But a by-product is radioactive waste, 'with a half-life of centuries' that must be dumped somewhere.

Other forms of pollution include noise which can be damaging to health, and litter—especially when made from plastics that cannot be broken down.

Even the warm water discharged by power stations can be a source of pollution as it may disrupt well established eco-systems. But new species may be able to make use of the new environment, so it is certainly not pollution to them. This is the important point about the term pollution. It is used to imply damage – but what is damaging to one species may be helpful to another. So always ask the question – pollution for whom?

MAN AS A CONSERVER Man has tremendous power over his environment. He can make it work for him, and produce the food that he needs. Or by mismanagement he can turn a fertile area into a desert. An appreciation of this has led to the concept of *ecological management* which attempts to regulate biological changes rather than submit to them. This involves keeping density of a species in a given environment at an economic level – trying to prevent our seas from being overfished for instance. The problem is not confined to food. As the affluence of the world increases so the demand for raw materials such as fossil fuels and timber increases. Without careful management we could completely destroy our forests.

Many man-made changes produce no rapid ill-effects, but they may have profound biological consequences, such as the changes in forest tree species that may follow the invention of chipboard.

Man can also conserve nature for reasons that are not solely economic. It may mean attempts to prevent a rare species from becoming extinct. It may also mean attempts to preserve the character of an ecosystem – as small as a nature reserve, or as large as 'the countryside' and protect it from violent, or ill-considered, or irreversible biological changes. In the case of the nature reserve, man tries to pretend that there he is not a biotic factor at all.

Fig 7.14 **Oil pollution as a threat to bird life**

8 Respiration

8.1 Introduction

Respiration is a term that can be used at three levels;
1 the *level of the complete organism*, where it means the inspiration/expiration of air — we refer to this as 'breathing';
2 at *tissue level*, where it refers to the chemical/osmotic processes involved in carbon(IV) oxide/oxygen exchange;
3 at *cell level*, where it is concerned with cell energetics, the way in which the living organism extracts energy from molecules.

Many reactions in the cell are endothermic, *e.g.* synthesis of proteins, and for the cell to perform this type of reaction, it must have not only the correct enzymes, but also an adequate supply of energy from another source; *i.e.* a series of exothermic reactions must run parallel with the endothermic ones. The endothermic and exothermic reactions are then linked so that the overall free energy change (ΔG) is either zero (thermodynamic equilibrium) or negative. If ΔG is negative there is liberation of energy overall and spontaneous reaction is possible. In animals and plants, the principal cellular reaction for releasing energy is the oxidation of glucose.

Probably the most important energy carrier in the cell is ATP or high energy phosphate. The structure of ATP or adenosine triphosphate(V) is given in Fig 8.1. The energy produced in respiration, or indeed photosynthesis (§9.2), is used to make ATP which carries the energy to where it is needed in the cell. To extract the stored energy ATP is converted to ADP (Fig 8.2) by hydrolysis.

$$\text{ATP} + H_2O \rightarrow \text{ADP} + \text{P} \quad \triangle H = -33{\cdot}6 \text{ kJ mol}^{-1}.$$

The presence of ATP in muscle had been demonstrated by 1929. The energy of ATP can be stored in other molecules, for example, in muscles it converts creatine to creatine phosphate(V).

The substrates which break down to release energy (coupled to ATP synthesis) arrive at the body as food (except in green plants). Food reaches the cell, as a rule, after it has been broken down to smaller molecules which the cell can handle. The breaking down is known as *digestion* in man; the resulting mixture of sugars, amino acids and fatty acids is absorbed through the intestine wall and enters the blood stream. Under normal conditions, the majority of the cell's energy requirements are met by breakdown of sugars, although both fats and proteins can also be broken down to aid synthesis of ATP.

8.2 Respiration of carbohydrates

The respiration of carbohydrates takes place in two distinct stages. The first stage is called *glycolysis*, which literally means 'sugar splitting'. In this stage C-6 sugar is oxidised to 2-oxopropanoic (pyruvic) acid. What happens in the second stage depends on whether or not oxygen is available. If oxygen is available the respiration is *aerobic*, and the 2-oxopropanoic acid enters the *Krebs' cycle* (p 125). If oxygen is not available then the respiration is anaerobic, and the 2-oxopropanoic acid

or A-Ⓟ-Ⓟ-Ⓟ

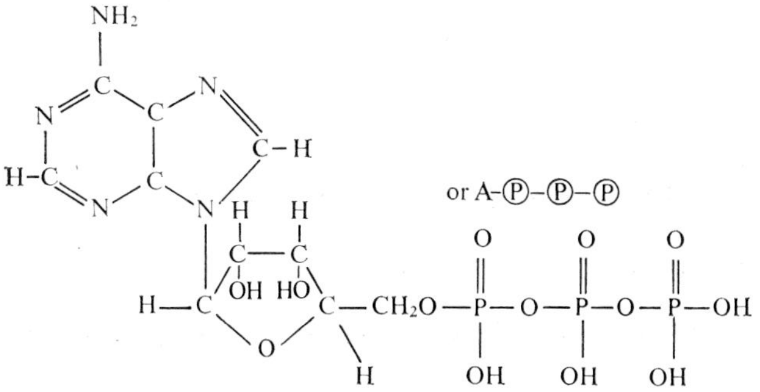

Fig 8.1 Adenosine triphosphate (ATP)

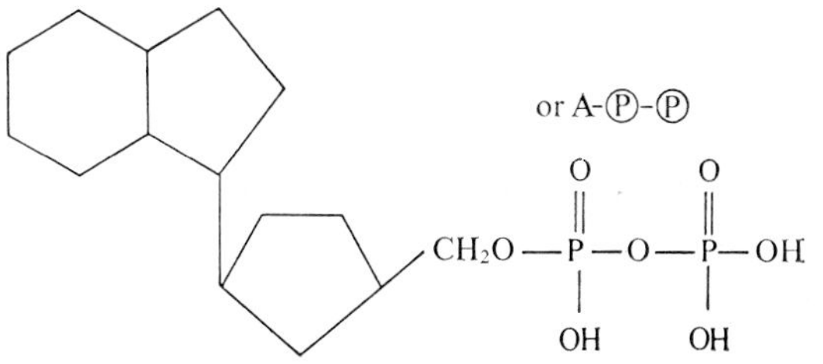

or A-Ⓟ-Ⓟ

Fig 8.2 Adenosine diphosphate (ADP)

is broken down into ethanol in plants or into 2-hydroxypropanoic (lactic) acid in animals. This is shown schematically in (Fig 8.3).

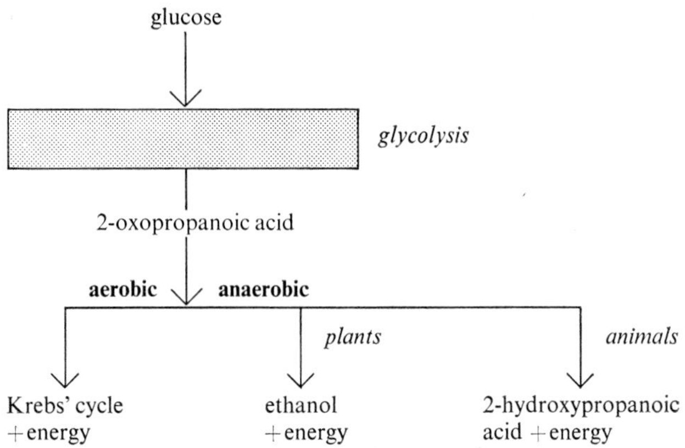

Fig 8.3 **Aerobic and anaerobic respiration**

GLYCOLYSIS Glucose can enter the animal cell either directly through the circulatory blood system or be present in some body cells as the storage polysaccharide *glycogen.* Glycolysis can itself be divided into two stages. Firstly, the C-6 sugar is *phosphorylated.* Phosphorylation is an endothermic process, the energy coming from the change of ATP into ADP. The terminal phosphate(V) group of ATP becomes attached to the sugar molecule, forming a sugar phosphate(V). Two phosphorylations take place so that a total of two phosphate(V) groups are added to each molecule of sugar (fructose 1, 6-diphosphate(V) is eventually formed).

sugar + 2 ATP→sugar-diphosphate(V) + 2 ADP

The phosphorylation process is vital because it gives energy to the sugar molecules that can be released later on.

The glycolysis sequence can start with larger molecules which do not necessarily have to go to glucose first, but can undergo the initial reaction of phosphorylation directly.

The second stage of glycolysis involves the breaking down of the phosphorylated sugar into two C-3 molecules (triose sugar), each having one phosphate(V) grouping. Each of these fragments is then oxidised to a C-3 acid by removal of hydrogen.

C—Ⓟ, C, C, C, C, C—Ⓟ (chain not ring just for convenience) → C—Ⓟ, C, C + C, C, C—Ⓟ

CH_2O—Ⓟ, CH(OH), CHO $\xrightarrow{H_2O}$ CH_2O—Ⓟ, CH(OH) + H_2, COOH
3-phosphato (V)-2-hydroxypropanoic acid

3-phosphato (V)-2-hydroxypropanal

Fig 8.4 **The breakdown of phosphorylated sugar and the glycolysis flow diagram**

This reaction is linked to the conversion of 1 ADP plus phosphate(V) to 1 ATP, an oxidation requiring the low relative molecular mass cofactor NAD (nicotinamide adenine dinucleotide). Thus, the above oxidation is coupled with the complementary reduction of NAD, in a REDOX reaction.

$NAD + [H_2] \rightleftharpoons NADH_2$

However, no useful energy is derived from the removal of the two hydrogen atoms by NAD as the process takes place in the cytoplasm, outside the mitochondria. The hydrogen atoms are used in reduction processes in the cytoplasm.

Four further sequences then convert the 3-phosphato (V)-2-hydroxypropanoic (phosphoglyceric) acid to 2-oxopropanoic acid, and in the process the phosphate(V) group from 3-phosphato(V)-2-hydroxypropanoic acid becomes attached to ADP forming ATP. Thus, so far we have 2 ATP from 1 mole of glucose (two ATP's are needed to phosphorylate each glucose molecule and each glucose molecule then yields two molecules of 2-oxopropanoic acid).

In the glycolysis flow diagram (Fig 8.4) remember that each step is enzyme-catalysed (and these come from the cell matrix, *not* the mitochondria).

ANAEROBIC RESPIRATION In the absence of oxygen the sequence stops here and the NAD is reformed by

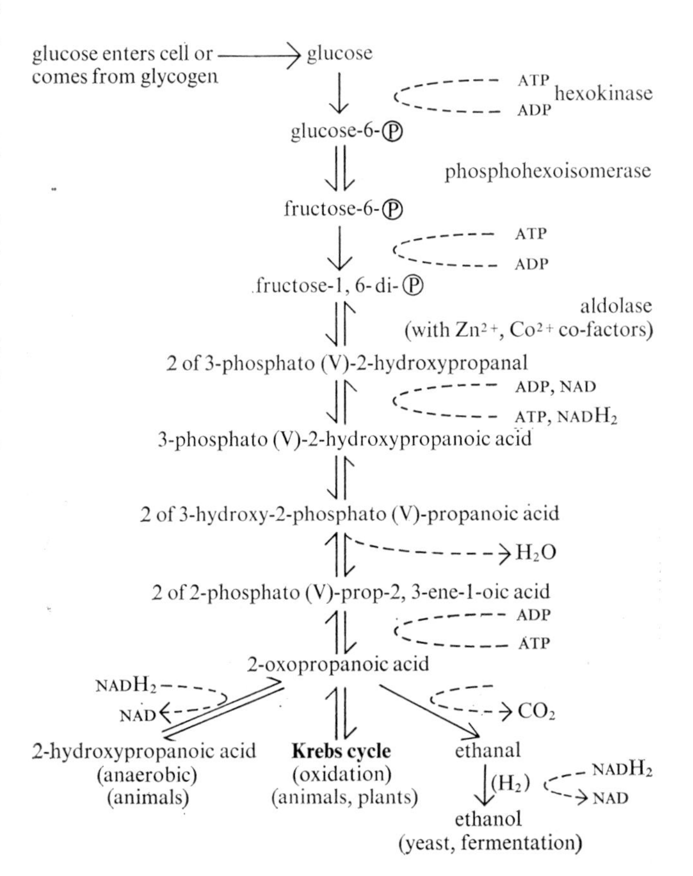

reduction of 2-oxopropanoic acid to 2-hydroxypropanoic acid.

$$\begin{array}{ll} CH_3 & CH_3 + \text{NAD} \\ | & | \\ C=O + \text{NADH}_2 \rightleftharpoons & CH(OH) \\ | & | \\ COOH & COOH \end{array}$$

2-oxopropanoic (pyruvic) acid	2-hydroxypropanoic (lactic) acid

In yeast an alternative reaction takes place, ethanol being formed together with carbon(IV) oxide. This is *fermentation*.

$$\begin{array}{ll} CH_3 & CH_3 \\ | & | \\ C=O + \text{NADH}_2 \rightarrow & CH_2 + CO_2 + \text{NAD} \\ | & | \\ COOH & OH \end{array}$$

ethanol

Some micro-organisms will generate 2-hydroxypropanoic acid. For example it is built up in the souring of milk, a fermentation which turns milk to *yoghurt*.

AEROBIC RESPIRATION The oxidative phase of respiration involves the sequence called the Krebs' cycle (Fig 8.5). The first stage in the degradation of 2-oxopropanoic acid is to oxidise it to ethanoic (acetic) acid.

$$2CH_3COCOOH + O_2 \rightarrow 2CH_3COOH + 2CO_2$$

ethanoic (acetic) acid

It involves several stages requiring co-enzymes, for example co-enzyme A, a B-vitamin named pantothenic acid. In fact although ethanoic acid is the active component, it enters the Krebs' cycle in the form of ethanoyl co-enzyme A. If we consider 2-oxopropanoic acid being converted to ethanoyl co-enzyme A, carbon(IV) oxide is evolved and the 2-oxopropanoic acid loses two hydrogen atoms which eventually cause the formation of three units of ATP. Ethanoyl co-enzyme A links glycolysis with the Krebs' cycle reactions. Although the oxidation has been written in a straight-forward chemical manner above, it really involves the use of NAD to accept hydrogen atoms, forming NADH_2 (cf oxidation of 3-phosphato(V)-2-hydroxypropanal). It is in oxidising the remaining two carbon atoms of ethanoic acid to carbon(IV) oxide that the Krebs' cycle reactions (Fig 8.5) are involved. A C-2 molecule combines with a C-4 molecule to form a C-6 molecule. This is then degraded stepwise forming a C-5 and then a C-4 molecule. Two carbon(IV) oxide molecules are evolved. The C-4 molecule is then converted into a form suitable for combination with ethanoic acid. Oxidation stages result in the production of NADH_2 from NAD and ATP from ADP. What the cell is really doing is to pick the ethanoic acid molecule to

2-oxopropanoic acid (3-C)
NAD
NADH_2
ethanoic acid (2-C)
(4-C)
2-hydroxypropane-1, 2, 3-tricarboxylic acid (6-C)
FPH_2 (a reduced co-factor like NADH_2)
CO_2
NADH_2
(5-C)
(4-C)
CO_2, ATP, NADH_2

Fig 8.5 **The Krebs' (citric acid) cycle**

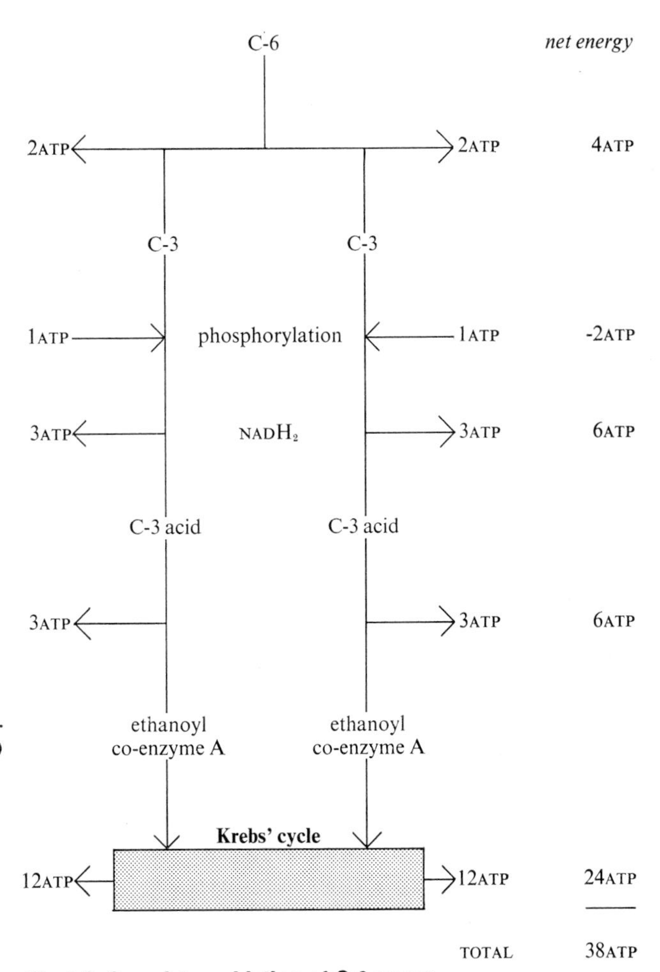

Fig 8.6 **Complete oxidation of C-6 sugar**

pieces very carefully so that ATP can be synthesised from ADP. The Krebs' cycle is the centre-piece of cell energetics.

If the cell needs energy quickly, the demand for oxygen may be so great that it exceeds the concentration in the cell. The cell then operates glycolysis without the oxidative part of the sequence, producing only a small amount of the energy which would normally be obtained but giving the cell a second line of attack should the oxygen run out. The 2-oxopropanoic acid formed is reduced to 2-hydroxypropanoic acid and this is released into the bloodstream for return to the liver, where glycogen can be resynthesised. Glycolysis provides much of the energy for a sprint run, and at the finishing line the athlete would have a high 2-hydroxypropanoic acid content in his blood. A build up of 2-hydroxypropanoic acid in muscle causes *cramp* pains.

The Krebs' cycle is the most important source of energy in the metabolic process. It can be shown that for every molecule of ethanoic acid passing into the cycle 12 ATP molecules are formed. What happens when one molecule of C-6 sugar is completely oxidised is shown in Fig 8.6. The complete oxidation results in the production of 38 ATP units.

By separating organelles of cells by centrifugation it has been shown that glycolysis takes place in the cell cytoplasm, whereas the Krebs' cycle (and the transfer of hydrogen atoms and electrons) takes place in the mitochondria. Mitochondria are the sites of ATP production; Krebs' cycle reactions take place in the matrix and ATP producing (electron transfer) reactions occur on the inner membranes and cristae.

8.3 Fats in cells

Fats are first hydrolysed to fatty (carboxylic) acids and propantriol (glycerol), and the fatty acids are then oxidised to smaller relative molecular mass acids, in the presence of co-enzyme A, to form ethanoic acid which enters the Krebs' cycle. This process continues until the fatty acids have been completely oxidised. A fatty acid with a 6-carbon skeleton yields energy equivalent to 44 ATP on oxidation. However the Krebs' cycle requires carbohydrate fuel for continued operation. The enzymes required to oxidise fatty acids occur in the cell mitochondria.

8.4 Proteins

Proteins are only degraded when there is great starvation and all carbohydrates and fats have been used up. When this happens the protein is first hydrolysed to amino-acids. These are deaminated with the eventual formation of a keto-acid RCOCOOH. This either reacts with co-enzyme A to form ethanoic acid, which enters the Krebs' cycle, or groups are stripped off it by successive oxidation reactions. Less energy is derived than for fats or carbohydrates.

8.5 Enzymes

The enzymes for anaerobic glycolysis are found in the cell matrix, but those for the Krebs' cycle and oxidation of fatty acids are found in the matrix of the *mitochondria.* These take in substrates and produce ATP which is released to the cell. The higher the rate of metabolism of a cell the more mitochondria it will have. The oxidases, phosphorylases and hydrogen accepting enzymes are in the cristae and membranes of the mitochondria.

8.6 Respiratory quotient

The rate of aerobic respiration can be determined by a measurement of the O_2 used, the organic material consumed, or the CO_2 formed. In most cases both O_2 consumed and CO_2 formed are measured and the rate of respiration is quoted as the quantity of gas consumed or produced per fixed mass of tissue per unit time. This rate is affected by the substrate used. The relationship between the O_2 and CO_2 is expressed as the *respiratory quotient* (RQ) which is given by

$$\text{RQ} = \frac{CO_2 \text{ formed}}{O_2 \text{ consumed}}.$$

It is of value when determining what food materials are being oxidised in respiration. For example carbohydrates have a RQ of 1·0, fats 0·7, proteins 0·85 (animals) and 1·0 (plants); organic acids have a RQ of 4·0 as they are rich in oxygen.

8.7 Animals

Whereas the ambient temperature affects the respiration of poikilothermic ('cold blooded') animals, the homoiotherms ('warm blooded') respire in accordance with the basal metabolic rate. Mass for mass, the homoiotherm has a much higher respiratory rate. For very small animals the rate of heat loss to the surroundings is high

Table 8.1 **Breathing organs in different species**

species	*breathing organ*
man	lungs
spider	lung books
insect	tracheae
fish	internal gills
worm (aquatic)	external gills

(large surface area compared with volume) and so is the respiratory rate.

Breathing is the inhalation/exhalation process which aids the interchange of gases between the blood and the surroundings. Structurally, the organs for breathing vary widely (Table 8.1).
All systems have the common property of providing a large surface area for gas exchange.

GAS EXCHANGE IN MAN In man the inhaled air passes through the nose to the nasal cavity, is warmed and moistened and dust is filtered off by hairs and mucus on the nasal membrance, and then directed downwards via the *pharynx* to the *larynx* (the enlarged upper part of the *trachea*). Inside the larynx are the vocal chords (connective tissue) and outside it are cartilage rings. From the larynx (Fig 8.7) the air goes to the trachea, lined with mucous membrane, covered with *cilia;* the motion of these keeps the mucous secreted by the trachea lining moving upwards to the back of the throat, where it is swallowed down again.

The trachea divides into two *bronchial tubes* (bronchi) and each *bronchus* divides into many small tubes called *bronchioles*. These have *alveoli* at the ends, each of which is a microscopic empty sphere surrounded by *capillaries.*

The lungs are in the *pleural cavity,* whose floor is the diaphragm, separating the pleural and abdominal cavities. The wall of the thorax makes up the sides of the pleural cavity and it is supported by ribs. Air is inspired and expired by increasing and decreasing the volume of the pleural cavity. As the volume increases, the air pressure is lowered and air rushes down the trachea to equalise the pressure, filling the lungs. The increase in volume is caused by *1* the lowering of the diaphragm, *2* the outward and upward movement of the ribs. A man at rest inhales and exhales about 0·5 dm^3 of air per cycle and this exchange (tidal air) is repeated about once every four seconds.

Table 8.2 **Differences in inhaled and exhaled air**

	inhaled	*exhaled*	*difference*
N_2	79%	79%	zero
O_2	20%	16%	−4%
CO_2	0·04%	4%	+4%
noble gases	trace	trace	
water vapour	varies	saturated	
temperature	varies	body temperature	

For transport to the body cells by the blood, oxygen must diffuse into the blood through the membrane of the alveoli and the capillary wall; carbon(IV) oxide moves in and out in the same manner but the process is more complex than diffusion. The tendency for a gas to diffuse into a liquid is measured by its partial pressure. Dalton's law states that each gas in a mixture of gases exerts a

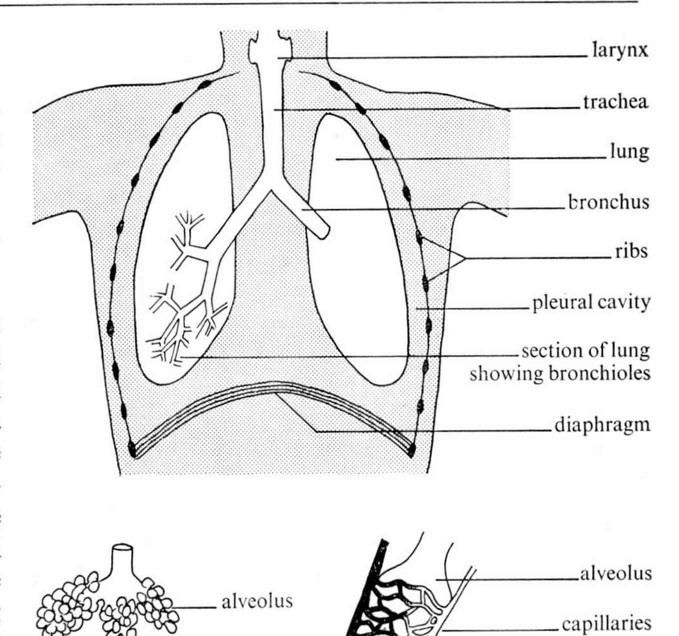

Fig 8.7 **Lungs in man**

pressure proportional to its percentage in the mixture disregarding the other gases present. The tendency of a gas to escape from a liquid is called gas tension, expressed in $N\ m^{-2}$ (or in mm of mercury). Oxygen passes into the blood and a state of equilibrium is reached when the number of molecules leaving the blood per second equals the number of molecules passing from the air into the blood per second, *i.e.* the oxygen pressure equals the partial pressure of oxygen in the air. This value is $21 \times 10^3\ N\ m^{-2}$ (standard atmospheric pressure is approximately $1 \times 10^5\ N\ m^{-2}$). If the partial pressure of oxygen increases there is a tendency for additional oxygen from the air to diffuse into the blood, whereas a reduction of oxygen pressure would cause oxygen to diffuse in the opposite direction.

Table 8.3 **Gas tension/partial pressure in $N\ m^{-2}$**

gas	*alveolar air*	*arterial blood*	*venous blood*
O_2	$13·3 \times 10^3$	$13·1 \times 10^3$	$5·3 \times 10^3$
CO_2	$4·6 \times 10^3$	$5·3 \times 10^3$	$6·2 \times 10^3$

The oxygen diffuses from the alveoli to the blood as it passes through the capillaries to the veins; the oxygen partial pressure in the alveoli is less than $21 \times 10^3\ N\ m^{-2}$ because the lungs are not filled with air completely at each breath. Carbon(IV) oxide diffuses from the venous blood to the alveoli.

Rate of respiration is measured by a *spirometer*; the person breathes into a mouthpiece linked to a metal piece supported in water. The metal piece is linked, via a pulley, to a recording pen which traces a graph on a revolving reel of paper.

Most oxygen is carried in the blood in combination with haemoglobin (Hb) in the erythrocytes; a small percentage is dissolved in the plasma. In many crustaceans and a few molluscs it is transported in combination with haemocyanin (a colourless or blue pigment containing copper). Haemoglobin contains iron and occurs in chordates, arthropods, annelids and molluscs. The haem (iron(II)-containing) part is attached to a protein called globin to form a molecule of high mass (about 7×10^4).

$Hb + O_2 \rightleftharpoons HbO_2$ (oxyhaemoglobin)

The globin part differs in different species. The amount of oxygen carried depends upon the partial pressure of the oxygen, a maximum of 4 'O_2' per haemoglobin molecule is possible in humans, the relevant partial pressure being just under 10^4 N m^{-2}.

The above reaction can proceed in either direction depending on the conditions in the tissue. In the lungs it goes from left to right because the concentration of oxygen is high. It remains as oxyhaemoglobin until the erythrocytes reach capillaries where the oxygen concentration is low (due to cell activity); here oxygen is released. Very active cells are all the time using up oxygen and so oxygen from the blood capillaries diffuses into them.

Since cell respiration releases carbon(IV) oxide, this concentration in the cells and tissue fluids is higher than in the blood stream; carbon(IV) oxide diffuses into the blood in the capillaries where it is converted to carbonic acid (enzymically, using carbonic anhydrase).

$$CO_2 + H_2O \rightarrow O{=}C(OH)_2$$

The carbonic acid combines with Na^+ and K^+ ions in plasma forming hydrogencarbonates which maintain the pH of the blood approximately constant (*buffering*). When the blood comes to the capillaries of the lungs the changes are reversed since the carbon(IV) oxide tension is low (lung action). The HCO_3^- ions break down mediated by carbonic anhydrase forming carbonic acid, which then gives carbon(IV) oxide and water. A little CO_2 is also carried in true solution in the blood plasma and a little in combination with haemoglobin as carbaminohaemoglobin. Buffering is necessary since H_2CO_3 might ionise and lower the pH, which is at a critical value.

GAS EXCHANGE IN OTHER ANIMALS Gas exchange in other animals depends on the size and complexity. *Simple diffusion* of gases, O_2 in, CO_2 out, occurs between protozoa, coelenterates and platyhelminths and their surrounding water. Oxygen also diffuses in through the epidermis of the earthworm, but there is a simple circulation to the tissues via the dorsal and ventral blood vessels.

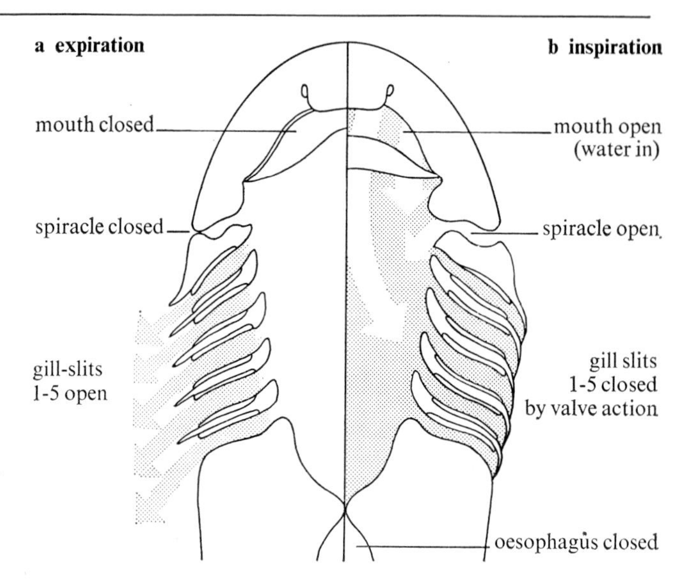

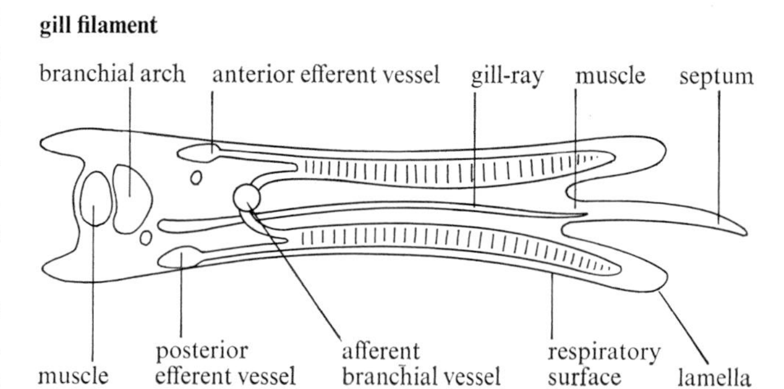

Fig 8.8 **Respiration in the fish**

In many arthropods air percolates through lateral openings called spiracles, whose opening is controlled by *occlusor* muscles. From here the air flows into the tracheae (the method is called *tracheal respiration*) which branch into tracheoles which closely surround the tissues and contain a fluid in which the oxygen dissolves. The passage of air into the tracheae is aided by flexing of the insect (cf flexing of leaves of a plant), and is very rapid, consistent with the high metabolic rate of the insect. There is an oxygen gradient along the respiratory pathway. When the metabolic rate is high the tracheolar fluid may be absorbed into the tissues by osmosis and the tracheoles oxygenate the blood directly. Insect blood does not have an oxygen carrier but some oxygen is dissolved in the blood. Part of the CO_2 may be removed by the reverse route but most combines with nitrogenous waste products to form uric acid which is excreted.

Gills are found in most chordates, at least in the embryo or early stages of development, and in many crustaceans (*e.g.* crayfish) and molluscs (*e.g.* mussel). In the dogfish there are five pairs of gill slits, lateral clefts in the pharynx which extend right down to the alimentary canal. The gills, one between each pair of

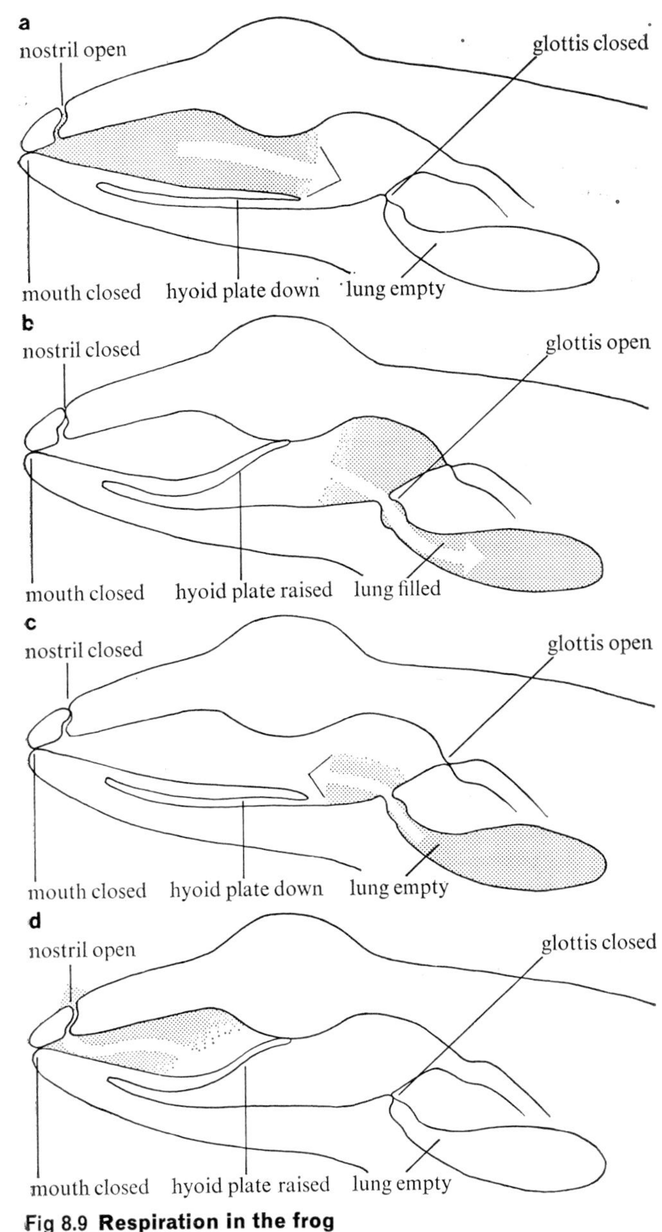

Fig 8.9 Respiration in the frog

slits, are filamentous and thus present a large surface area for the exchange of gases between blood and water. The method is thus a specialisation associated with aquatic life. Fig 8.8 shows the structure and details of the breathing cycle.

The gills are red owing to their high blood supply, and are protected. Deoxygenated blood is brought to the gill via afferent branchial vessels from the ventral aorta and flows across the many capillaries in the filament. Water is pumped across the filament by suction of the parabranchial cavity or by raising the floor of the branchial cavity. Removal of oxygen from the water takes place and the blood, containing oxyhaemoglobin, passes to the efferent epibranchial vessels, thence to a dorsal vessel and hence around to various tissues.

The frog exhibits several gaseous exchange mechanisms during its life. Newly-hatched tadpoles use external gills, which have thin walls and allow gases to diffuse freely in and out of the blood being pumped through them. They are soon replaced by internal gills which closely resemble those of the fish. Adult frogs carry out gaseous exchange by three methods.

1 The moist skin is well supplied with blood vessels; this method is used when the frog is on land or under water.

2 The lining of the mouth cavity is also well supplied with blood vessels. The frog moves its mouth up and down and air passes in and out through the nostrils. This method is used on land.

3 Lungs, which are used in emergencies. They are simpler than mammal's lungs, with no ribs or diaphragm. The frog pumps air into the lungs with its mouth (Fig 8.9).

Animals have widely differing oxygen consumptions. For example a resting butterfly consumes 0·17mm^{3}s^{-1} (150 times more when in flight), whereas a resting *Lumbricus terrestris* consumes only 0·015mm^{3}s^{-1}.

8.8 Plants

In lower plants, simple diffusion over the tissue surfaces is effective (*e.g.* algae, fungi) aided by the fact that the habitats tend to be damp or aquatic. In higher plants, the *stomata* (non-woody) and *lenticels* (woody) are used (Fig 8.10). The stomata (singular stoma) are small pores in the leaf epidermis bordered by two *guard cells* which cause the stoma to open and close. The ability to do this depends mainly on the peculiar thickening of the walls of the guard cells. The wall remote from the pore is thin and the other walls are somewhat thicker. When the guard cell absorbs water, its volume increases owing to turgidity, the increase in volume being rendered possible by a change in shape of the cell. The thin wall of the cell bulges out, the cell moves a small amount, and the pore opens. When the turgor falls, the opposite movement occurs and the pore closes. The mechanism of guard cell operation is complex. The traditional explanation revolving around sugar concentration and osmotic pressure does not explain the rapidity of the response observed. Most plants (*e.g.* ordinary herbaceous plants) have stomata only on the lower surface of the leaf, but in some (*e.g.* grasses) they are found on both upper and lower surfaces. On the floating leaves of water plants they are on the top side only. Xerophytes (desert plants) have very small stomata. Their number varies from about

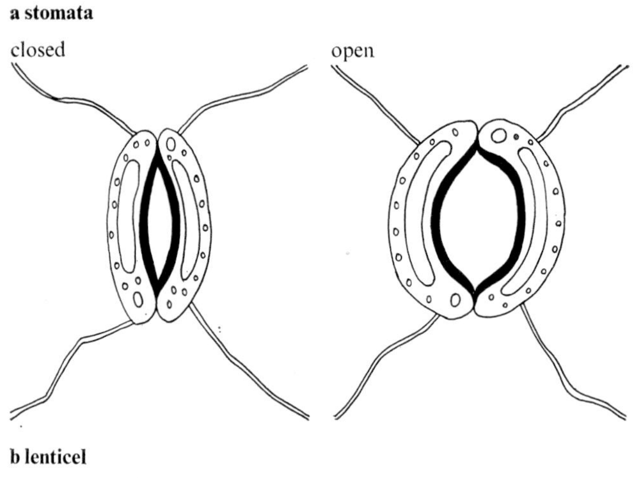

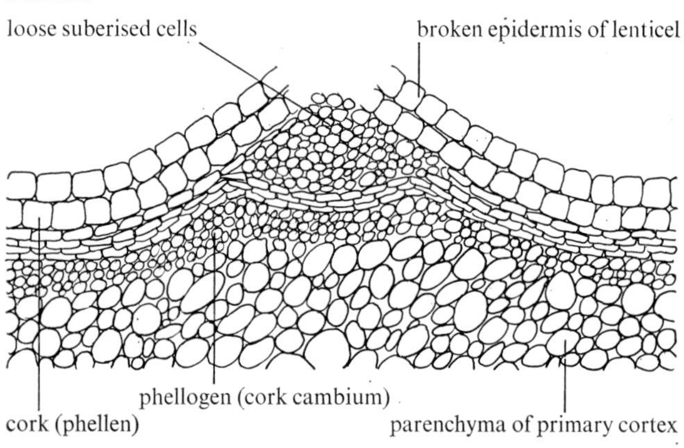

Fig 8.10 **Stomata and lenticels**

25 to more than 1 000 per mm^2.

When the stomata are open oxygen and carbon(IV) oxide can pass through. Under normal conditions, a plant will use up much of the oxygen it forms during photosynthesis.

Examination of the surface of the cork covering a young stem shows that slits are present, filled with a brown powder. These structures are called *lenticels*. They are formed in the phellogen to allow gaseous exchange. They arise below the stomata in the original epidermis. The suberised cells are not tightly packed, leaving intercellular spaces and it is through these spaces that slow gaseous diffusion can take place.

8.9 **Conclusion**

Rates of respiration (*e.g.* in m^3 or mm^3 of CO_2 per kg of dry mass tissue per ks) can be measured by observation of mass changes (difficult) or volume changes. Thus, the air that has passed over the organism/tissue can be collected and passed through barium hydroxide solution, precipitating barium carbonate; a back titration can be carried out to determine how much of the alkali has been used (Fig 8.11). Alternatively, the volume of gas can be measured manometrically, *e.g.* using Warburg's manometer.

Gas exchange in the dark should be studied. In the light, photosynthesis exceeds respiration and so carbon(IV) oxide formed is used again in the nutrition. In photosynthesis (dark reaction), 3-phosphato(V)-2-hydroxypropanoic acid (phosphoglyceric) acid is formed and some of this must go into the aerobic phase of respiration to generate energy, via 2-oxopropanoic (pyruvic) acid and the Krebs' cycle.

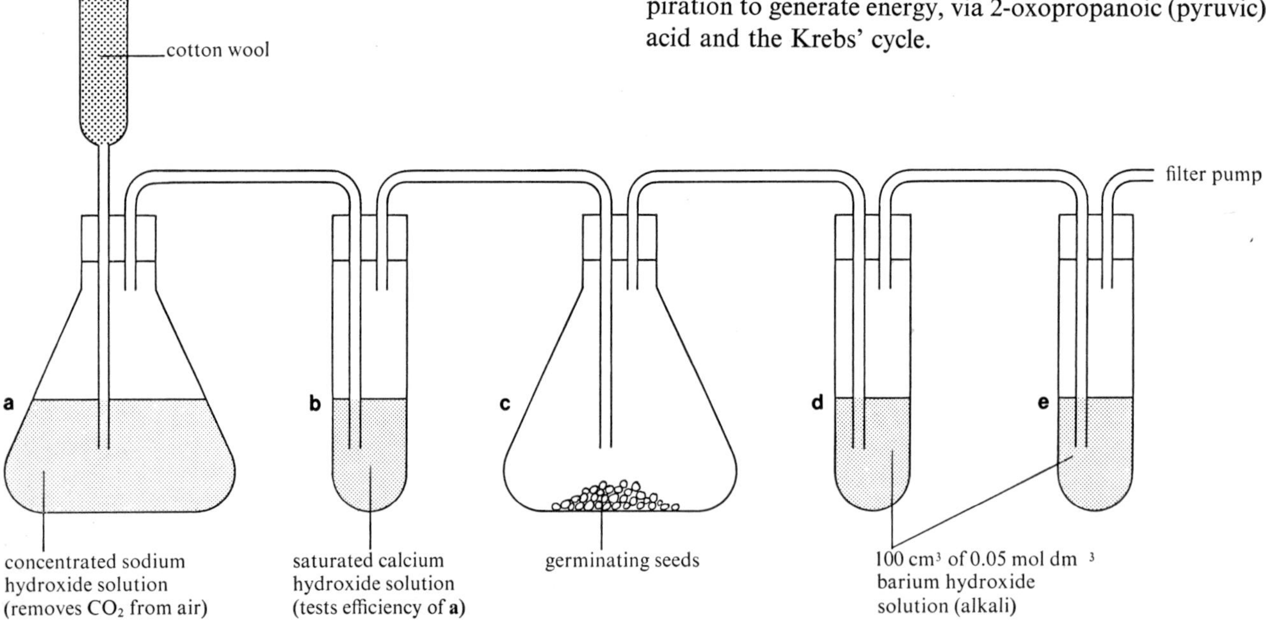

Fig 8.11 **Uptake of oxygen and evolution of carbon dioxide in germinating seeds**

9 Nutrition

9.1 Introduction

Nutrition is the process by which an organism makes or obtains its food. In general animals are said to feed *holozoically* and plants *holophytically*. Nutritional methods can be broken down as shown in Fig 9.1.

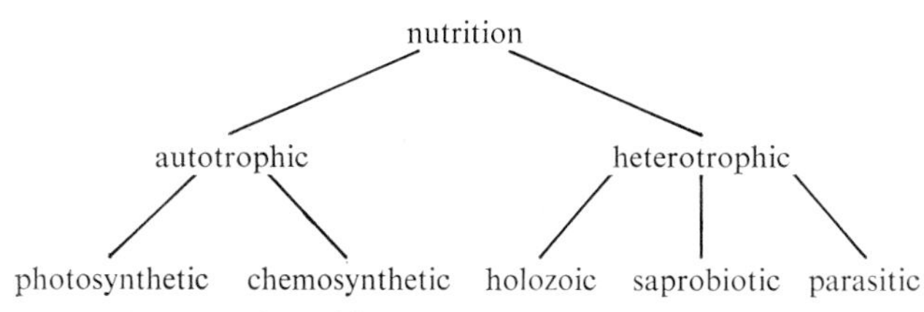

Fig 9.1 **Modes of nutrition**

AUTOTROPHIC NUTRITION This process involves the manufacture of food from simple inorganic compounds. In photosynthesis (§ 9.2 & 9.3) the energy of the sun is used to synthesise organic compounds from carbon(IV) oxide and water. This process is used by all green plants. In chemosynthesis (§ 9.4) the energy used is that derived from the oxidation of terrestrial materials.

HETEROTROPHIC NUTRITION Heterotrophs are unable to synthesise organic from inorganic compounds and therefore have to obtain their food already in organic form. They thus depend on the autotrophs to build up the organic compounds which are then broken down to provide nutrition.

Holozoic nutrition, typically the taking in of food at the mouth, is essentially an animal method of feeding. Saprobiotic nutrition is used by both animals (saprophagous) and plants (saprophytic). Saprophytic organisms such as fungi usually take in food over the whole surface of the body.

Parasitic nutrition is the method used by parasites. Parasitism as a way of life will be discussed in § 9.9 along with commensalism and symbiosis.

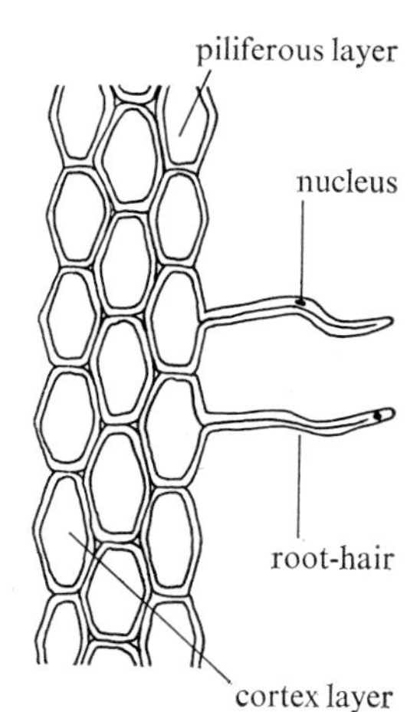

Fig 9.2 **Root hairs**

9.2 Photosynthesis

Plants which contain the green pigment chlorophyll, whether land or water plants, are examples of photosynthetic autotrophs. The essential reaction in photosynthesis is between carbon(IV) oxide and water in the presence of energy (sunlight) and chlorophyll to produce foodstuffs (carbohydrates) and oxygen.

WATER Although water may be absorbed through lenticels and cracks in the surface of suberised roots, maximum uptake is in a region near the root tip where *root hairs* develop. The hairs develop as an extended part of an epidermal cell (Fig 9.2). Each hair may be 0·5 to 3·0 mm long. There are many hairs per unit area of root surface and a very large surface area is therefore exposed to the surrounding soil. In moist soil, each particle is covered with a film of water; this is really not water but a dilute solution of mineral salts. How does the soil water get into the roots? To understand this we must understand *osmosis*.

OSMOSIS When a solution is separated from a solvent by a semipermeable membrane (which permits the passage of solvent molecules but not those of solute) there is a tendency for solvent to pass into the solution, this being called *osmosis*, *i.e.* osmosis is diffusion of water through a membrane. The stronger the solution, the greater is the tendency for solvent to cross the membrane. The osmotic pressure is the pressure which would have to be applied to the solution side of the membrane to prevent osmosis.

The normal plant cell has a vacuole, surrounded by at least two membranes, the cytoplasm and the cellulose cell-wall. The cytoplasm may well be formed of more than one membrane. The vacuole contains cell sap which is a solution of organic and inorganic substances in water. The cellulose cell-wall is permeable and water and dissolved substances can pass freely through it. The layer of cytoplasm which lines the cell wall on the inside, however, is semipermeable (at least, when alive). If such a cell is immersed in water, there is a fairly concentrated solution (sap) inside the vacuole separated from the water by the semipermeable membrane; under these conditions water will enter the cell by osmosis. Water will enter the cell until finally stopped by the tension in the cellulose wall, the so-called *wall pressure* (or turgor pressure). At a given instant, the pressure causing entry of water into the cell (*water diffusion potential* or suction pressure) is equal to the difference between the effective osmotic pressure (the difference between the osmotic pressure of the cell contents and the external solution) and the wall pressure.

$$\text{water diffusion potential} = \text{effective (net) osmotic pressure} - \text{wall pressure}$$

If the cell is placed in a solution more concentrated than the sap, water will be withdrawn from the vacuole (this is often called exosmosis to distinguish it from the entry of water, endosmosis). The cytoplasm is now no longer pressed tightly against the cell wall but shrinks away from it; it may form a spherical mass still enclosing the greatly shrunken vacuole. When the cell has drawn in as much water as it can and the cytoplasm is tightly pressed against the cell wall, it is said to be *turgid* (Fig 9.3). As water starts to be removed from the cell and the shrinking of the cytoplasm can just be observed, the cell is said to be in a state of incipient plasmolysis; when the cytoplasm has completely shrunk away from the cell wall it is said to be fully *plasmolysed*.

There is a dilute solution (low osmotic pressure) outside the root hair and a more concentrated solution of sap inside it (sucrose and other osmotically active compounds reach the root from upper parts of the plant). The water diffusion potential tends to cause a movement of water into the root hair, since the osmotic pressure set up (about $5 \times 10^5\,N\,m^{-2}$) is greater than the capillary forces retaining the water film to the soil particles (about $5 \times 10^4\,N\,m^{-2}$). Transport through the cortical cells is also caused by gradients in potential.

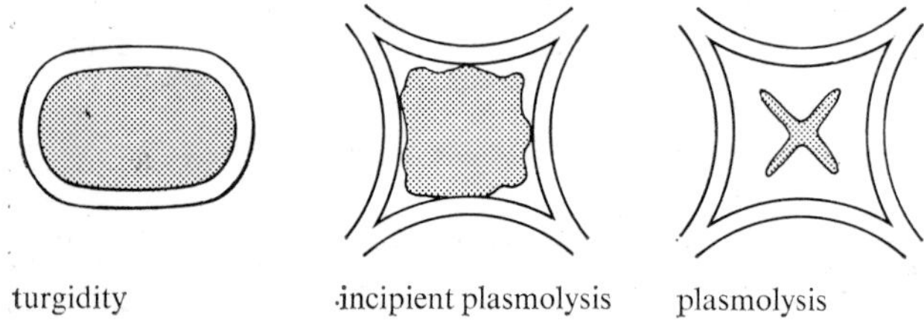

Fig 9.3 **Stages in plasmolysis**

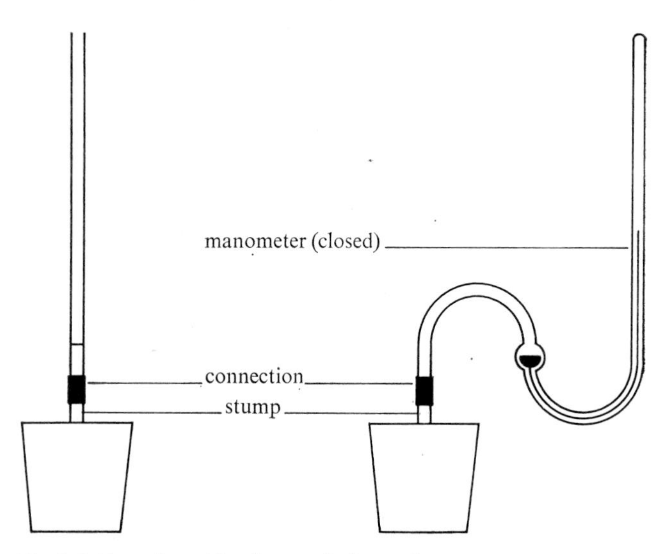

Fig 9.4 **Experiment to demonstrate root pressure**

The roots of a plant from which the main stem and leaves have been removed still force sap upward and therefore must exert pressure – called *root pressure*. This can be demonstrated with a potted plant (e.g. *Pelargonium*). The shoot is cut off about 30 mm from soil level and the stump connected by a piece of polythene tubing to a vertical glass tube (Fig 9.4). The liquid exuded from the stump will soon appear in the tube and, provided the diameter of the tube is known, the volume of sap can be deduced. Alternatively, instead of connecting a vertical tube, a mercury manometer can be used; the root pressure is then measured in mm of mercury and expressed in $N\,m^{-2}$.

The simple observation that cut flowers remain fresh and 'drink' water shows that water moves up a stem in the absence of roots. The ascent of water, particularly in trees, which may grow up to 110 m high (*Sequoia*), is not fully understood. *The cohesion theory* suggests that water vapour is withdrawn from the leaves by transpiration (p 133) and is replaced from the adjacent air spaces. The water lost from the air spaces is compensated for by movement of water from the leaf cells, and thence from *xylem vessels* and *tracheids* at the ends of the veins. The pressure created by the escaping tendency of water from the leaves supplies the pressure needed to raise water through the thin passages (average diameter 0·5 mm) of the tracheids and xylem vessels, against gravity.

TRANSPIRATION The content of the turgid mesophyll cells of the leaf (Fig 9.5) is a solution in which water is the solvent. The cellulose cell walls hold much water. The air spaces in the mesophyll layer in direct contact with these cells are therefore completely saturated with water. The air surrounding the leaf will contain water vapour but the relative humidity will normally be much less than the 100% relative humidity of the cell air spaces.

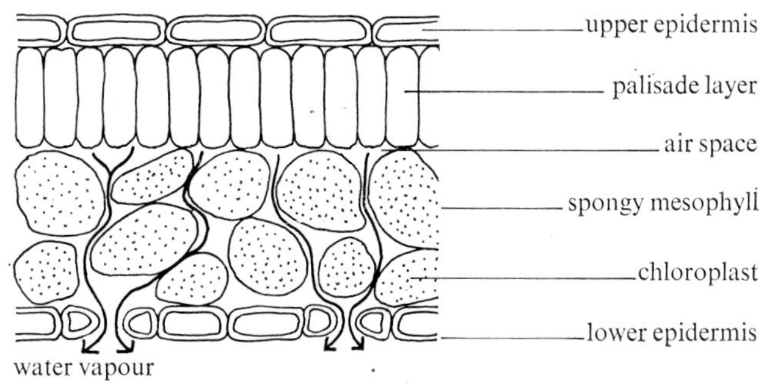

Fig 9.5 **T. s. leaf (diagrammatic) to show water movement**

Under these circumstances, the cells of the leaf will tend to lose water vapour to the air outside when the stomata are open. The bigger the difference in the relative humidity between inside the leaf and outside, the bigger will be the water diffusion potential causing water to move up the xylem and out of the leaves.

$$\text{The relative humidity of the atmosphere} \propto \frac{1}{\text{diffusion potential}}$$

e.g. if the humidity is 50% the diffusion potential $= 10^8 \text{ N m}^{-2}$. Transpiration can be demonstrated with a potted plant, the pot being well watered and placed in a polythene bag to stop evaporation from it before it is placed under a glass jar. In a short time the inside of the jar will be covered with moisture. The actual water loss can be determined by weighing the plant at the start and after a period of time of about 2 hours (if this time is short the gain due to photosynthesis will be small and can be neglected). Loss of water by transpiration takes place mainly through the stomata of the leaves and occurs even when the stomata are only slightly open. Since the tendency for transpiration to occur is the difference between the relative humidity inside the leaf cells and the value in the atmosphere, anything which lowers the relative humidity of the atmosphere (wind, warmth) increases the transpiration, whereas factors which increase the relative humidity of the atmosphere (rain, fog, cold) lower the rate. Even when external conditions are the same, different plants transpire at different rates because of anatomical differences. The *potometer* (Fig 9.6) is an instrument for studying the effects of external conditions on transpiration. It measures uptake of water by the plant; it does not measure transpiration rate.

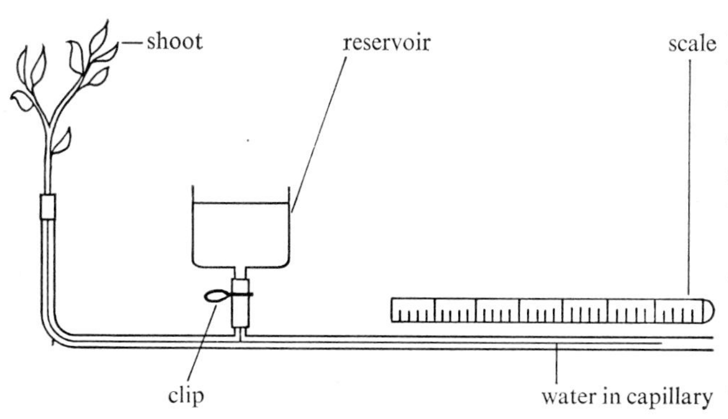

Fig 9.6 **Potometer**

Plants with thin leaf blades, showing no protection against excessive transpiration, are called *mesophytes;* they grow where the soil is neither waterlogged nor excessively dry. The rate of flow of water through the xylem of mesophytes can be greater than 20 m ks^{-1} (m^{-3}).* *Hydrophytes* live where water is plentiful and *xerophytes* are plants which live in dry places (*e.g.* desert); for the latter a lavish expenditure of water is undesirable if desiccation is to be avoided; the rate of flow through the xylem might well be less than $0{\cdot}1 \text{ m ks}^{-1}$ (m^{-3}).*

The water vapour will have to traverse a path out of the leaf in which molecules of it will be resisted by molecules of air. If this resisting force is too large there will be no transpiration. The opposite, wilting, will occur if the transpiration rate is so high that it is too fast for the replacement of water in the leaves up the xylem. Under some conditions, water can ooze out of leaves, *e.g.* when there is free water absorption, a great root pressure and atmospheric conditions are humid—so that transpiration proceeds slowly, if at all. Many plants will then *guttate* water; guttation is prevalent in glasshouses and tropical regions. Early morning dew, after a humid night, is often accompanied by water droplets guttating from the leaves, from hydathodes at the leaf tips.

Transpiration involves evaporation which cools the plant; it also provides a mechanism for the movement of water and dissolved minerals to the leaves.

CARBON(IV) OXIDE Carbon(IV) oxide enters the leaves via the stomata, diffuses across the air spaces, and dissolves in the aqueous solution on the surface of the cells capable of photosynthesising (usually the mesophyll cells). The CO_2 molecules in solution diffuse through the plasma membrane into the cytoplasm and reach the *chloroplasts*, the site of the photosynthetic reaction. With the stomata closed, some CO_2 can be taken up through the cells of the epidermis; CO_2 is fat-soluble and can dissolve in the thin layers of lipid material which comprise the cuticle.

*metres per kilosecond (per cubic metre of xylem tissue).

9.3 The chemistry of photosynthesis

The chlorophyll in all green plants has two components, chlorophyll *a* and chlorophyll *b*. In the purple and green synthetic bacteria, bacteriochlorophyll is present. In addition, yellow carotenes and yellow to orange/brown xanthophylls are present. This can easily be seen by grinding up and extracting green leaves with propanone (acetone) to form a dark solution, and separating the components on a thin-layer silica gel *chromatoplate* run in petroleum/ethoxyethane (ether) solvent. Three distinct bands are formed, a slow green one (containing chlorophylls) and faster yellow (xanthophyll) and orange (carotene) bands.

Chlorophyll *a* has a formula $C_{55}H_{72}O_5N_4Mg$ and chlorophyll *b* has a formula $C_{55}H_{70}O_6N_4Mg$. Pyrrole rings are present. Fig 9.7 shows the structure of chlorophyll.

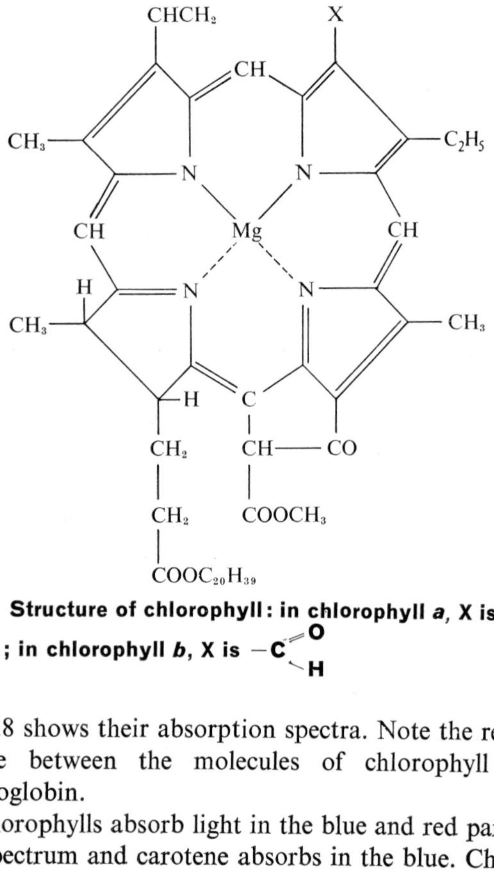

Fig 9.7 Structure of chlorophyll: in chlorophyll *a*, X is $-CH_3$; in chlorophyll *b*, X is $-C(=O)H$

Fig 9.8 shows their absorption spectra. Note the resemblance between the molecules of chlorophyll and haemoglobin.

Chlorophylls absorb light in the blue and red parts of the spectrum and carotene absorbs in the blue. Chlorophyll pigments are photosensitisers, in that the energy they absorb is passed on to the carbon(IV) oxide and water for the endothermic reactions of photosynthesis.

Experiments in which CO_2 concentration, temperature, and light conditions were varied have shown

1 Variations within a normal range of temperatures at low light intensity produce no changes in amount of photosynthesis, even though CO_2 concentration is optimal.

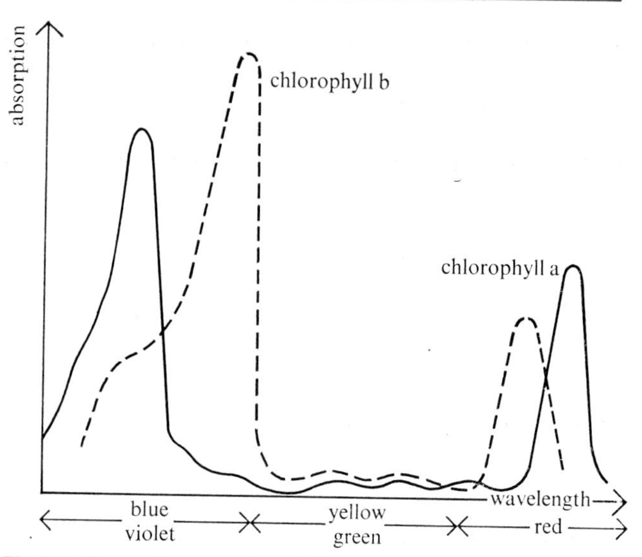

Fig 9.8 **Absorption spectra of the components in chlorophyll**

2 At higher light intensity, increase in temperature gives a marked increase in photosynthesis, particularly if CO_2 is in low concentration.

3 Increasing CO_2 concentration increases photosynthesis (within limits).

4 Interrupted light regimes produce more photosynthesis than continuous light.

The amount of photosynthesis can be measured by measuring the carbon(IV) oxide taken up, the oxygen evolved or the organic compounds formed. Since respiration is proceeding as well, the true photosynthesis is the net value plus the respiration correction, assessed by subjecting the plant to alternate periods of light (respiration and photosynthesis) and dark (respiration only). Carbon(IV) oxide uptake can be measured using ^{14}C labelled CO_2 or by use of a manometer and passing a known concentration of the gas over the plant, determining how much is removed. Oxygen produced can be measured with a photosynthometer (*e.g.* for aquatic plants); the O_2 is collected and measured along a graduated scale.

If the CO_2 concentration is varied and other quantities kept constant, the results would be as shown in Fig 9.9. The flat parts of the curve are caused by the fact that the light intensity is the limiting factor.

These observations show that there are two phases in photosynthesis, the *light reaction*, a photochemical chain type of reaction, and the *dark reaction*, a thermochemical type of the ordinary sort.

In total darkness, a green plant respires only (Fig 9.9). In light of low intensity, the amount of photosynthesis may be insufficient to use up all the carbon(IV) oxide from respiration. As the light intensity increases, the CO_2 uptake in photosynthesis eventually equals the CO_2 output from respiration; this is called the *compensation*

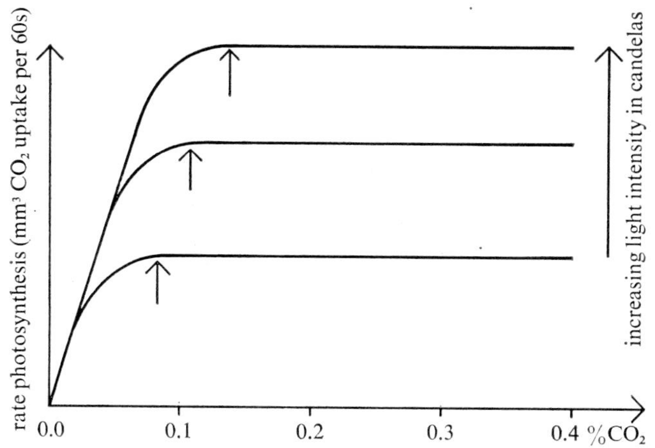

Fig 9.9 **Rate of photosynthesis and illumination**

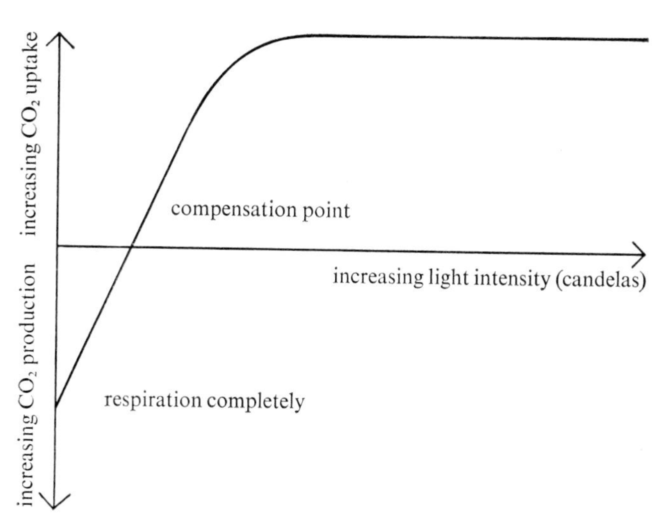

Fig 9.10 **Compensation point**

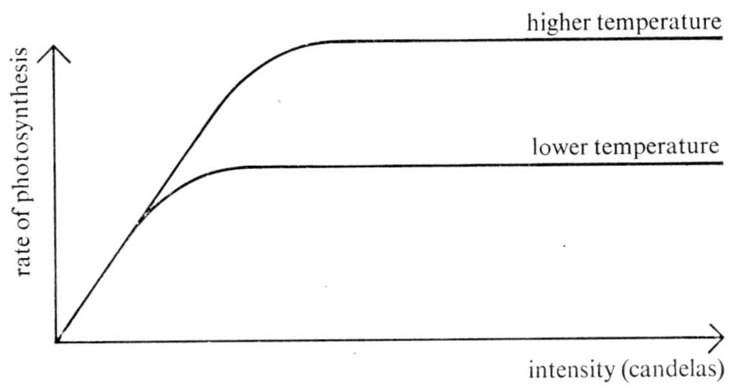

Fig 9.11 **Rate of photosynthesis and temperature**

point (Fig 9.10). The compensation point can be determined with a suitable indicator, *e.g.* distilled water containing $NaHCO_3$ and 0·1% thymol blue turns a yellow colour; this is adjusted to *just* blue by Na_2CO_3 (very dilute, 0·00001 mol dm^{-3}).

When the intensity of the light is increased above the compensation point, oxygen is liberated. Normally, maximum photosynthesis is at 10% of the full intensity of sunlight. In shade, the rate can be limited by the light intensity. Plants can photosynthesise in artificial light of the right colour and intensity. The length of the daily period of light is also important; the longer this is (all other factors being optimum) the more they photosynthesise.

We have already seen that, as temperature affects the rate of photosynthesis, at least two stages must be involved. If light intensity is varied (with other factors constant) and the temperature is kept quite low, the rate varies as shown in Fig 9.11. A higher curve is obtained at a higher temperature.

Radioactive tracer ^{14}C techniques have been used to study fixation of carbon. Labelled carbon(IV) oxide is generated in a small test-tube by reacting together ^{14}C labelled barium carbonate and dilute sulphuric(VI) acid. The tube is held against the basal surface of a green leaf, *e.g.* broad bean, and the leaf kept in bright light; the whole arrangement can be enclosed in a polythene bag. A control experiment is carried out in the dark. After exposure for about 900 seconds, the leaf is detached and taken to a dark room where it is sealed between two sheets of X-ray film and kept in a refrigerator for two weeks. The lowering of the temperature is necessary to keep the solutes stationary. The film is developed and the one in light should show clearly the uptake of labelled CO_2, producing an autoradiograph. The control should be almost free from dark grains of silver. After exposure to the labelled carbon(IV) oxide, the leaf can be processed for chromatography and the chromatogram exposed to film to give an autoradiograph (p 21).

Another interesting experiment is to study carbon(IV) oxide exchange in, say, *Chlorella*, using labelled carbon(IV) oxide, following the exchange with a Geiger-Müller tube (geiger counter).

When *Chlorella* photosynthesises in water containing the heavy isotope ^{18}O, this tracer appears in the oxygen evolved; if ^{18}O labelled carbon(IV) oxide is used instead, no ^{18}O appears in the oxygen evolved. This shows that the oxygen evolved is derived from water. The ^{18}O is determined by use of a mass spectrometer.

THE LIGHT REACTION In this, the first part of the process, chlorophyll absorbs some of the energy from the sunlight. This is used to catalyse the formation of ATP from ADP (photophosphorylation) and the energy formed is used in the dark stage. Light energy is also used to split water into oxygen (evolved) and hydrogen (which is transferred to an acceptor such as NAD). The $NADH_2$ passes on the hydrogen to the dark reaction and the actual fixation of carbon; some, however, is oxidised via the cytochrome mechanism to give ATP.

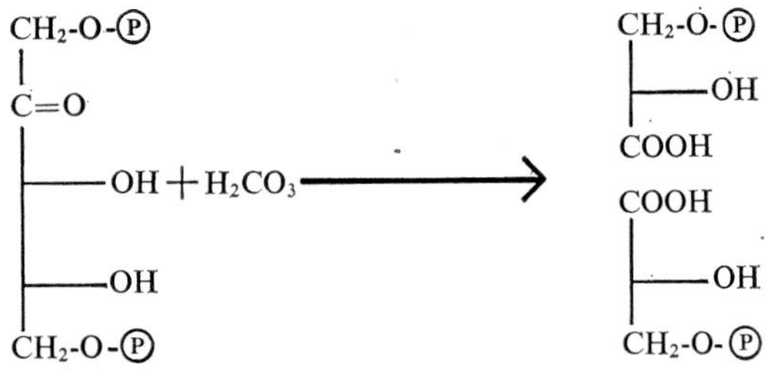

CH₂-O-Ⓟ
C=O
OH + H₂CO₃
OH
CH₂-O-Ⓟ
CH₂-O-Ⓟ
OH
COOH
COOH
OH
CH₂-O-Ⓟ

Fig 9.12 **Conversion of ribulose diphosphate into phosphoglyceric acid**

THE DARK REACTION Calvin determined the fate of the carbon from the CO_2 by chromatography and radioactive tracer methods. The separated materials were tested for radioactivity on the autoradiograph and those which originated from the radioactive CO_2 were labelled. Radio-carbon was found in sugars, simple organic acids and amino acids if photosynthesis had proceeded for some time. But if it lasted only seconds most of the radioactivity was in one compound, 3-phosphoglyceric acid (3-C atoms). Calvin concluded that carbon(IV) oxide (as H_2CO_3) first becomes attached in the chloroplasts to a 5-C compound called ribulose 1,5-diphosphate (enzyme activity in the chloroplasts (Fig 9.19) causes the attachment); two molecules of 3-phosphoglyceric acid are then formed (Fig 9.12).

The 3-phosphoglyceric acid (using hydrogen from the light reaction) becomes reduced to 3-phosphoglyceraldehyde (Fig 9.13); that the hydrogen is 'active' is evident because –COOH is not normally reduced direct to –CHO.

The 3-phosphoglyceraldehyde is a phosphorylated, 3-C triose sugar. Its formation needs ATP from photophosphorylation in stage 1. Some of the triose is used to form glucose (6-C skeleton) and some is needed to regenerate ribulose 1,5-diphosphate, via two intermediates which have 4-C and 7-C respectively. This resynthesis cycle is called the *Calvin Cycle*.

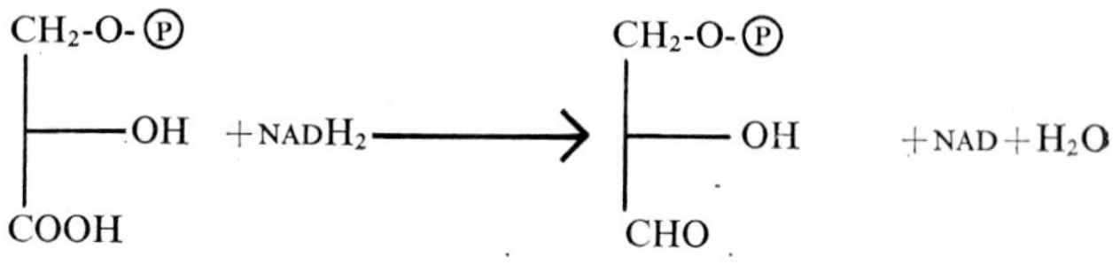

CH₂-O-Ⓟ
OH + NADH₂
COOH
CH₂-O-Ⓟ
OH
CHO
+ NAD + H₂O

Fig 9.13 **Conversion of phosphoglyceric acid into phosphoglyceraldehyde**

By combining the light and dark reactions, the mechanism of photosynthesis is as shown in Fig 9.14.

The resynthesis cycle, in which 5 C–3 molecules give 3 C–5 molecules, can be set out as in Fig 9.15, the C–4 compound is erythrose phosphate, the C–7 compound septoheptulose phosphate, the C–6 compound fructose phosphate (from which glucose can be obtained) and the C–5 compound xylulose phosphate (from which ribulose phosphate is obtained).

From the glucose formed, various other types of carbohydrate can be synthesised. The polysaccharide starch is an immediate product and grains of this appear

C-3
C-3
C-3
C-3
C-3
C-6
C-4
C-7
C-5
C-5
C-5

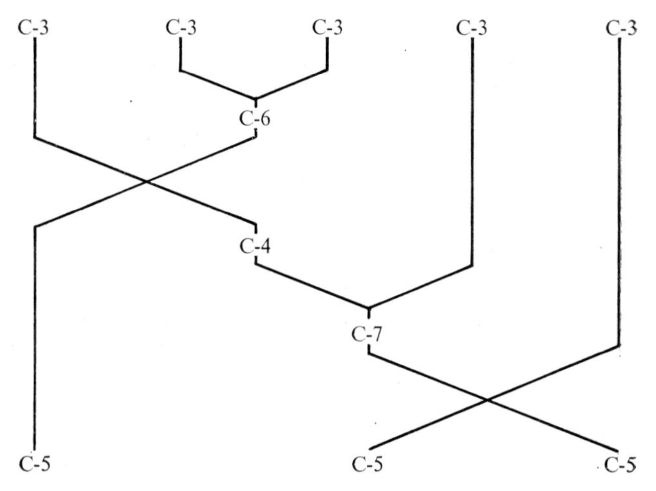

Fig 9.15 **Resynthesis cycle**

light
6 O₂
12 H₂O
12 H₂
ATP
12 NADH₂
dark
6 C-5 + 6 C-1
12 C-3 acid
12 C-3 sugar (PGA)
10 C-3
2 C-3
1 C-6

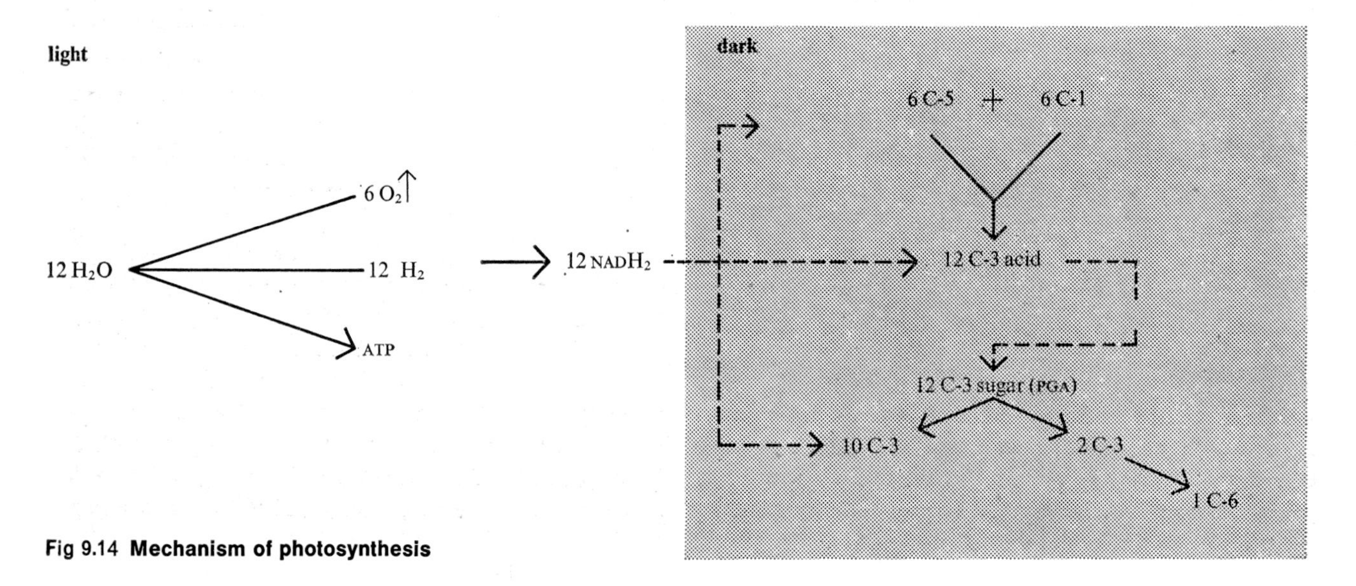

Fig 9.14 **Mechanism of photosynthesis**

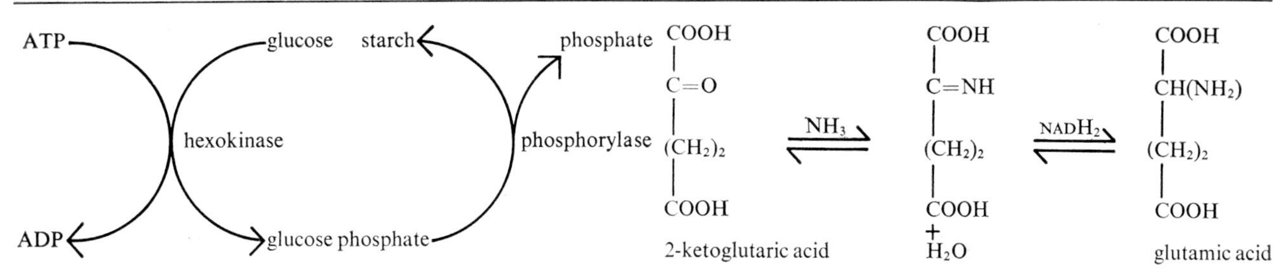

Fig 9.16 **Carbohydrates formed in photosynthesis**

Fig 9.17 **Formation of glutamic acid**

in all photosynthesising cells. Plant enzymes are necessary for these syntheses (Fig 9.16).

A part of the phosphoglyceraldehyde (PGA) enters the Krebs' cycle, forming at one stage the compound 2-ketoglutaric acid which can react with ammonia present in the plant to form glutamic acid, an amino acid (Fig 9.17). Other amino-acids can then be formed by transaminase enzymes that transfer the NH_2 group from glutamic acid to other organic acids. An example is the formation of alanine (Fig 9.18).

$$\begin{array}{c} COOH \\ | \\ C{=}O \\ | \\ CH_3 \end{array} + \begin{array}{c} COOH \\ | \\ CH(NH_2) \\ | \\ (CH_2)_2 \\ | \\ COOH \end{array} \rightleftharpoons \begin{array}{c} COOH \\ | \\ CH(NH_2) \\ | \\ CH_3 \end{array} + \begin{array}{c} COOH \\ | \\ C{=}O \\ | \\ (CH_2)_2 \\ | \\ COOH \\ \text{2-ketoglutaric acid} \end{array}$$

Fig 9.18 **Formation of alanine**

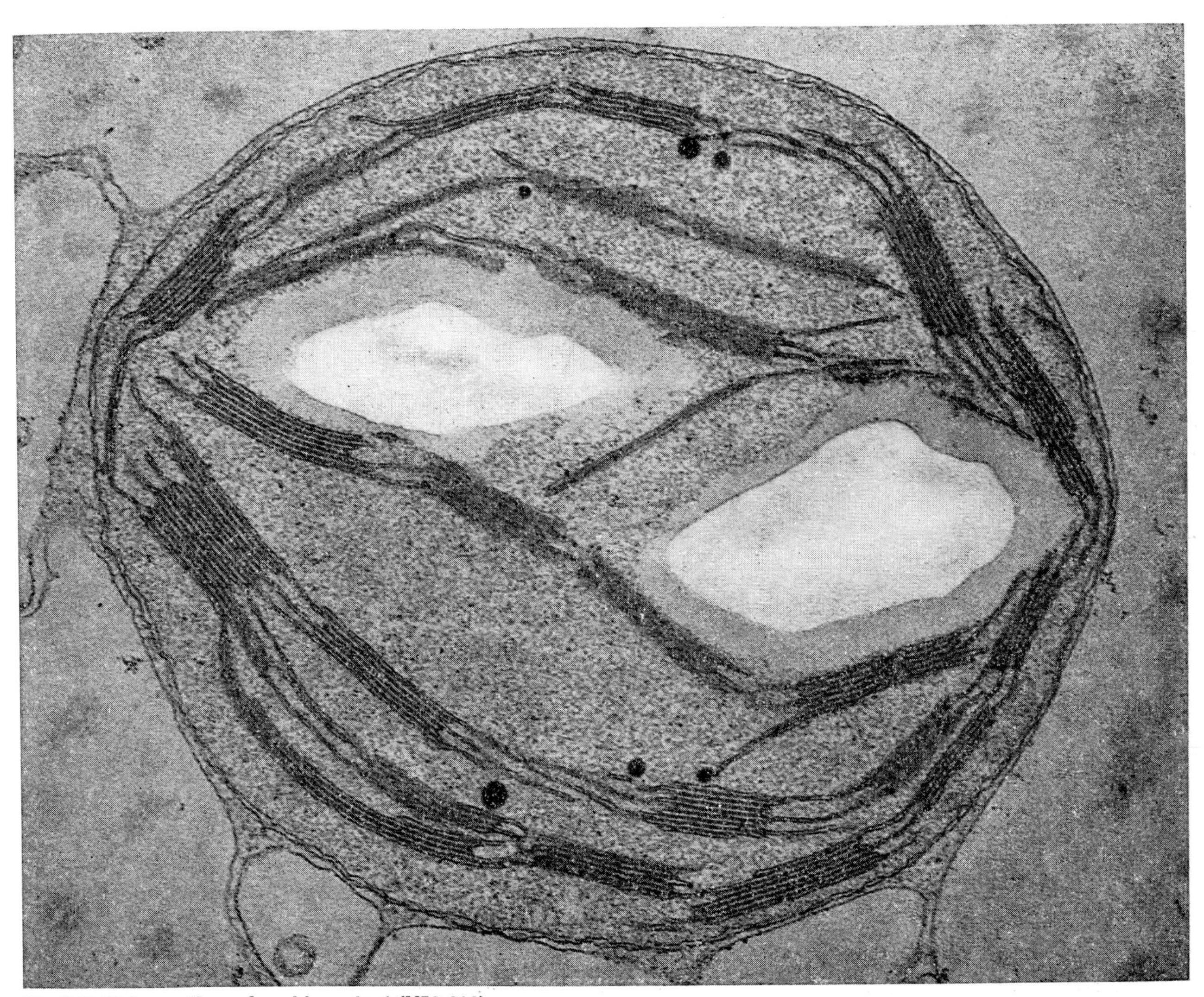

Fig 9.19 **Thin section of a chloroplast (X50 000)**

Instead of amino–acids, some amides ($RCONH_2$) are formed and these are used in the meristems for synthesis of proteins.

Among the products from the PGA are glycerol and carboxylic (fatty) acids which can combine by esterification to form lipids. The path followed through the photosynthetic and respiratory reactions is shown in Fig 9.20.

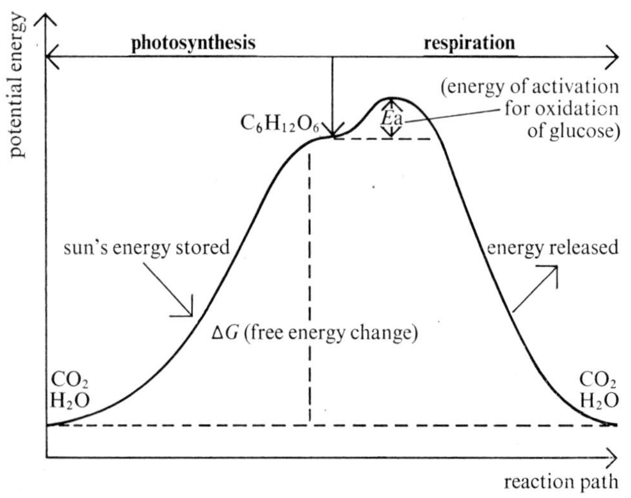

Fig 9.20 **Relationship between photosynthesis and respiration**

TRANSLOCATION Although many carbon compounds are built up in the leaves, most are moved elsewhere – either stored or other compounds synthesised from them. Carbohydrates are stored in roots, stems and leaves as starch (normally). Large amounts of proteins occur in seeds, particularly of legumes (peas, beans, etc.). Organic acids such as malic, citric and tartaric are often found in plant cells. The resins in pines are also carbon compounds. Fats and oils are stored in many seeds and nuts.

Experiments using dyestuffs and ^{14}C tracers have shown that the *phloem* is the organ of translocation. Transport is active because it can take place quite easily in any direction. The rate is often high, *e.g.* 0·1 m ks^{-1} (a factor at least 10^4 higher than diffusion). Dissolved compounds pass along the phloem sieve tubes, simultaneously with circulation of water up the xylem.

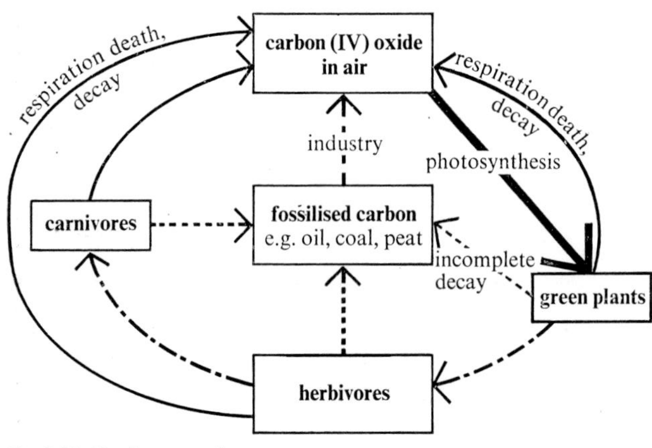

Fig 9.21 **Carbon cycle**

CARBON CYCLE The place of photosynthesis in the carbon cycle can be seen from Fig 9.21. Its role is vital for most heterotrophic animals.

Practical work

What practical work can be done on photosynthesis?

1 Extract, using propanone (acetone) the pigments from chopped-up green leaves and run a *chromatogram* as indicated on page 20 using ethoxyethene (ether) as solvent. Alternatively, just place a spot of the extract in the centre of a circular filter paper, allow to dry, and, using a glass tube, drop a small volume of propanone (acetone) on to the spot. Notice that a green disc appears surrounded by a yellow ring.

2 Show the evolution of oxygen from a suitable water weed by use of the inverted funnel/test-tube method described in any good elementary textbook of biology.

3 Radioactivity experiments can be carried out, using carbon–14, as outlined in the text, provided the person in charge is qualified in this field*.

4 Boil pieces of onion with distilled water, filter, and test the filtrate for reducing sugar.

9.4 Chemosynthesis

Chemosynthetic organisms synthesise organic from inorganic materials (carbon(IV) oxide and water) using energy derived from the oxidation of other simple compounds. Particularly important chemosynthetic organisms are the *nitrifying bacteria.* For example *Nitrosomonas* obtains the energy necessary to fix carbon(IV) oxide as follows: during the decay of plant and animal remains the nitrogen of the amino groups is liberated as ammonia. This is oxidised, $NH_4^+ \rightarrow NO_2^- +$ energy.
The *Nitrobacter* group of bacteria obtain energy by oxidising these nitrate(III) ions, $NO_2^- \rightarrow NO_3^- +$ energy. The nitrate(V) ions are available to plant roots (see p 131).

* No radioactivity experiments should be undertaken until the person in charge has attended a proper course and is proficient. This is a legal requirement.

9.5 Transport of minerals and essential ions

The ions taken up by plant roots come from the soil water and from the adsorbing surfaces of colloidal clay particles (pp 24 and 112). Clay particles are also reactive and ions like K^+, Na^+, Ca^{2+} and Mg^{2+} may be associated with them. Clay soils are thus richer in nutrients than sandy soils. Saprophytes in the soil render mineral salts available by degrading protoplasmic material.

Much of the nitrogen in the soil is bound up in the bodies of protozoa, mites, nematodes, insects and worms; in their faeces, in the remains of these and other animals, in roots, in decaying plant material, and in bacteria and fungi. The conversion of this nitrogen into forms which plants can assimilate (NO_3^-, NH_4^+) is the basis of the nitrogen cycle.

Experiments with fungi and bacteria in petri dishes, in which different elements are in turn deleted from the agar culture medium, show (e.g. *Mucor*) that the elements N, P, K, Ca, Mg and S are needed in quite large amounts and that small amounts of the minor or trace elements Fe, Cl, Mn, Cu, Zn, Mo and B are also required. Similar deletion experiments with higher plants with their roots in inert containers in which nutrient solutions are supplied in a solution or as a mist show that their needs are similar, a trace of cobalt (Co) ions being necessary for some.

Ions may enter root cells by diffusion but they mainly enter by an exchange process called active transport.

ACTIVE TRANSPORT OF IONS IN CELLS Active transport of ions across cell membranes has been studied using the fresh-water algae *Nitella* and *Valonia* whose cell sap is easily removed. By comparing the sap from within the vacuole and from outside it with the pond water in which the algae grow, it has been shown that Na, K, Ca, Mg and Cl are all in higher concentration in the vacuole than in the pond water. The ions are found to be absorbed to different extents and it has been concluded that the cytoplasmic membrane must operate a mechanism which both selectively absorbs ions and also provides the energy (*i.e.* ATP) needed to drive the ions by diffusion against the concentration gradient, *i.e.* to drive them into a more concentrated internal solution.

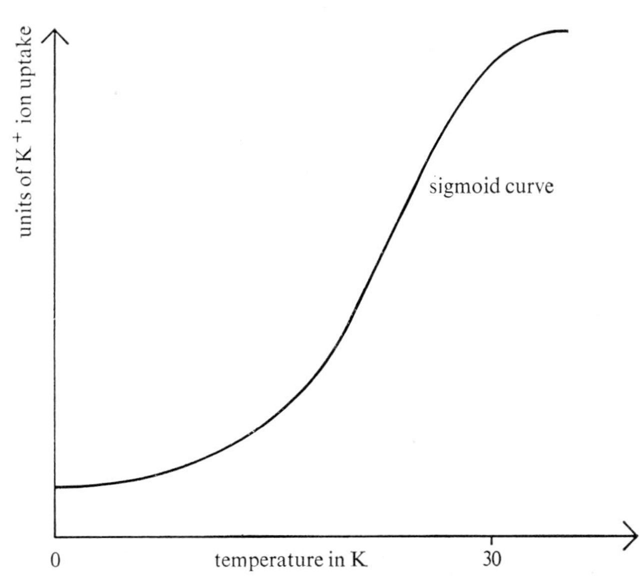

Fig 9.22 **Variation of uptake of potassium ions with temperature in barley root tips**

It was proposed that the thermal energy of the ions could do this and that once the ions reached a vacuole they were in some way adsorbed, lowering the concentration inside and thus allowing more ions to pass into the vacuole. Conductivity experiments have ruled this out, however, because it has been found that free ions remain in solution without ever becoming adsorbed.

It is now thought that the mineral ion uptake is linked to the process of respiration. A study has been made of the uptake of K^+ ions by young barley root tips as a function of temperature. The graph obtained (Fig 9.22) resembles that for an enzyme-controlled process such as respiration. If simple diffusion, a purely physical process, was the mechanism involved, its rate would not increase so steeply in response to rising temperature. Furthermore, the uptake of Br^- ions by young barley root tips has been shown to be greatly reduced in an atmosphere of nitrogen, *i.e.* oxygen is important, as would be required by respiration. Other inhibitors have also been used and radioactive rubidium (like potassium) has also proved useful. Using rubidium ions (Rb^+), it has been found that ions enter the cell cytoplasm rapidly but subsequent entry into the vacuole is slow. Also, ions often leave the cytoplasm while ions of the same sort are entering it. It is clear then that the *tonoplast* (cytoplasm/vacuole membrane) is the energy barrier to ion passage in the cell.

Some process linked to respiration drives ions across the tonoplast and into the cell vacuole. Perhaps, it is argued, a *carrier molecule* in the cytoplasm picks up the ion which has entered there by diffusion and transfers it to the tonoplast so that it can be released into the vacuole. Although the evidence is not yet to hand, it could be that the formation and regeneration of the carrier are linked to the redox (oxidation and reduction) processes of respiration.

An excellent account of ions in cells is given in W. M. M. Baron *Organisation in Plants* (Edward Arnold, London 1967).

Whatever the mechanism of active transport, there is a large concentration of ions built up in the cell vacuole. By active transport the ions move from cell to cell across the cortex, eventually entering the xylem for transport up the stem; there is a slower movement in the phloem which is more local. It is worth remembering that electrical neutrality must be preserved, *i.e.* cations of charge n^+ must be taken up with anions bearing a total charge of n^-.

Acidity of the soil can have a direct effect on plant growth and it can also increase the solubility of certain salts (*e.g.* Al(III) and Mn(II)); except in low concentration, these are toxic. Also, acid soils often prevent the activity of nitrifying and nitrogen fixing bacteria.

Chemical analysis of plants indicates that at least traces of most elements are found in some species under certain conditions. Not all elements are needed for nutrition. Which ones are needed for any species can be established by deletion experiments. The growth is compared with healthy plants and deficiency symptoms are noted. These are very similar in different plants; some characteristic ones are given in Table 9.1

Table 9.1 **Nutrient deficiency symptoms**

element	*enters as the ion*	*deficiency symptom*
N	NO_3^-	general chlorosis.* Poor growth (N is essential part of chlorophyll and amino-acids)
P	$H_2PO_4^-$	abnormal red colours. Delayed maturity. (transport and respiration impaired)
K	K^+	marginal necrosis† of older leaves. (Enzyme and membrane function suffers)
Ca	Ca^{2+}	stunting, wilting (cell wall structure affected)
Mg	Mg^{2+}	interveinal necrosis after chlorosis in basal leaves. (Mg of chlorophyll removed to older leaves)
Fe	Fe^{2+}	tip of shoot chlorotic, stunting; chlorophyll and cytochrome systems affected. Fe is immobile unlike Mg. In strongly alkaline soils iron(II) hydroxide is formed and Fe is not available as this has low solubility.
S	SO_4^{2-}	chlorosis, stunting except roots. Needed for cystine and cysteine.

* chlorosis lack of green colour, turning yellow
† necrosis dying, death.

An excess of some ions gives toxicity symptoms. An imbalance of ions may be caused by excessive applications of certain fertilisers by farmers (*e.g.* lime). Applying the right amount of fertiliser at the right time is thus important in agriculture.

9.6 Animal nutrition

Vertebrate animals select different kinds of material on which to feed.

Feeding can be classified as follows:

Herbivores feed on plants; *e.g.* cow, pigeons, aphids.

Carnivores feed on flesh of other animals; *e.g.* tiger, wolf, eagle, ladybird.

Omnivores feed on both plants and animals *e.g.* man, seagull, cockroach.

Micro-feeders can be divided into the following classes:

Fluid suckers feed by piercing/sucking, *e.g.* aphids.

Deposition feeders feed by taking in lots of material and extracting the beneficial food, *e.g.* earthworm.

Filtering feeders feed by straining off food particles from surroundings, *e.g. Paramecium.*

Components of a complete diet are set out in Table 9.2. Most holozoic organisms require such a diet, although quantities and vitamins (Table 9.3) needed will vary.

9.7 Mammalian digestion

Mammals alone have well-differentiated teeth. *Ingestion* begins when the food is taken in at the mouth (buccal) cavity. Solid food is chewed into small pieces by teeth, which are of four kinds: *incisors* (cutting teeth in front

Table 9.2 **A complete diet for most holozoic animals**

component	*occurs in*	*examples*	*role*
water	water	water	as solvent in cell, transport of food and waste temperature control
minerals	many foods	chlorides and sulphates(VI) of sodium and potassium carbonates and phosphates(V) of calcium, magnesium, potassium and sodium many salts of iron	enzyme action, as cofactors
carbohydrates	sugars, starches (such as bread and potatoes) cellulose (usually digested by bacterial symbionts)	glucose, fructose	main source of carbon for herbivores and carnivores
protein	milk cheese, eggs, meat, fish	essential amino-acids: others also synthesised by organism	large proportion of solids in cells, muscle, hair and blood
fats	animal and vegetable oils, eggs, milk, most meats	stearates also built up from deaminated proteins *e.g.* insects, mammals, protozoa	alternative to carbohydrates, storage
vitamins	*see Table* 9.3		

Table 9·3 **Vitamins**

vitamin	*name*	*formula*	*occurs in*	*deficiency symptons in man*
A		$C_{20}H_{30}O$	fish, liver oils, milk, butter, carrots, eggs	poor growth, night blindness
B_1	thiamine	$C_{12}H_{18}ON_4SCl_2$	wheat germ, eggs, nuts, liver	beri-beri
B_2	riboflavin	$C_{17}H_{20}N_4O_6$	lean meat, liver, eggs, cheese, green vegetables	general ill-health
B_{12}	cobalamin	pyrrolidine ring containing cobalt	most green plants	pernicious anaemia
C	ascorbic acid	$C_6H_8O_6$	fresh vegetables, fruit	scurvy, bleeding
D	calciferol	$C_{28}H_{44}O$	fish, liver oils, milk, cheese, butter, eggs	rickets, poor bones
E	tocopherol	$C_{29}H_{50}O_2$	salads, seeds	muscle decay, low fertility
K			most foods synthesised by intestinal bacteria	poor clotting of blood
Other B vitamins	pyridoxal			
	nicotinamide			grey hair, pellagra
	pantothenic acid			
	biotin			
	folic acid			various types of anaemia

used for gnawing and scraping, chisel-shaped), *canines* (pointed, used for tearing), *premolars* and *molars* (Fig 9.23) (at the back, used for grinding and chewing). For carnivores the canines are well-developed and used for tearing at flesh and the premolars and molars are specialised for tearing also. For herbivores, the incisors are well developed for cropping plants. The cow has a horn-like pad instead of upper incisors. No canines are needed but the *diastema* gap allows good circulation/mixing of food. The lower jaw has a loose joint to allow free rolling motion.

Some dental formulae are as follows.

	I	C	Pm	M	
MAN	$\frac{2}{2}$	$\frac{1}{1}$	$\frac{2}{2}$	$\frac{3}{3}$ (one-side only)	= 32 teeth
RABBIT	$\frac{2}{1}$	$\frac{0}{0}$	$\frac{3}{2}$	$\frac{3}{3}$	= 28 teeth
SHEEP	$\frac{0}{3}$	$\frac{0}{1}$	$\frac{3}{3}$	$\frac{3}{3}$	= 32 teeth
DOG	$\frac{3}{3}$	$\frac{1}{1}$	$\frac{4}{4}$	$\frac{2}{3}$	= 42 teeth

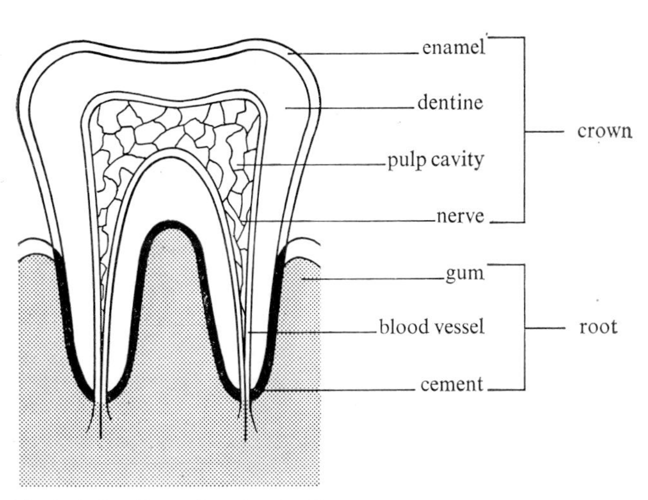

Fig 9.23 Section of molar tooth

The tongue aids the chewing and *masticating* of food. The food comes into contact with taste buds and this stimulates glands to secrete enzymes. *Saliva* is mixed in from the salivary glands. The saliva has mucus which acts as a lubricant and the saliva itself acts as a solvent. It provides (in a few mammals, *e.g.* man) two enzymes, salivary amylase (ptyalin), a polysaccharase, and maltase, an oligosaccharase. The former hydrolyses starch and glycogen to disaccharides and the latter hydrolyses maltose to glucose. The tongue becomes arched to force food into the pharynx, voluntary control ends and the food travels by reflex actions to the stomach. The soft palate (*uvula*) is elevated when the food is in the pharynx so that food cannot get into the nasal region. The *epiglottis* stops food from entering the larynx. Food in the oesophagus starts a wave motion of contractions called *peristalsis*, the rhythmic squeezing of the muscular walls. Mucus is added and the food moves ahead of the wave.

STOMACH The *stomach* (sac) is divided into upper (cardiac) and lower (pyloric) regions. The *fundus* is the small chamber where the oesophagus enters. The stomach is bounded by a membrane of connective tissue, the *visceral peritoneum*, which is continuous with the body wall (*parietal*) *peritoneum*.

In *ruminants* (*e.g.* cow) a rumen lies at the end of the oesophagus and from it food is passed back to the buccal cavity for extra chewing. While the food is in the rumen, symbiotic bacteria/protozoa digest materials like cellulose forming fatty acids and proteins.

The stomach prepares the food for digestion in the intestine. During churning, the muscular ring (the *pyloric sphincter*) separating the stomach and duodenum – the initial part of the intestine – stays contracted so that food cannot leave the stomach. In the stomach the food becomes almost fluid (called *chyme*). When the chyme reaches the pyloric region, the pyloric valve opens and the chyme is squeezed into the duodenum. Vomiting is a reflex action that can be initiated by nerve endings in the stomach. Vitamin B_{12} is absorbed in the stomach.

GASTRIC JUICE (the release of which is stimulated by the hormone gastrin) is made up of hydrochloric acid, pepsin, rennin, lipase and mucin. It is secreted from tiny glands in the walls of the stomach. Rennin changes casein (milk protein) to a curdled form, to remain longer in the stomach, allowing protein digestion to be initiated. Pepsin hydrolyses proteins into smaller molecules (but not amino-acids); in fact, pepsin comes from hydrolysis (by hydrochloric acid) of the endopepsidase pepsinogen, and it hydrolyses peptide bonds adjacent to aromatic rings. When the stomach is empty, a small amount of alkali passes through the pyloric valve to neutralise excess acid.

SMALL INTESTINE The long *small intestine* (Figs 9.24, 9.25) is divided into the *duodenum* (digestion occurs here) and the *ileum* (assimilation occurs here); the *jejunum* separates these regions. The small (and large) intestine is attached to the body wall by the *mesentery*. The lining of the small intestine contains thousands of glands which secrete enzymes (intestinal juice controlled by the hormone enterocrinin). The pancreas produces enzymes and hormones (Table 9.4). Pancreatic secretion is stimulated by the hormone *secretin* and pancreozymin; food in the duodenum causes glands to produce proscretin (inactive) and this is activated by hydrolysis with hydrochloric acid, then transported via the blood to the pancreas which is thereby stimulated to produce enzymes. The *liver* has many functions, one being the production of bile. This is produced owing to the action of secretin, and is collected by ducts that lead to the *gall bladder*. This concentrates the bile (removal of water) and it is then emptied, by contraction of the gall bladder, controlled by the hormone cholecystokinin, into the duodenum. The dark colour of bile is due to broken down red blood cells. It aids digestion of fats (by bile salts) by emulsifying them, thereby causing a greater surface area to be presented to the fat-hydrolysing enzymes.

In the ileum assimilation of inorganic and organic materials takes place. The surface area is greatly increased by *villi* with a fine surface structure of *microvilli* (Fig 9.26). The villi are 'laced' with a network of blood capillaries and fine lymph vessels. The surface area of the human small intestine is about 5×10^4 m^2. Absorption of digested food needs energy which comes from respiration of the cells of the villi. Monosaccharides and amino-acids, including dipeptides hydrolysed in cells of the villi, are absorbed into blood capillaries and transported by veins to the portal vein, thence to the liver where some of the glucose is converted to glycogen. Propantriol, fatty acids and glycerides enter the villi by *pinocytosis* as small droplets called *chylomicrons*. Then they are absorbed into lacteals of the lymphatic system, where recombination to fat takes place; this then passes into the large veins before entering the right atrium (heart). Vitamins are absorbed directly (fat-soluble K and D need bile salts). Mineral ions are absorbed under the control of mineralocorticoid hormones.

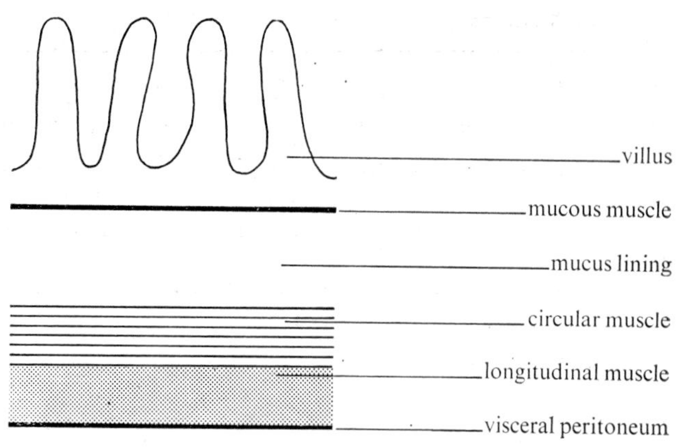

Fig 9.24 **Diagramatic section of mammalian small intestine**

Table 9.4 **Digestive enzymes**

enzyme	*in*	*pH optimum*	*acts on*	*gives*
trypsin	pancreas	alkaline	proteins	peptides
chymotrypsin	pancreas	alkaline	proteins	peptides
lipase	pancreas	alkaline	fats	propantriol, fatty acids
amylase	pancreas	alkaline	glycosides	maltose
ribonuclease	pancreas	alkaline	RNA	nucleotides
deoxyribonuclease	pancreas	alkaline	DNA	nucleotides
carboxypeptidase	intestine	alkaline	peptides	amino-acids
aminopeptidase	intestine	alkaline	peptides	amino-acids
enterokinase	intestine	alkaline	trypsinogen	trypsin
maltase	intestine	alkaline	maltose	glucose
sucrase	intestine	alkaline	sucrose	glucose, fructose
lactase	intestine	alkaline	lactose	glucose, galactose

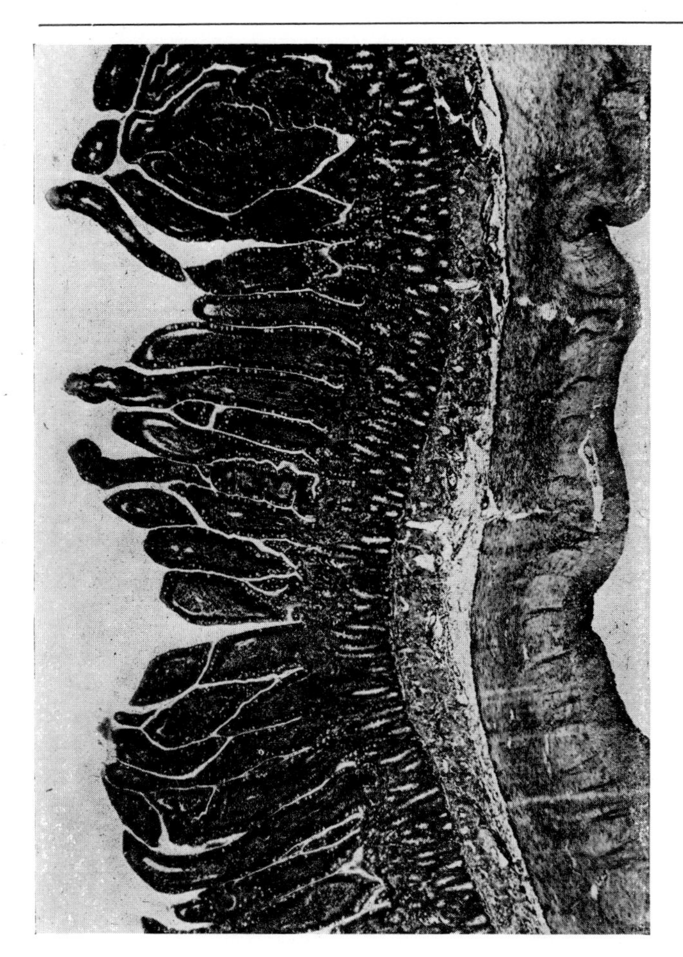

Fig 9.25 **Microscopical structure of small intestine (low power t.s.)**

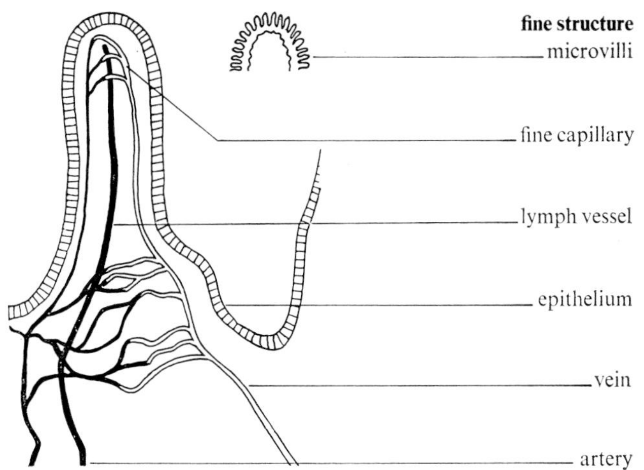

Fig 9.26 **Structure of villi**

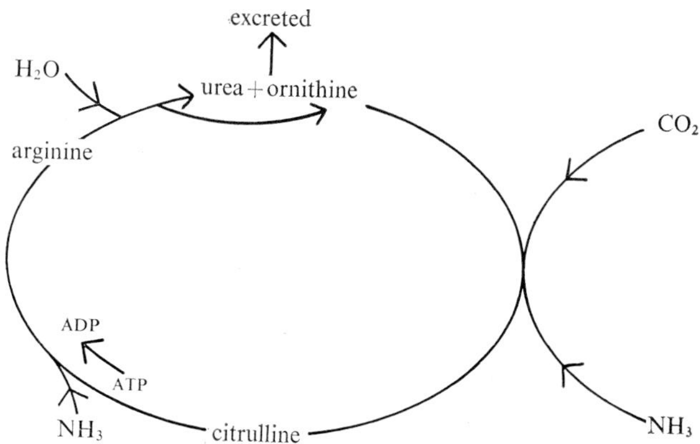

Fig 9.27 **Ornithine cycle**

LARGE INTESTINE The *large intestine* (colon) is responsible for egestion (elimination of unabsorbed materials). At the end of the ileum is the *ileocolic valve* to keep liquid food in the ileum. The *sacculus rotundus* sends food into the diverticulum of the caecum and appendix. The latter has little use in man but in some animals such as the horse it helps digestion of cellulose. It contains many symbiotic bacteria/protozoans. There are no enzymes and so this region is less effective than the rumen.

The colon removes water from the contents and delivers unabsorbed material to the rectum to be eliminated from the anus (*defaecation*). Faeces are made up of undigested food fragments, unabsorbed materials, bile pigments plus huge colonies of bacteria.

FOODS ABSORBED INTO THE BLOODSTREAM Monosaccharides and amino-acids are conveyed via the hepatic portal vein to the liver, where glycogen is synthesised and stored, controlled by the hormone *insulin* (*adrenalin* is responsible for release of sugars from the liver).

Amino acids are either deaminated and the keto-acids formed pushed into the Krebs' cycle for glycogen syntheses, or returned to the body if they are needed. The ammonia from the deamination unites with carbon(IV) oxide to form urea (carbamide), excreted in due course.

This combination takes place in the *ornithine cycle* (Fig 9.27). Urea is much less toxic to the body than ammonia and levels between 18 and 38 mg per 100 cm^3 are normal. Shark hearts will not beat unless urea is present. Energy is needed in the synthesis and urea is formed by hydrolysis of the guanido grouping of arginine (Fig 9.28). In the liver CO_2 and NH_3 form carbamyl phosphate. Then citrulline is formed. Citrulline then acquires an NH_2 group from aspartic acid.

arginine ($H_2N-C(=NH)-NH-(CH_2)_3-CH(NH_2)-COOH$) $+ H_2O \rightarrow$ urea ($O=C(NH_2)_2$) $+$ ornithine ($H_2N-CH(COOH)-(CH_2)_3-NH_2$)

$\downarrow$

$CO_2 + NH_3 + 2$ ADP $\rightarrow H_2N-C(=O)-O-P + 2$ ADP $+ HO-P$

$\downarrow$

ornithine $+$ carbamyl phosphate $\longrightarrow$ citrulline ($NH_2-C(=O)-NH-(CH_2)_3-CH(NH_2)-COOH$)

$\downarrow$

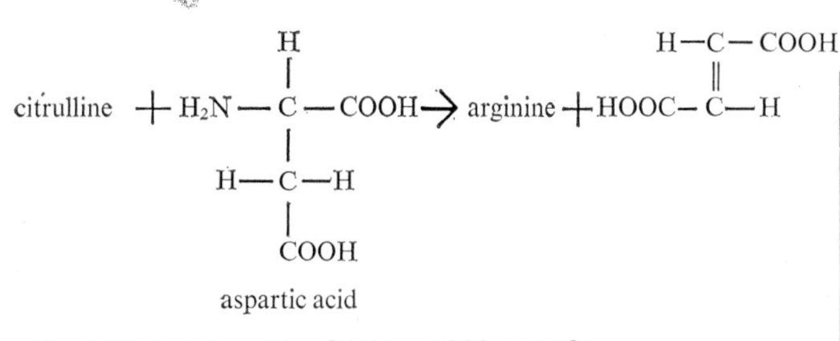

Fig 9.28 **Relationships in the ornithine cycle**

ALIMENTARY CANALS IN VERTEBRATES Fig 9.30 shows simplified drawings of the alimentary canal of the dogfish, the frog, a bird and a mammal. They form an example of adaptive radiation (p 102). The dogfish has a large liver, gall bladder and intestine with spinal valve which increases the surface area available for absorption. The frog has a smaller liver, a gall bladder and a convoluted ileum. The bird has developed the specialised crop, but has no gall bladder. The mammal masticates in the buccal area, while the dogfish and frog swallow their food whole. But similar regions of the alimentary canal are definable in each case.

9.8 Digestion and nutrition in other orders

PROTOZOA *Amoeba* envelops its food particles by flowing around them (Fig 9.29). During digestion and absorption, the food remains in food vacuoles and when digestion is complete, the amoeba flows away from it thereby egesting it. Food may also be absorbed from

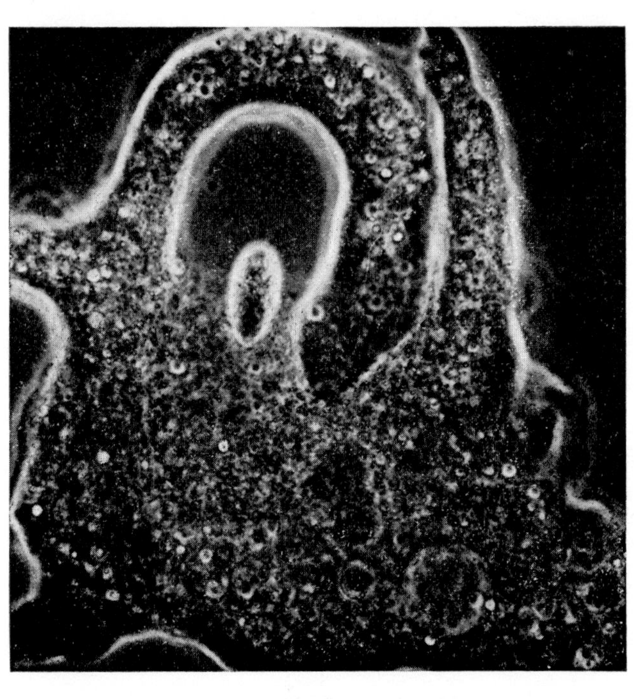

Amoeba engulfing food (a ciliate in this case) by flowing around it

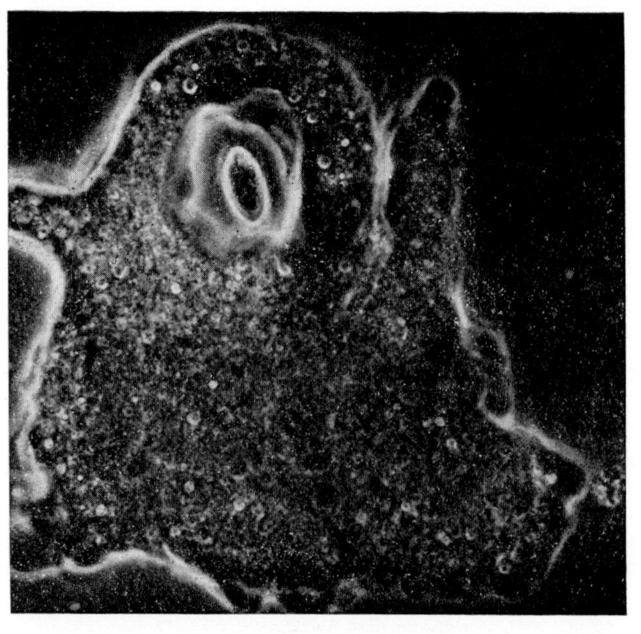

Fig 9.29 **Digestion and nutrition in Amoeba**

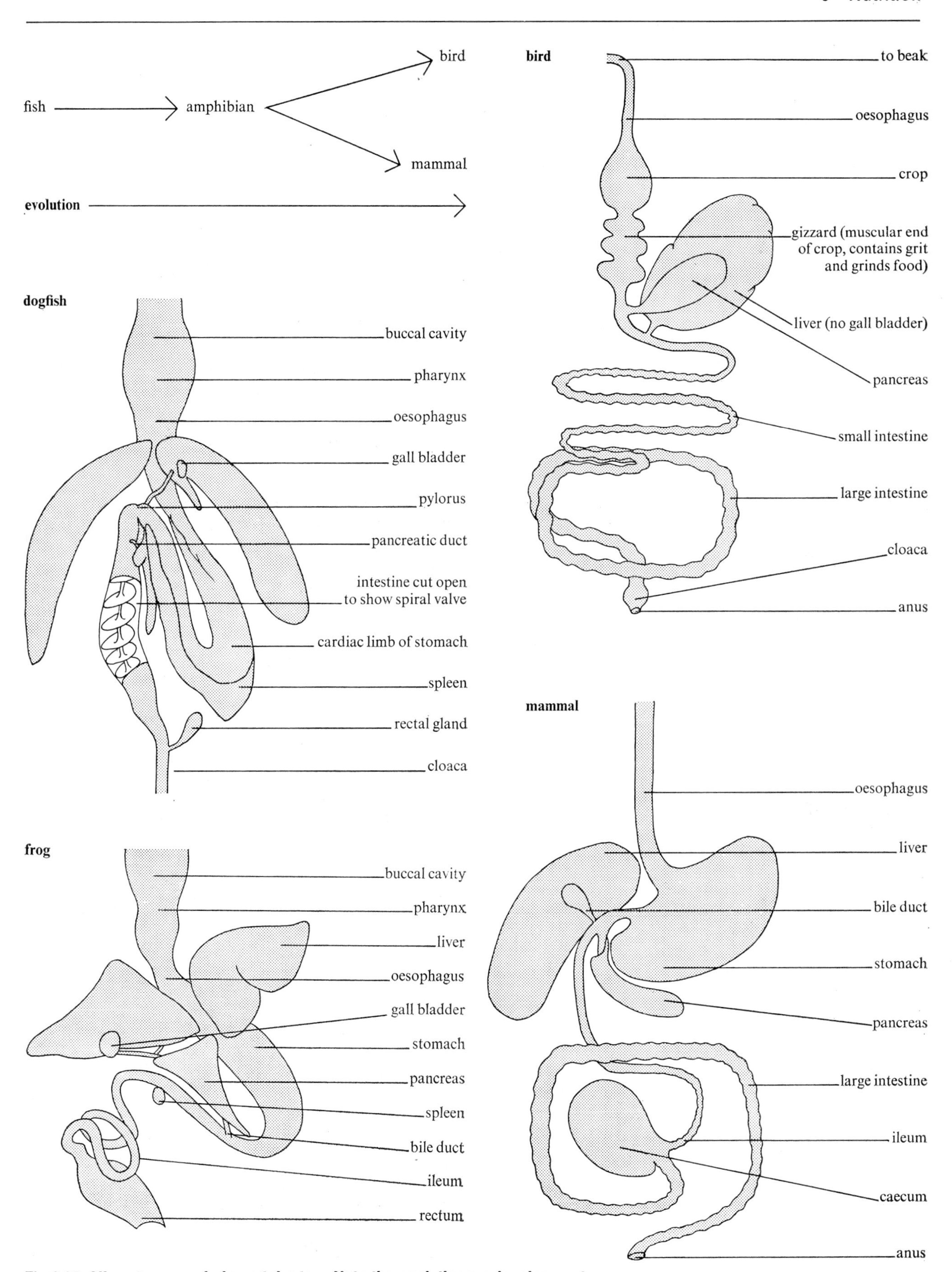

Fig 9.30 **Alimentary canals in vertebrates. Note the evolutionary development**

small vacuoles on the tips of the pseudopodia. *Paramecium* draws bacteria and small algae, using the cilia, into the oral groove and through the cytostome into the cell cytoplasm. The undigested residue is passed out through the cytoproct or anal spot. *Monocystis* appears to absorb food over all its surface and thus probably exudes digestive enzymes.

COELENTERATES *Hydra* captures its protein food by the specially adapted nematocysts which pass the food to the mouth. Glands in the endothelium of the enteron secrete enzymes, the products circulate through the enteron and digestive cells remove soluble materials. Large particles are then ejected through the mouth. Excretion products in the enteron are digested by symbiont algae, and these algae may form a source of food. There are also digestive cells in the enteron capable of ingesting and digesting food like protozoa. Notice that in protozoa digestion is typically intracellular but in coelenterates we have both intra and extracellular digestion.

PLATYHELMINTHS Planarians shown progression towards an organised digestive tract, with a mouth, a muscular pharynx, and intestine. The food is drawn in through the mouth by eversion and inversion of the pharynx and digestion may be both intra and extracellular.

NEMATODES These have a mouth, pharyngeal pump, and an intestine containing specialised brush border cells for absorption.

ANNELIDA The Polychaete worm, *Nereis*, has jaws, pharynx, oesophagus, intestine and anus, the pharynx being everted to trap the food. The oesophagus and intestine secrete digestive juices and absorb digested food. *Lumbricus* has a gizzard. Both worms excrete nitrogen via nephridia.

ARTHROPODS These have a still more specialised alimentary canal. The crayfish *Astacus* has a mouth with associated mouth parts, oesophagus, gizzard and filter chamber, separated from the gizzard by a pyloric valve, a mid-gut with digestive diverticula in the pouch-like midgut, the caecum, which secrete enzymes used in the filter chamber. Food is milled in the gizzard and after digestion is absorbed via the digestive diverticula.

The insect mouth parts (Fig 9.31) enable food to be identified (maxillary and labial palps) and passed into the area between the mandibles which have cutting and grinding surfaces. Saliva is added, and the pulp passes into the crop (Fig 9.32) where it is churned with enzymes secreted by the mesenteric caeca. The food is assimilated via the mesenteric caeca, and in the stomach and intestine.

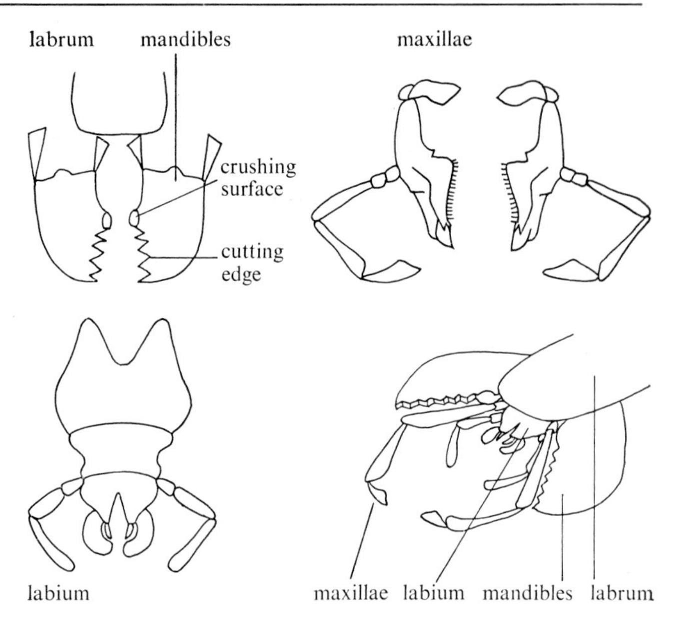

Fig 9.31 **Mouthparts of insect**

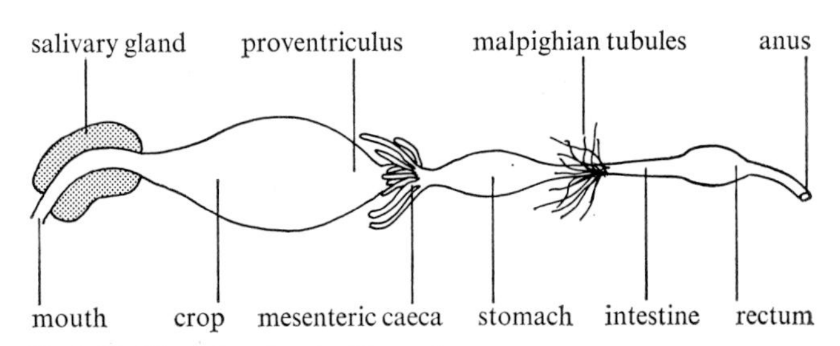

Fig 9.32 **Digestive tract of insect**

9.9 Specialised methods of nutrition

PARASITISM This is a relationship between two organisms, in which one (the *parasite*) lives and feeds off the other (the *host*) causing damage to it. It can be temporary or permanent. Parasites have developed different specialities according to the type of parasitism, but they have in common a need to invade the host, not to be destroyed by its enzymes, wound reactions or antibodies, and to have an effective form of dispersal which may include a resting stage like a spore, cyst or egg. Parasites have usually evolved so as not to kill the host, and have frequently lost powers of movement and co-ordination (animals) or photosynthetic capacity (plants). Specialised structures associated with parasitism are hooks as on the body of the tape worm and the legs of lice, suckers, sucking mouth parts and degeneration of legs, wings and other un-needed organs.

Parasitism is essentially passive, a characteristic which distinguishes it from predation where one animal actively preys on another. Examples of animal on animal, animal on plant, plant on animal and plant on plant are given in Tables 9.5, 9.6, 9.7 and 9.8.

Animal parasites can be divided into two classes, ectoparasites which live on the outside of their hosts, and endoparasites which live inside the host's body. Endoparasites are generally more modified for life in a particular region of a particular host. They obviously cannot move from host to host as ectoparasites sometimes do.

Table 9.5 **Some animal parasites of animals**

host	*parasite*	*classification*	*notes*
man	*Trypanasoma*	protozoa	cause of sleeping sickness
	Plasmodium		cause of malaria
	Entamoeba		cause of dysentery
sheep	*Taenia*	platyhelminth	tapeworm
	Fasciola		liver fluke
pig	*Ascaris*	nematode	pig nematode
man	*Filaria*		cause of elephantiasis
	Ancylostoma		hookworm – causes debility
	Hirudo	annelid	leech
	Pediculus	arthropod	louse
	Pulex		flea

Table 9.6 **Some animal parasites of plants**

host	*parasite*	*classification*	*notes*
potato	*Heterodera*	nematode	root eelworm
various plants	aphids thrips hoppers mites	arthropods	many animals in this group are not specific to a single host plant e.g. the aphid *Myzus persicae* has many recorded plants

Table 9.7 **Some other parasites of plants**

host	*parasite*	*classification*	*notes*
potato	virus		mosaics, leaf curl
sugar beet	virus		'yellows'
barley	virus		yellow dwarf
pear	*Erwinia amylovora*	bacteria	fire-blight disease
potato	*Phytophthora infestans*	fungus	potato blight
nettle	*Cuscuta* (dodder)	Angiosperm	see below
broom, gorse	*Orobanche*	Angiosperm	broom-rape, obtains food from roots of host on which grows

Table 9.8 **Some other parasites of animals**

host	*parasite*	*classification*	*notes*
man	virus		influenza
	bacteria		tuberculosis, typhoid
aphids	*Entomophthora*	fungus	
man	*Tinia*		ringworm

Of special interest is the head louse known as 'super louse' which has developed immunity to DDT. The human body louse, *Pediculus*, is a carrier of typhus. The head louse lives in human hair, laying its eggs on individual hairs. The body louse is much larger and lays its eggs in clothing or on body hairs. The pubic louse ('crab') is even larger and infests the genitalia, producing great itching.

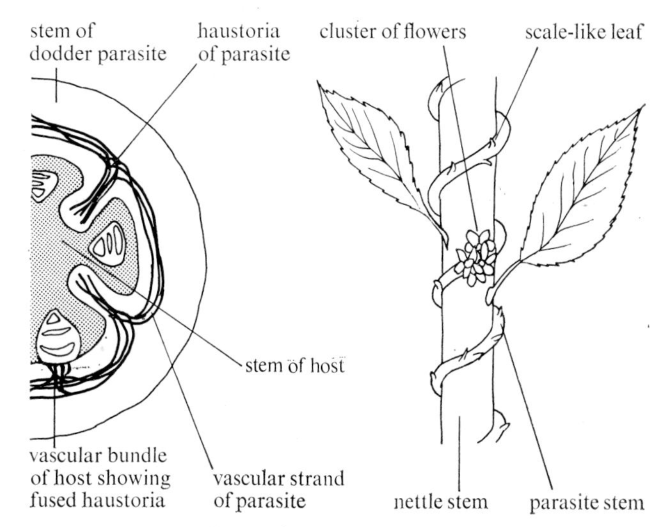

Fig 9.33 **Dodder (of Nettle)**

Dodder (*Cuscuta europaea*) is a parasite on plants like nettles. It belongs to the family Convolvulaceae. Another common species is the clover dodder (*C. trifolii*). The nettle dodder (Fig 9.33) is seen most markedly on nettles growing on river banks. The seed, which has a thread-like embryo embedded in endosperm, starts to germinate late in spring when the nettles have already produced shoots. The seedlings send out roots into the ground and as the shoot gets longer it eventually meets the stem of the nettle, twines around it, and develops haustoria (suckers) which penetrate the host tissue. The xylem and phloem of the dodder fuse with those of the host so that the parasite can obtain its water, organic materials and mineral salts; the root of the dodder then dies. The thin red stem, like a piece of pink cotton, spreads and forms small scaly leaves and small flowers (in clusters).

PARTIAL PARASITISM Such animals as the mosquito do not feed on animals as larvae, but the adult females require a blood meal to lay eggs and thereby complete the life cycle. Mistletoe (*Viscum album*) is also a partial parasite in that it absorbs only water and mineral salts from its host (*e.g.* apple, poplar, oak, hawthorn) and it photosynthesises to obtain organic food (Fig 9.34).

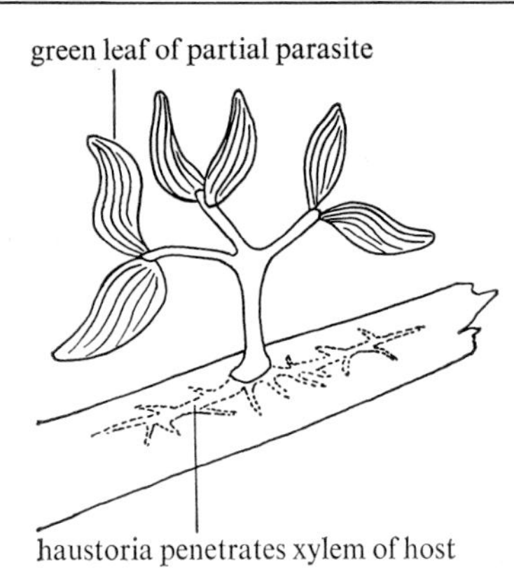

Fig 9.34 **Mistletoe**

FACULTATIVE PARASITISM In this form parasites do not necessarily depend on the host for food and shelter (in contrast to *obligate* parasites). *Lampetra fluviatilis* has a life cycle shown in Fig 9.35. It can live on its own – it swims well and can feed on small creatures on the bed of the river – but it succeeds best as a parasite. Another example is the fungus *Botrytis* which feeds on organic debris but in moist weather may invade soft live tissue like strawberries.

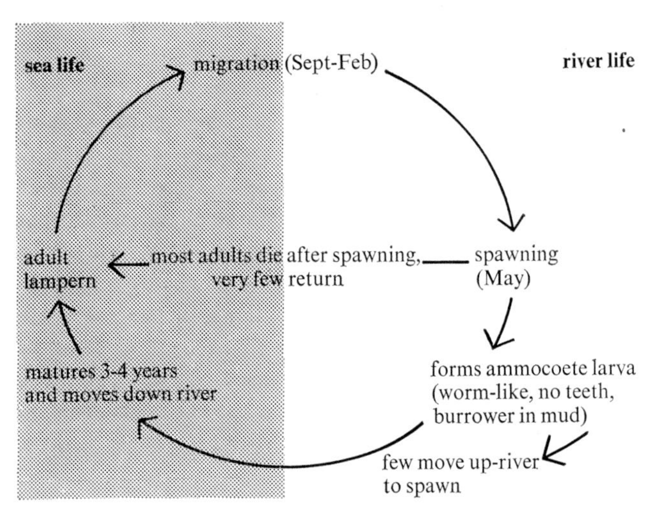

Fig 9.35 **Facultative parasitism**

PARASITE CHAINS Parasite relationships can be extremely complex. Starting with a single host the parasites become progressively greater in number and smaller in size at each stage. A typical parasite chain might run as shown in Fig 9.36.

COMMENSALISM What distinguishes this relationship from parasitism is that the commensal is neither beneficial nor harmful to the host. Epiphytes (*e.g.* orchids) grow on the surface of other plants. The cuckoo, in as much as it ejects other eggs from the host nest, is a parasite; but while it induces the parents to feed it with-

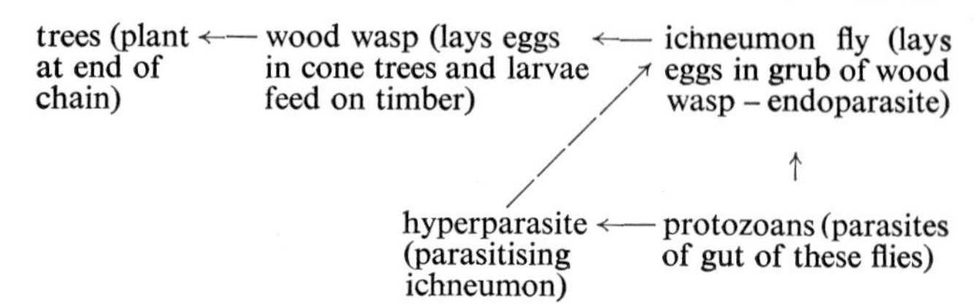

Fig 9.36 **Parasite chain**

out harming them, it is a commensal. The barnacle attaches itself to the whale or the mite to the aphid to gain transport.

SYMBIOSIS In this relationship the symbiont is actually beneficial to the host. Examples of symbiosis are the photosynthetic flagellates which live in the tissues of animals; the host animal uses the oxygen released by photosynthesis and the flagellate uses the carbon(IV) oxide from respiration of the animal. True symbiosis occurs in lichens, algae and fungi which are so closely related that they appear to be a single plant. The fungus (not photosynthetic) derives food from the alga and maintains the water supply for the alga.

The freshwater coelenterate *Chlorohydra viridissima* has in its endoderm cells many small unicellular green algae, *Chlorella*. The latter are often called zoo-chlorellae and they give to the animal its green hue. The animal benefits in obtaining its oxygen from photosynthesis of the plant. The animal is a carnivore and so forms much nitrogeneous/phosphatic material by metabolism of animal protein, and the algae helps assimilation of these, *i.e.* the algae form auxiliary excretory organs to the animal. The plant uses the metabolic waste from the animal by combining it with carbohydrates which it synthesises to form protein. It also receives carbon(IV) oxide from respiration of the animal to assist this synthesis.

SAPROBIOSIS This is a mode of life in which an organism feeds on dead or dying material. Saprophagous larvae of many species of flies perform a useful function in degrading organic matter, e.g. *Fannia* sewage fly. Near habitations of course flies may be a nuisance and spread infections (*Musca* – house fly).

Saprophytes can build their protoplasm from the chemical products of other living organisms or from their dead bodies. They take in dissolved food over the surface of their body. Saprophytes usually have the ability to respire anaerobically, since their bodies become covered with food and the air is restricted. Saprophytes must be able to multiply and grow rapidly, firstly because there are bound to be huge losses when the organism seeks out new territory and secondly because the food may be eaten by other organisms if the saprophyte does not absorb it quickly (and hence grow quickly). Aligned to the fact that there will be large wastages of offspring, saprophytes tend to reproduce sexually *and* asexually, and also form resistant spores.

Saprophytic bacteria in the soil, as well as saprophytic fungi, aid the breakdown of plant and animal materials to form humus.

Some parasites kill their host and then live off them saprobiotically.

The life history of the saprophytic fungus *Saccharomyces* (yeast) has been discussed on p 61.

Practical work

What practical work can be done on digestion?

TO DIGEST STARCH Mash some raw potato with a little water to give a pulp and allow saliva to stand with this pulp in a test-tube for about 2 ks. Add a small volume of distilled water and filter. Test the filtrate for sugars. Check whether reducing sugars are present – by seeing if the tests work before or only after boiling with dilute mineral acid.

INTESTINE Examine a transverse section of the intestine of a frog; note the villi.

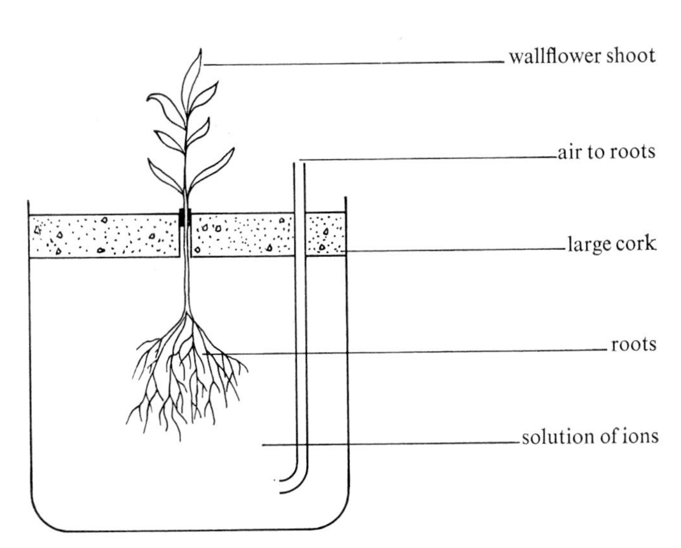

Fig 9.37 **Experiment to illustrate the influence of ions on the rate of growth of plants**

NUTRITION Prepare a dilute solution A containing calcium nitrate(V), potassium phosphate(V), potassium nitrate(V), magnesium sulphate(VI). Make another solution B like A but with calcium sulphate(VI) instead of calcium nitrate(V) and potassium chloride instead of the nitrate(V) (*i.e.* no NO_3^- ions). Make a solution C like A but without potassium phosphate(V) (*i.e.* no PO_4^{3-} ions). Make a solution D like A but with sodium nitrate(V) and phosphate(V), replacing the potassium salts (*i.e.* no K^+ ions). Make a solution E like A but containing no calcium nitrate(V) (*i.e.* no Ca^{2+} ions). Grow plants in these solutions (Fig 9.37).

TRANSPIRATION Demonstrate transpiration by covering a pot containing a leafy plant with tinfoil, and also cover the soil surface. Place the plant in a bell-jar in sunlight. Notice that water droplets eventually collect on the glass – test with anhydrous copper(II) sulphate(VI). Measure the rate of transpiration by use of a potometer.

Fig 10.1 **Athlete at the start of an important race**

10 Response and Locomotion

10.1 Co-ordination in animals and plants

In every organism there is some system of co-ordination. If a set of specialised cells co-ordinate the organism by receiving and transmitting impulses, we say that it has *nervous co-ordination*. In many instances the nervous system operates in conjunction with the *endocrine system* (§10.5), a set of secretory cells or *glands* which pass their secretions (called *hormones*, from the Greek, meaning to excite) directly into the blood plasma. The hormones act as chemical messengers, acting on specific types of tissue.

Plants do not have a nervous system, co-ordination being achieved by chemical messengers called *auxins* (§10.8). Chemical messengers are not capable of the speed of response of the nervous system.

Organisms react to changes in their environment, which means that they have to be able to *detect* such changes and then *respond* to them. In animals, detector and response organs are communicated by means of a nervous system aided by hormones: sometimes response is associated with complex behaviour and/or movement in the animal. Plants respond to changes in their environment by growth movements. In this chapter we discuss *why* organisms move, and also *how* they move.

10.2 Nervous systems

All multicellular organisms of the animal phyla except Porifera have some form of nervous system which, for convenience, can in most animals be divided into a *central nervous system* (CNS) and *peripheral nerves*. The latter connect the CNS to *effectors* and *receptors* (sensory cells).

NEURONE The unit of the synaptic nervous system is the neurone. These are usually associated with cells for support and protection, which as a whole make up the brain, the spinal cord and nerves.

There are three types of neurones, *sensory neurones*, *motor neurones* and *interneurones*. All motor and sensory neurones are basically similar in structure, but interneurones vary considerably. Fig 10.2 shows the structure of motor and sensory neurones. Each has a cell body consisting of a nucleus surrounded by cytoplasm enclosed within a membrane. The cytoplasm contains granules of RNA, called Nissl granules. Extending from the cell body

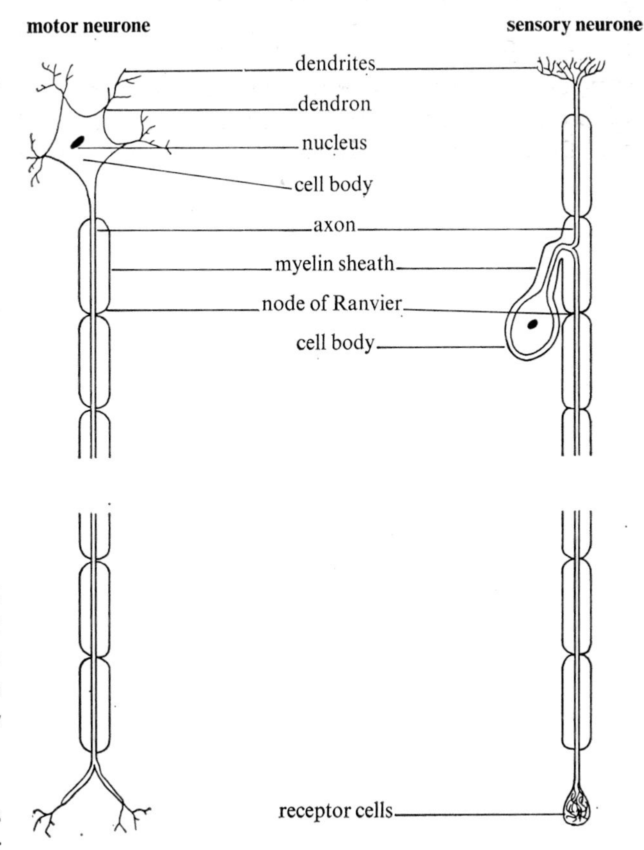

Fig 10.2 **Motor and sensory neurones**

are a series of thin fibres or processes called *axons* and *dendrons*. Axons carry information away from the cell body, and dendrons, terminating in fine *dendrites*, carry information towards the cell body. In the motor neurone the cell body is at the end of the axon, whereas in the sensory neurone it is to the side and has no branching fibres. Fig 10.3 shows the structure of interneurones. These are found only in the CNS.

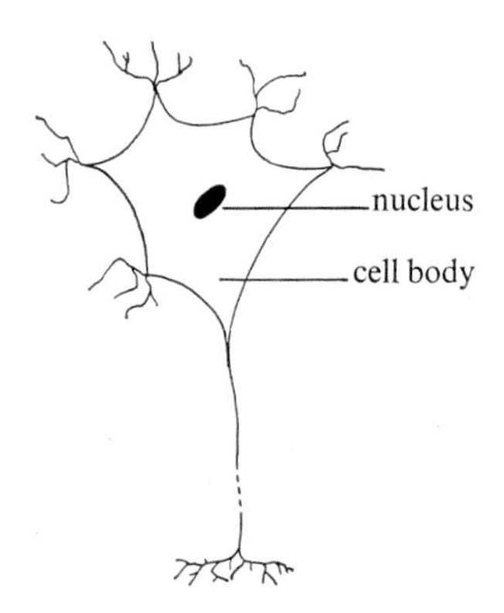

Fig 10.3 **Interneurones**

NERVOUS SYSTEM The vertebrate nervous system comprises the brain, the spinal cord and the nerves (Fig 10.4). In the brain there are *centres* (groups of interneurones) connected with each other by tracts. Different centres are associated with different senses; some centres are associated with responses (*motor areas*) while others serve to correlate information. In the spinal cord there are *ascending tracts* connecting receptors with sensory areas in the brain, *descending tracts* connecting motor areas with effectors in the body, plus interneurones. Motor and sensory nerves are sometimes connected directly to the brain, but more often they connect effectors and sense organs with the spinal cord.

The nervous system is divided into:

1 *the somatic system*, concerned with voluntary actions, *i.e.* response to external conditions,
2 *the autonomic system*, concerned with responses of internal organs, *i.e.* actions that are not as a rule voluntary,
3 *brain centres.*

TRANSMISSION OF AN IMPULSE The transmission of an impulse from receptor to effector is made up of two parts, the impulse along the fibre and the impulse across the gap between one nerve cell and another, called the *synapse*.

Du Bois-Raymond (1840) showed that when an impulse passes along a fibre, it is accompanied by some change in the electrical state. The impulse is a wave of activity whose rapid motion is accompanied by a change in potential in the nerve, this wave travelling along the

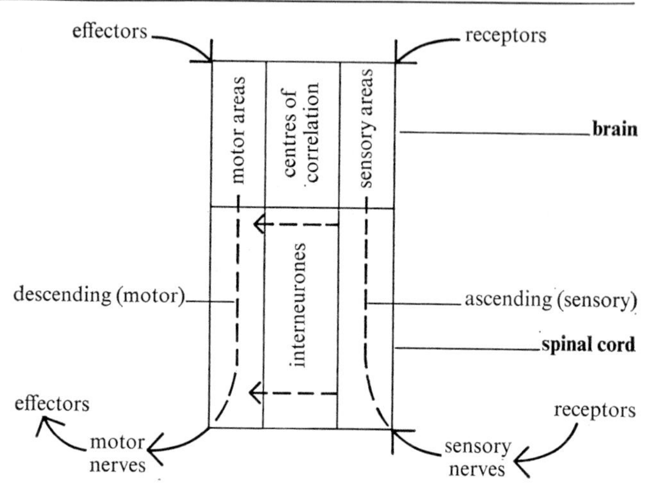

Fig 10.4 **The vertebrate nervous system**

sensory fibre into the cell body, and thence along the axon. We use the term *action potential* to denote the way in which the potential changes when the nerve is transmitting an impulse. If an impulse is strong enough to be propagated at all, it will pass regardless of the strength of the stimulus. This is known as the '*all or nothing law*'. If a stimulus is below a certain intensity known as the threshold intensity no impulse is passed. But above the threshold intensity the same impulse is passed whatever the strength of the stimulus.

After an impulse has passed there is an *absolute refractory period* (about 1 ms) during which no impulse can pass, however strong the stimulus. The speed of the impulse is *independent* of the strength of the stimulus. The speed does however vary with the type of fibre. A fibres (large, somatic, myelinated) can pass messages faster than B fibres (small, visceral, thinly myelinated), which in turn can pass messages faster than C fibres (small non-myelinated). For A fibres, the speed is directly proportional to the diameter of the axon. Transmission speeds vary from less than 0.5 m s^{-1} in certain invertebrate neurones, up to 160 m s^{-1} in some mammals.

In a resting nerve fibre (Fig 10.5), the axon membrane separates two solutions of about the same electrical conductivity. The types of ion differ. Outside more than 90% of the ions are Na^+ and Cl^- whereas inside less than 10% are of this sort; the main cation is K^+ and there are many organic anions such as aminoethanoate ($NH_2CH_2COO^-$) which are too large to diffuse to the outside. The concentration of K^+ is 30 times greater inside and the concentration of Na^+ is 10 times larger outside. The unequal distribution of ions results in potential difference of about 50mV across the membrane with the inside negative with respect to the outside. This probably involves active transport of sodium ions, the *sodium pump mechanism*, where Na^+ is pumped one way making room for K^+.

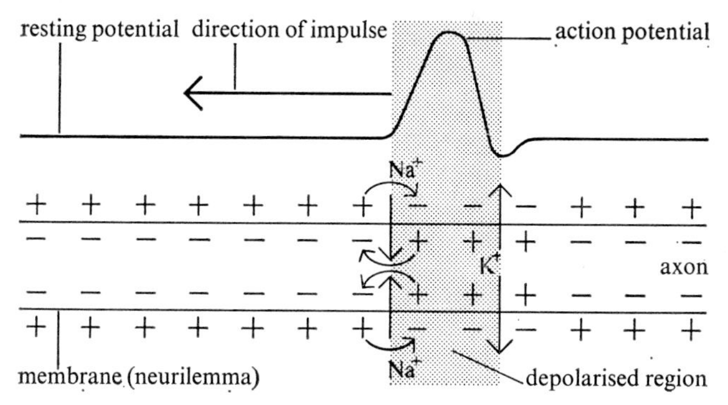

Fig 10.5 **Transmission of an impulse**

Passage of an impulse is due to a temporary alteration in the permeability of the membrane. The permeability to Na^+ and K^+ alters so that Na^+ enter and K^+ leave by diffusion, a process which is very rapid owing to the high concentration. The overall result is that the outside becomes temporarily negative and the inside positive. This is know as depolarisation. Very soon afterwards the sodium pump mechanism gets going again (Na^+ pumped out, K^+ pumped in) and repolarisation takes place. Thus the impulse passes along by this process, continuing down the length of the neurone. The actual number of ions involved is small so that the composition of the solution inside and outside is hardly affected.

In nerves covered with a myelin sheath the conduction is greatly enhanced. It is thought that the myelin sheath causes the action potential to jump between the nodes of Ranvier where depolarisation takes place as normal, thus increasing transmission speeds.

The refractory period is caused by the fact that another impulse cannot be transmitted until repolarisation has occurred. The '*all or nothing*' law is caused by the fact that the stimulus must be strong enough to overcome the resting potential.

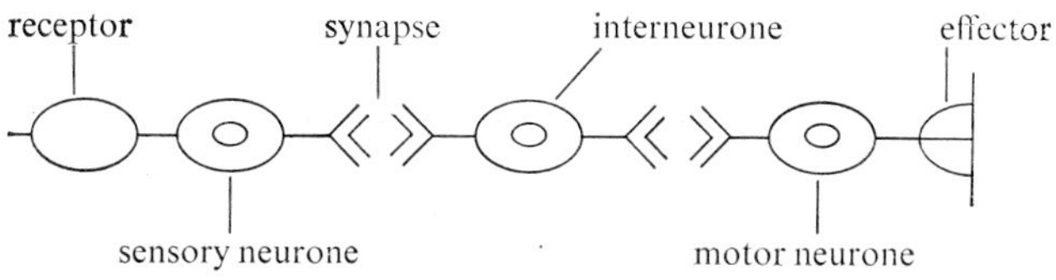

Fig 10.6 **Synapse**

SYNAPSE When a message reaches this point *acetylcholine*, a chemical transmitting agent, is liberated (Fig 10.6). This diffuses from between nerve endings of the axons (or nerve endings and muscle), changes the degree of permeability of the tissue that it enters, which sets up another wave. Note that although diffusion, a relatively slow process, is involved here, it can be fast enough in operation if the distance is small. After carrying out its task, the acetylcholine is degraded by the enzyme cholinesterase which is present in high concentration at the synapse.

REFLEX ARCS If the passage of an impulse from receptor to effector is direct, *i.e.* if the receptor is stimulated the effector *has* to come into play, the response to stimulation is called a *reflex action.* The neurone arrangement is the *reflex arc* (Fig 10.7).

The simplest reflex arc is when a sensory neurone connects directly with a motor neurone. But in general the sensory and motor neurones are connected by one or more interneurones. When the afferent fibres of a neurone pass into the brain or spinal cord they make many connections with other neurones. Some of these are able to relay the impulse to other effectors and some are capable of carrying it to other parts of the brain or spinal cord where it can link up with other efferent neurones. A more accurate picture of a reflex arc would be as illustrated in Fig 10.8.

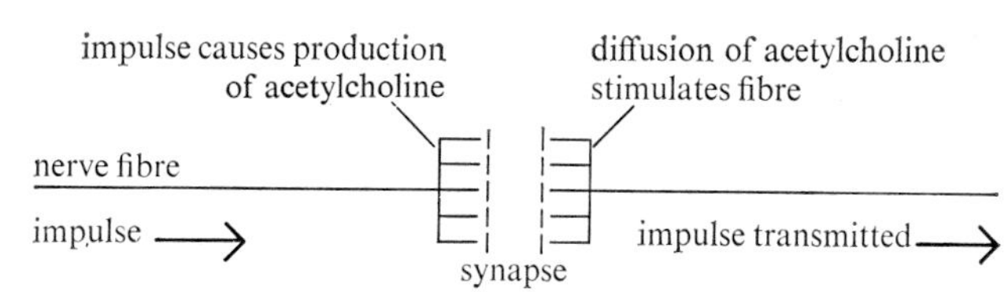

Fig 10.7 **Simple reflex arc**

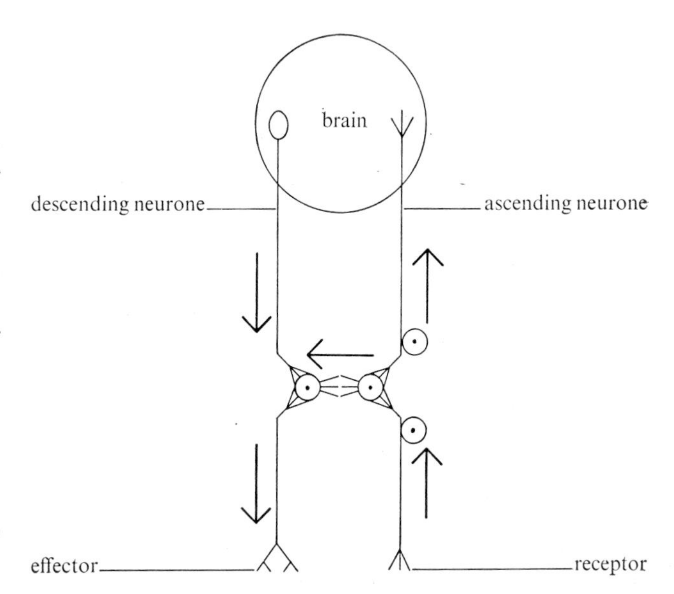

Fig 10.8 **Reflex arc**

If the nervous system of an animal was constructed only of reflexes, its activity would be unco-ordinated; in fact, in higher animals reflex is the exception.

AUTONOMIC NERVOUS SYSTEM This is made up of visceral reflexes and consists of two sub-divisions, the *sympathetic* and *parasympathetic* system. For both there is a complex set of synapses called a *ganglion* outside the central nervous system where a neurone leading from CNS to the ganglion (*i.e.* a pre-ganglionic neurone) synapses with a neurone that is post-ganglionic and medullated; the latter leads from the ganglion to the organ served by the nerve. In the sympathetic system the ganglion is near the CNS but in the parasympathetic system (*e.g.* the vagus nerve) the ganglion lies near the organ served (Fig 10.9). The effects of parasympathetic stimulation are *localised* whereas sympathetic stimulation covers a wider area over the material served by nerve. In both, the preganglionic fibre liberates acetylcholine as chemical receptor. At the ending of the muscle or gland, however, the parasympathetic neurone gives ethanoylcholine whereas the sympathetic post-ganglionic neurone liberates *noradrenaline*.

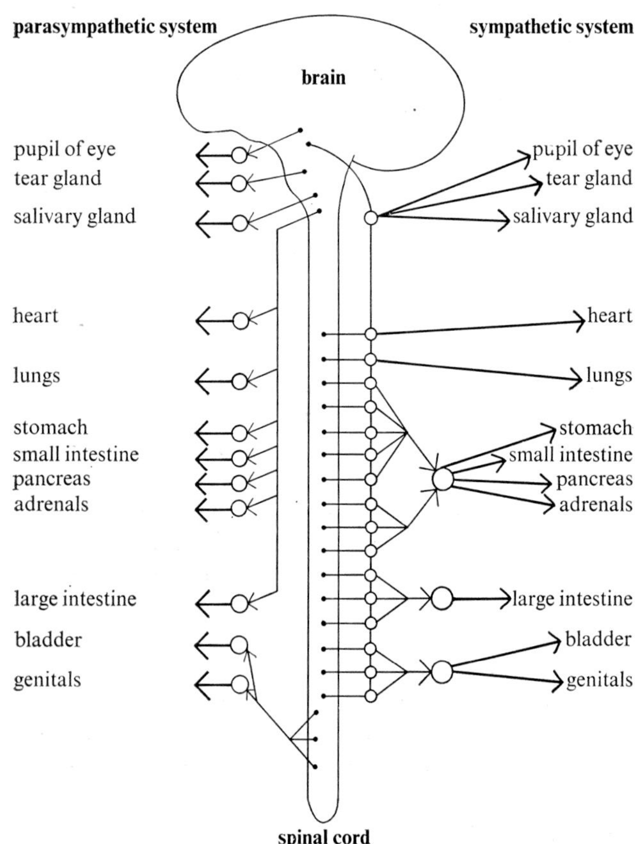

Fig 10.9 **Sympathetic and parasympathetic system**

BRAIN CENTRES The brain is the region of action between receptors and effectors in which there are centres for different activities (Table 10.1). The centres show different degrees of development in different species which correlate well with the behavioural needs of the species.

Table 10.1 **Centres of activity in the human brain**

region	*centre*	*function controlled*
hindbrain	medulla	respiration circulation salivation vomiting
	cerebellum	balance fine control of movement
	pons	? dreams
midbrain	roof	eye movement ear function
	floor	regulates muscle action
forebrain	thalamus	see text
	hypothalamus	temperature control osmoregulation hunger and thirst sleeping speech
	cerebral cortex	sight hearing smell taste touch initiation of movement intelligence

Electrical stimulation of the brain areas has helped to show the areas responsible for initiating movement in different parts of the body. The brain exhibits natural electrical activity and the study of people's brain waves (called electroencephalograms or EEG's) provides visual evidence of normal and disturbed mental states and are frequently used in psychiatric hospitals as routine measurements on patients.

HINDBRAIN The medulla is vitally important since it is concerned with visceral functions such as respiration, circulation etc. Most of the cranial nerves originate here; it is a nerve centre connecting the spinal cord with the fore-regions. The cerebellum has sensory neurones from the eyes as well as the neurones from the inner ear, controlling balance It also receives information from muscle proprioceptors and information from the cerebral cortex.

Removal of the cerebellum in a dog makes the creature unable to stand or to make voluntary movements. In man, damage to the cerebellum makes his hand tremor when he tries to follow an object through space. This shows that the cerebellum is concerned with posture and voluntary movements.

MIDBRAIN In fish the midbrain is all-important in the correlation of sensory information, but in the mammal it is far less important. The roof has four optic lobes, the superior colliculi and inferior colliculi. The former are for the control of eye movements while the latter are concerned with auditory functions and reflexes. A cat from which the inferior colliculi have been removed cannot differentiate between tones and cannot detect low intensity noises. The crura cerebri of the floor of the mid-brain joins the spinal cord and the anterior regions of the brain. Many nuclei control eye muscles. Regulation of posture and movement is the province of the red nucleus.

FOREBRAIN In man this is the most important area of the brain. It consists of three parts, the thalamus, the hypothalamus and the cerebral cortex.

1 *The thalamus* is the posterior part of the forebrain. Its main function is that of a relay station between the cerebral cortex and other regions of the brain. It receives, modifies and passes on information from sensory organs.
2 *The hypothalamus* lies under the thalamus. Its functions are to regulate the unconscious activities of the body, and to maintain the internal environment. It is for instance the site of the thermostat of homoiotherms. Action is taken via the autonomic nervous system and endocrine system. It is also connected with the action of the pituitary gland (p 160). Some hormones, from the posterior pituitary such as the one regulating urine flow, are produced in the hypothalamus.
3 *The cerebral cortex* is extensive only in the mammals. Its size and complexity which increases throughout the mammals culminating in man represents a major evolutionary advance in the development of the CNS. It lies in the roof of the forebrain. It is the 'thinking' area of the brain and consists of millions of nerve cells connected together in an extremely complex pattern.

INVERTEBRATE BRAIN The brain in an arthropod or annelid plays little part in co-ordination of locomotion; if the head of a locust is removed the animal still moves in a co-ordinated manner. Co-ordination of locomotion is spread throughout all the nervous system. The brain serves as a nerve centre, receiving impulses from various receptor organs in the head and sending them via the nerve cord to effector organs in far removed parts of the body. *Lumbricus*, a burrowing vegetarian animal, has a simpler head than the carnivorous more active ragworm, *Nereis*, which has palps, sensory tentacles, eyes and cirri. Development of the head in evolution – *cephalisation* – is less advanced in *Lumbricus* than in *Nereis*.

10.3 Primitive response systems

AMOEBA If a pseudopodium of *Amoeba* comes into contact with, for example, a grain of sand, it retracts and another pseudopodium pushes itself out. Thus we can say that the protoplasm of *Amoeba* can detect nervous messages over its whole surface. The fact that another pseudopodium is pushed out means that the protoplasm must also be able to conduct impulses. The co-ordination is not very pronounced and it is sometimes found that a particle of sand sticks into a pseudopodium when *Amoeba* is moving fast.

Amoeba is able to respond to the stimulus of light. If it is suddenly illuminated it will move away from the source of light: pseudopodia are not formed on the illuminated side.

In general the co-ordination shown is of a very simple nature. The ectoderm layer which covers the entire surface of the animal is responsible for the limited ability to perceive.

PARAMECIUM If *Paramecium* is travelling in a straight line its cilia beat with definite co-ordination. When the animal approaches an unfavourable situation, such as a pool of acid placed in its path, it takes action. All the cilia reverse their action in such a way that the direction of travel is at right angles to the previous one; all the cilia on one side propel that particular side forward. The cilia of *Paramecium* are mechanically co-ordinated and not by a fibre (neurofibrillar) system as once thought – see Gliddon *School Science Review* 1966. *Vol.* 48 *p* 482.

LUMBRICUS The earthworm has quite a complex nervous system. It has a two-lobed brain, comprising two cerebral ganglia, above the pharynx, connected to the first, subpharyngeal ganglion of a ventral nerve cord by two thin connective fibres called the circumpharyngeal commissures. There are nerves from this subpharyngeal ganglion and the brain passing to the anterior regions of the animal. A ganglion is formed in each segment, and nerves from each ganglion serve the tissues and organs in that segment. The ventral nerve cord is really made up of two parallel cords, close together. In the dorsal part of the nerve cord there are 'giant fibres' capable of conducting impulses at about 10 m s^{-1}. The sudden contraction of the earthworm will be familiar.

The earthworm reacts to stimuli in a manner similar to higher animals. The ganglia serve as centres for simple reflex actions, neurones taking sensory impulses from the surface to the ganglia. The impulses have to cross synapses and then pass along motor neurones to muscles or glands, which are triggered into action. It has been shown that acetylcholine and adrenaline, substances which regulate muscle action, are present in the earthworm. There is a co-ordinated action because there are fibres in the cord which transmits impulses between the ganglia. In locomotion, the initiation comes from the 'brain' and there are waves of contraction of the circular/longitudinal muscles all down the body. It is known that proprioceptors in the muscles of one segment are linked to the muscles in an adjacent segment. The ventral solid nerve cord is characteristic of many invertebrates. In vertebrates the brain and spinal cord are hollow and are dorsal.

LOCUST The response system of the locust is typical of many insects. There is a central, solid, double ventral nerve cord with many ganglionic enlargements, and cerebral ganglia in the head. These are connected via circum-oesophageal connectives round the oesophagus to the ventral sub-oesophageal ganglia. From here double connectives pass to the large pro-, meso-, and metathoracic paired ganglia, and thence via smaller pairs in the first five abdominal segments to a larger, compound terminal pair. The mouthparts are controlled by the sub-oesophageal ganglia, the legs and wings by the appropriate thoracic ganglia; abdominal movement and reproductive organs are controlled by the abdominal ganglia. Within each segment, simple reflex actions are initiated by proprioceptors in the muscles which transmit information via neurones to the ganglion; impulses return via motor neurones. Respiration is controlled in this way. The brain exerts a regulatory influence by integrating stimuli and activating the appropriate ganglia for more complex activity.

Insects have compound eyes, made up of ommatidia (p 157) each of which is a unit capable of perceiving light and initiating impulses. Colour and general form are recognised and changes in light cause strong avoiding reactions. Many insects perceive polarised light, and the colour sense of some includes ultra-violet. Audio-reception in locusts is via the tympanum, a membrane stretched across a cavity which vibrates in response to the stridulatory noise made by the male. Some insects perceive higher-frequency supersonic signals. Chemo-reception is through the labial palps, tarsi and antennae, for identifying food, for instance. But the antennae of many insects also detect very low concentrations of chemical attractants used to bring the sexes together – the pheromones. Proprioceptors in the muscles and on the body surface control such features as co-ordination of posture, balance and flying attitude. Frequently, receptors of external stimuli are specialised sensillae, modified epidermal cells. These include temperature sensors. Touch is experienced by the bending of bristles called setae, with neurones at their bases.

10.4 Sense organs

MAMMALIAN EYE The eye detects the intensity, frequency, polarisation and direction of light in the visible range (400-750 nm). It is hence able to mediate the sensing of images in the brain. A diagram of the human eye is given in Fig. 10.10.

Light reception results in the passage of impulses along nerve fibres to the CNS, these impulses coming from endings in the back of the retina (Fig 10.11). Pigments in the rods and cones aid the change from light energy to electrical energy. The rods/cones are specialised cells. By electron microscopy, it has been shown that the outer surface of rods is a cilium connected by a stalk to the inner region. Discs in the rods contain the pigment *visual purple* (*rhodopsin*), which is changed by the light into a protein and retinene (via an unstable intermediate), the latter being the alkanal (aldehyde) of vitamin A. If there is a lack of this vitamin the ability to detect light is reduced ('night blindness'). Light continuously reforms rhodopsin. The cones contain three pigments in man again made up from retinene and a protein. They are not nearly so sensitive as rhodopsin. Vertebrates can detect a wide range of intensities owing to the presence of rods and cones in different ratios.

Diurnal (day) animals have $\frac{C}{R}$ = large

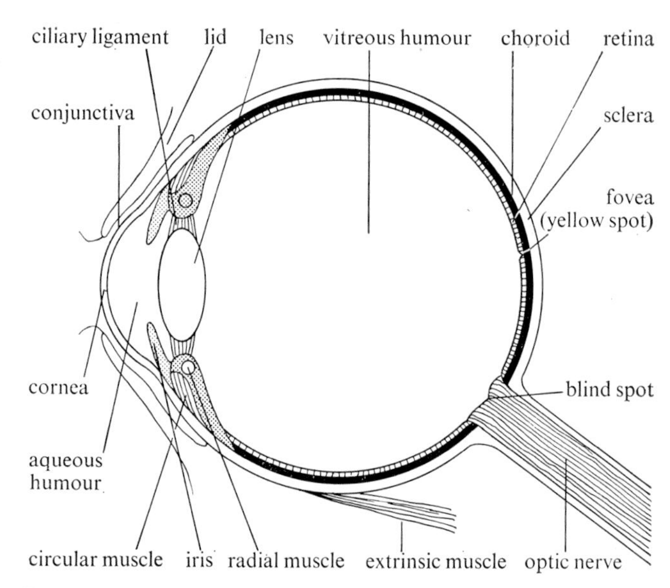

Fig 10.10 Mammalian eye

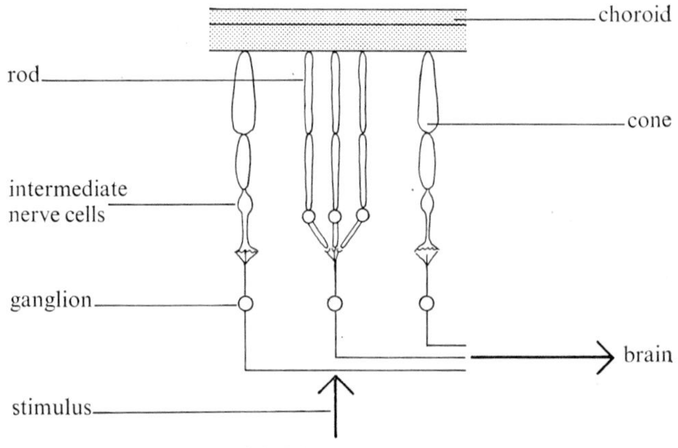

Fig 10.11 Reception of light

Nocturnal (night) animals have $\frac{C}{R}$ = small

where C is the number of cones, and R the number of rods. When rods and cones are excited, a nerve impulse passes along a whole network of fibres and synapses to the optic nerve which takes the message to the brain. At the fovea (yellow spot) there is a connection to a single neurone chain which goes straight to the brain; and so if light falls here the information passed to the brain will be very accurate. At other parts of the retina, several connections have to be made from visual elements to some intermediate neurone and so less detail is achieved when the message finally gets through to the brain. In order to reach the rods and cones the light has to travel through all the layers of the retina and reflections would tend to blur the image; the dark choroid helps to reduce this.

The chemical reactions taking place can be summarised

Rods (rhodopsin pigment) $\underset{\text{dark}}{\overset{\text{light}}{\rightleftharpoons}}$ intermediate unstable complex $\rightleftharpoons$ protein + retinase (vitamin A)

Absorption by rhodospin is at maximum at 500nm wavelength.

The trichromatic theory of colour vision is now generally accepted and depends on the ideas of colour mixing – blue, red and green primary colours mixed in various amounts give other colours. In the retina are three types of cone sensitive to the blue, red and green wavelengths, and stimulation of any one of these gives a single sensation of colour which is communicated to the brain. Any particular colour arises because the three types of cone are stimulated to varying extents. For evidence that there are three sorts of cone we can send a fine beam of light of a known wavelength through an eye retina placed under a microscope – a photomultiplier gives us the intensity of the light emerging from the eye lens. Control must be used by sending light through eye tissues lacking photoreceptor cells. These experiments, due to Wald and co-workers, show that there are three cones having absorption maxima at red (550 nm), green (525 nm) and blue(450 nm) respectively. Light *intensity* as opposed to colour, is determined by the frequency at which light pulses leave the three sorts of cone.

The shape of the lens can be changed to focus objects at different distances from the eye, this being called *accommodation*.

The eye can suffer from several defects. One is *far sightedness* (hypermetropia), in which rays focus behind the retina. This is corrected with a converging lens in spectacles if it is bad. In *near sightedness* (myopia) the rays focus in front of the retina, and this is corrected with a diverging lens in spectacles. In near sightedness there is a definite far point (rather than infinity) beyond which vision is not acute. Thus the eye cannot see distant objects. In *astigmatism* there are errors in the refracting surface of the lens system, generally an oblong cornea or lens. The eye focuses in one plane but not in another; correction is made with cylindrical lenses in spectacles.

COMPOUND EYE The compound eye (Fig 10.12) of insects (Fig 10.13) and other arthropods is different from the mammalian eye in that it consists of *ommatidia* instead of rods and cones. But because ommatidia are far larger than rods and cones the density of receptors in the compound eye is much lower than in the mammalian eye. Insects thus have less acute vision. However, the recovery time of an ommatidium (the time taken to receive an impulse, transmit it, and be ready for the next impulse) is far shorter than for a rod. If the frequency of a stimulus is too large, we get a continuous effect (cf looking at a film on a cinema screen). In man *flicker fusion* as it is called happens at 50 images per second, whereas in some insects it is as high as 200 images per second. The compound eye can therefore detect much more minute movements than the mammalian eye. To demonstrate this, try and trap the next fly that lands on your hand.

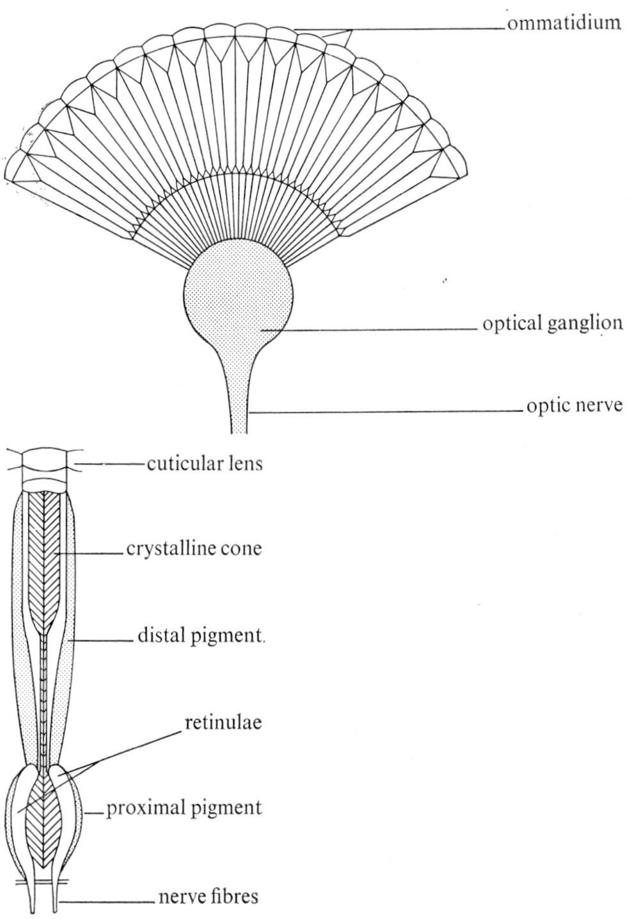

Fig 10.12 **Compound eye**

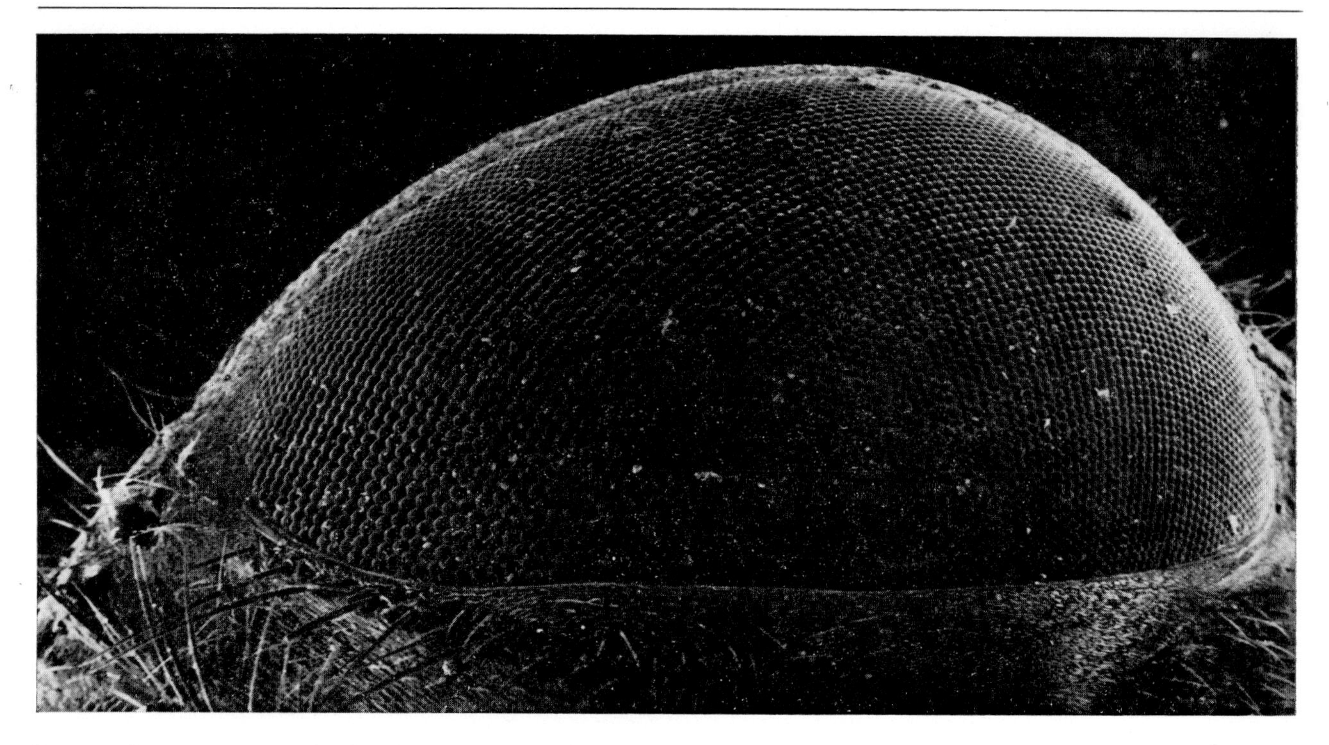

Fig 10.13 **Compound eye of a fly (Scanning Electron Micrograph)**

MAMMALIAN EAR This has three parts, the outer ear, the middle ear and the inner ear. The outer ear has a *pinna*, a skin-covered cartilaginous organ plus a connecting *auditory canal*, lined with cells that secrete wax. At the end of the auditory canal there is a *tympanic membrane* (eardrum) which vibrates in response to air pressure changes. The pinna collects *longitudinal sound waves* and concentrates them on to the tympanic membrane, making it vibrate (Fig 10.14). The middle ear is filled with air, and connected to the pharynx by the *Eustachian tube*. This tube has a valve to prevent the sound of the person's own voice from entering the middle ear.

There are three small bones in the middle ear, the *malleus* (hammer), attached to the tympanic membrane and to the *incus* (anvil); the incus is connected to the *stipes* (stirrup). The stipes are in contact with the *fenestra ovalis* (oval window). When the tympanic membrane vibrates, these vibrations are transmitted (with smaller amplitude) to the fenestra ovalis and so to the inner ear.

The inner ear is made up of two parts, the *cochlea* (Fig 10.14) where the *auditory nerve* starts and the *semicircular canals*. The cochlea is spiral-shaped and set in bone; it is narrower at the apex than at the base. It is made up of three tubes. At the apex, the vestibular canal connects with the tympanic canal, both being filled with *perilymph fluid*. The cochlear canal contains *endolymph fluid*. In the cochlear canal is the hearing part, the *organ of Corti*, and it extends the full length of the cochlea, being made up of row after row of hair cells with fibres projecting out.

The fenestra ovalis transmits the vibrations to the perilymph. The *fenestra rotunda* (round window), with its elastic membrane, makes possible the transmission of vibrations to the fluid in the vestibular canal. Movements of the fluid cause the *basilar membrane* to oscillate so that hair cells rub against the *tectorial membrane*; the motion stimulates the hair cells to start an impulse flowing via the auditory nerve to the brain. The loudness is proportional to the number of hair cells stimulated. It is known that the hairs protruding from the hair cells vary in length, being short at the base of the cochlea and longer at the apex. A given frequency causes stimulation of only those hair cells which respond to that frequency and nerves from each part of the cochlea go to different parts of the brain. It is therefore possible for a person to distinguish notes of different frequency (pitch). A man can detect sounds ranging in frequency from 20Hz to 20 kHz, but is most sensitive in the region 1 to 2 kHz.

The three semicircular canals (in Fig 10.14) are concerned with equilibrium and balance; (each canal lies in a plane at right angles to the planes of the other two). Equilibrium is concerned with the *sacculus* and *utriculus*, which maintain the head in a definite position with respect to the downward force of gravity acting on it.

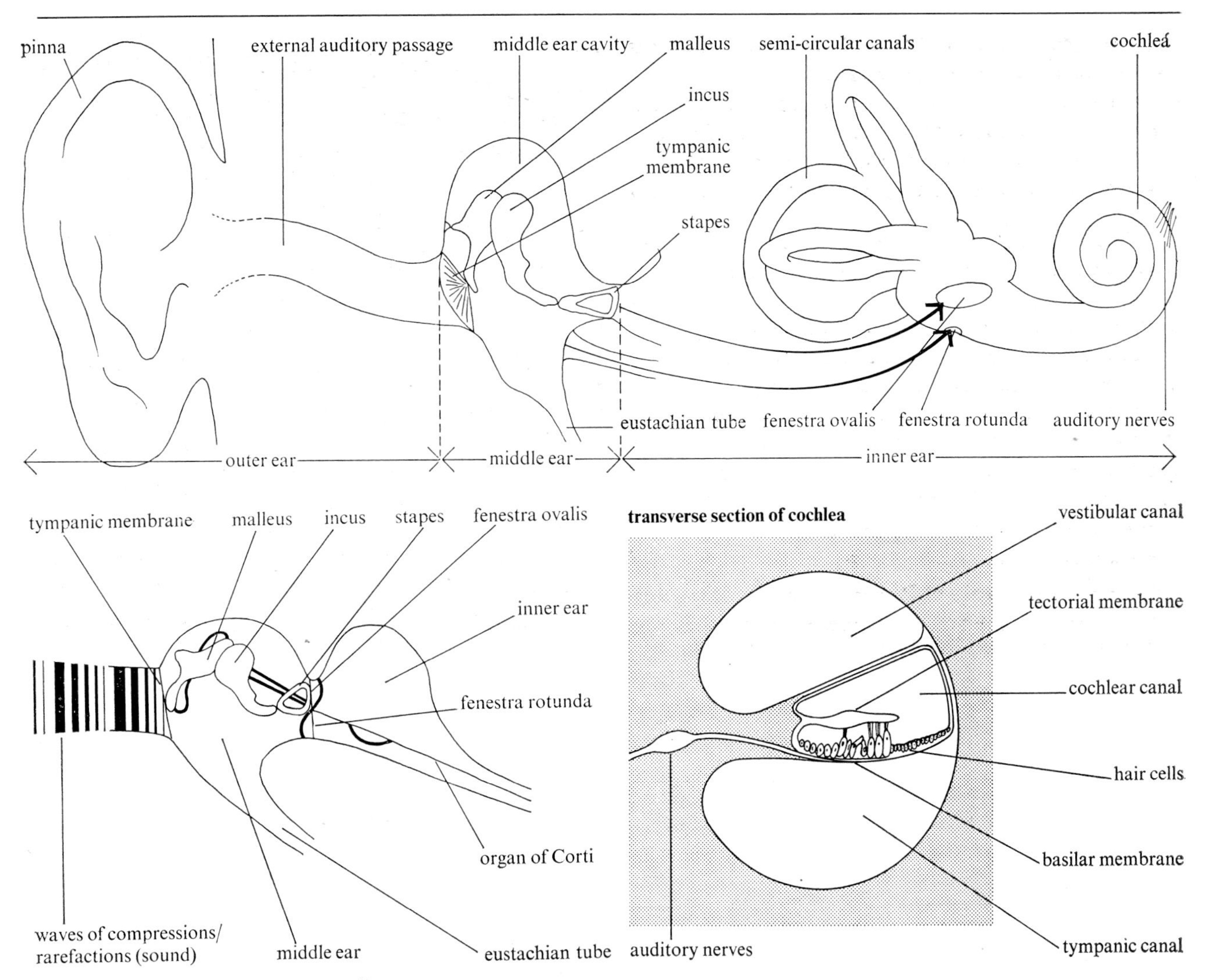

Fig 10.14 **Diagrams of the mammalian ear**

Both sacculus and utriculus have hair cells and an *otolith* of calcium carbonate (a small round 'stone') resting on the hair cells. If the person moves his head, the 'stone' moves, different hair cells respond, and different messages pass along nerve fibres to the brain. The interior of the semicircular canals has endolymph fluid and receptor nerve fibres. A movement of the head causes the endolymph to move and this stimulates the nerve endings. The combination of such stimulations by the brain enables a person to have a sensation of motion.

TONGUE This is used for *tasting* food by responding to the molecular geometry of the substrate; this causes stimulation of salivary glands which secrete juices containing enzymes for digestion. The four *true* tastes are sweet and salt (both detected by taste buds (Fig 10.15) at the front of the tongue), bitter (detected with the back of the tongue) and sour (detected with the front of the tongue). Taste detects dissolved substances.

NOSE The olfactory epithelium detects odours by detecting differences in molecular geometry, and conveys this information to the brain. Smells are substances in gaseous form and the number of smells that can be distinguished is very large. Smell is important in the

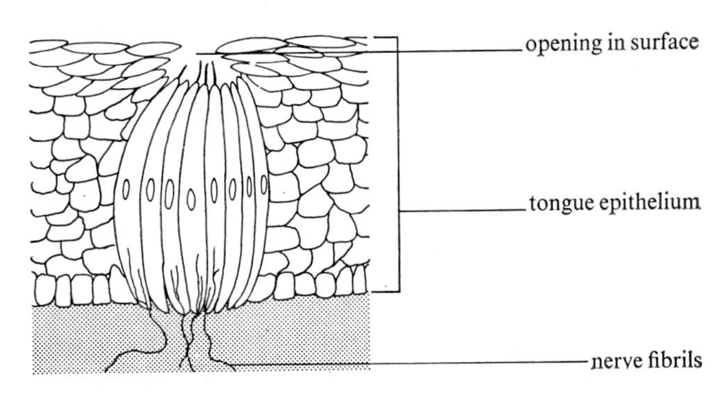

Fig 10.15 **Taste bud**

selection of food. It is now known that the *primary odours* are camphoraceous, musky, floral, pepperminty, ethereal, pungent and putrid, each being detected by a different receptor in the nose. Odours are comprised of several primaries in different proportions, and each primary has a different shaped molecule that enables them to fit onto different receptors.

SKIN The skin can detect touch (mechanical pressure) and thermal energy (hot and cold). The receptor organs are in the dermis and are very localised in certain regions such as the fingers. There are two types of touch receptor, the surface superficial receptor and the deep-lying pressure receptor.

10.5 Endocrines in man

Endocrine glands (Table 10.2) secrete hormones that pass via the bloodstream to the site in the body where they are needed. They form a system of communication with the body independent of the nervous system, except in so far as the secreting activity of some endocrine glands is controlled by the nervous system. Some hormones are secreted by nervous tissue, *e.g.* post-pituitary and adrenal medulla.

Table 10.2 **Endocrine glands**

endocrine gland	*situation*
pituitary gland	base of brain, just behind optic chiasma
thyroid gland	in neck
parathyroid gland	two pairs of glands on dorsal side of thyroid
sex glands	ovary and testis
adrenal glands	top of kidneys
islets of Langerhans	pancreas-region of duodenum

PITUITARY GLAND This is a small round body connected to the underpart of the brain below the hypothalamus just behind where the optic nerves cross as they go from eye to brain. It exercises central government over the endocrine system. The anterior lobe (hypophysis) secretes the growth promoting hormone *pituitrin*, *prolactin*-which controls the development of milk in the mammary glands, and the promulgative hormones which stimulate other gland centres; these are known collectively as the *trophins*. They are protein or glucoprotein in nature and vary in relative molecular mass between 1×10^4 and 1×10^5. The thyroid gland is stimulated by TSH (*thyrotrophin*) and the adrenal cortex by ACTH (*adrenocorticotrophin*). The gonads (see also p 161) are stimulated by the *gonadotrophins* FSH (follicle stimulating hormone) which stimulates the germinal cells of the ovary or testis to produce gametes, ICSH (interstitial cell stimulating hormone) which causes these cells to secrete their hormones, and LH (luteinising hormone) which causes the formation of a corpus luteum. The posterior lobe of the pituitary secretes ADH (antidiuretic hormone) which is concerned with assimilation of water and ions by the kidneys, vasopressin which increases blood pressure by causing contractions of the muscular walls of the vascular system, and oxytocin which causes uterine contractions and final release of milk through the nipples of the mammary glands.

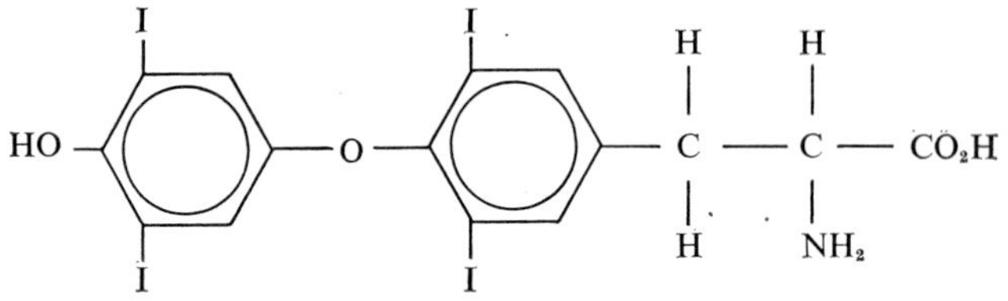

Fig 10.16 **Structure of thyroxine**

THYROID GLAND This is a gland with two lobes, one on either side of the windpipe, joined just below the voice-box. It secretes (from large glandular cells) a lyophilic colloid which contains thyroxine (Fig 10.16), which regulates the entire body metabolism by stimulating enzymes involved in respiration. In excess (thyrotoxicosis) it causes Graves' disease, an abnormal increase in the basal metabolic rate which can lead to heart failure if prolonged. Deficiency in children causes cretinism – failure to mature mentally and sexually, and in adults goitre, slowing of metabolism and obesity, the whole condition being known as myxoedema. The radioactive iodine–131 isotope is extremely useful in studying thyroid diseases.

PARATHYROID GLAND Situated in the posterior part of the lateral lobes of the thyroid, this secretes *parathormone* which regulates the amount of Ca^{2+} and PO_4^{3-} in the bones and the blood. Deficiency leads to a lack of calcium ions in the blood causing tetanus (irreversible contractions of the muscles); over-activity leads to excessive resorption of ions from the bones.

ADRENALS These are near the upper pole of the kidneys. The *cortex* (outer yellow layer) originates from epithelial cells near the embryonic gonads and the *medulla* (inner grey layer) from spinal cord cells. The adrenals secrete the *glucocorticoid* hormones (*e.g.* cortisol, corticosterone, deoxycorticosterone) which conserve sugar levels, the mineralocorticoids(*e.g.* aldosterone) which conserve mineral levels, and the *sex hormones*.

The adrenal cortex is absolutely vital being needed for

1 metabolism of fats, carbohydrates and proteins,
2 mineral ion and water balance,
3 function of the kidney,
4 maintenance of blood pressure,

5 resistance to stress,
6 inflammation and repair.
Destruction of the adrenal cortex, as for example by tuberculosis leads to a general disturbance of metabolism (weakness of muscles and loss of ions); stress situations can prove fatal. The treatment of Addison's disease is to administer corticosterone or deoxycorticosterone. If too much cortical hormone is present Cushing's disease results, in which there is an excessive breaking down of protein to glucose (overactive glucocorticoids) with bone/muscle weaknesses. The adrenals also secrete adrenaline. This comes from the adrenal medulla and, together with some noradrenaline, enters the blood during periods of stress. The effect is to stimulate the sympathetic system, making the body prepared for anything. Adrenaline has a powerful effect on the heart and blood vessels and it also breaks down some of the stored glycogen, making the level of blood sugar rise. It has a stimulating effect on the anterior pituitary, making it produce more ACTH.

GONADS AND OVARY The hormones are
1 oestrogen (♀),
2 testosterone (♂),
3 progesterone (from the corpus luteum)
The effects of these hormones have already been given (p 78). Deficiency of the hormones leads to sterility. Excess causes abnormal personality effects in the male and female.

ISLETS OF LANGERHANS These are endocrine cells in the pancreas, which is in the duodenum (p 142). *Insulin* and *glucagon* are secreted, which by exerting opposite effects regulate the synthesis/degradation of glucose into/from fats and amino acids, or for energy release. Deficiency causes *Diabetes mellitus* (detected by large amounts of sugar in blood and urine, and nervous disorders). Fats need energy from glucose for oxidation and so, in the absence of this, toxic degradation products of fats build up. Dehydration can result since the osmotic equilibrium of the body is broken. If insulin is in excess muscle/nerve operation is affected (too little sugar in blood).

FEEDBACK This is a term, also used in the physical sciences, to describe systems in which there is input and output, and part of the output is 'fed back' into the input, where it exercises a control effect upon the system. Fig 10.17 is a flow diagram illustrating the control of the supply of thyroxine to the body cells. It can be seen that blood levels of TSH and thyroxine exert control over the entire system; the level of thyroxine is self-regulating. Such feedback systems are very common in organisms and are mainly responsible for their physiological stability or

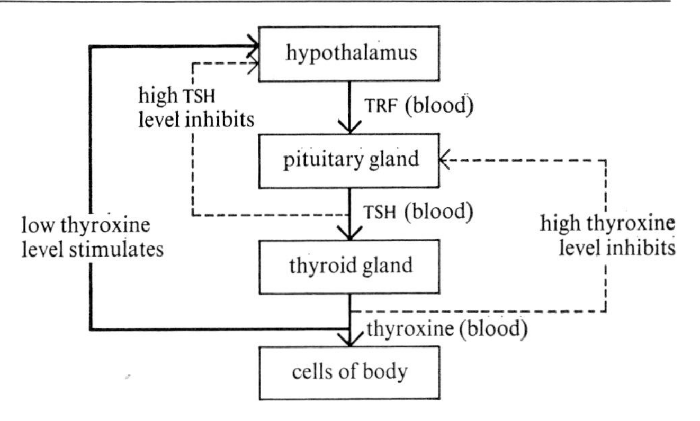

Fig 10.17 **Feed back controlling supply of thyroxine; TSH=thyroid stimulating hormone (thyrotrophin) TRF=thyroid releasing factor (environmental factor)**

condition of *homeostasis*. Feedback of this sort in which the trend to excess or deficiency leads to an opposing, correcting trend, is *negative feedback*; positive feedback operates to reinforce and accelerate an existing trend. Can you think of any examples of positive feedback?

COMBINED NERVOUS AND ENDOCRINE RESPONSE The reaction of the body to a stress situation can be represented as in Fig 10.18.

10.6 Endocrines in insects

The endocrine system in insects controls such things as reproduction, rate of metabolism, etc, as in vertebrates, but also metamorphosis/ecdysis. In metamorphosis, at least three hormones are involved. Cells in the brain secrete the brain hormone. The prothoracic glands in the thorax secrete a growth and differentiation hormone (GDH) called ecdysin. A gland known as the corpus

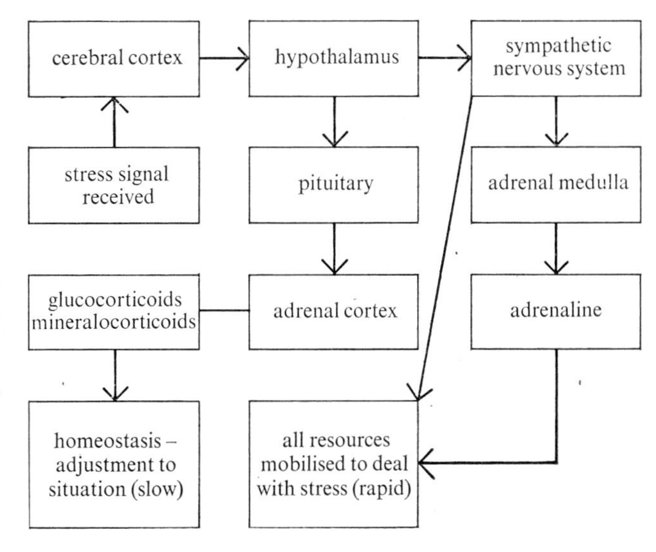

Fig 10.18 **Reaction of the body to a stress situation**

allatum (in the head near the brain) also secretes a hormone called the juvenile hormone.

Ecdysis in an insect with simple metamorphosis (such as the locust) begins when a stimulus is received by the cells of the brain. Brain hormone is secreted, which stimulates the prothoracic gland to secrete ecdysin. This acts on the embryonic tissues, causing differentiation, and the exoskeleton of the nymph is shed. The corpus allatum secretes juvenile hormone, which prevents the transformation to the imago before the last moult. When this last moult is due to take place, the juvenile hormone is no longer secreted and the imago emerges.

When there is a complete metamorphosis (as in the butterfly) the transformation from pupa to imago is controlled by the brain and prothoracic gland; the juvenile hormone is cut down very suddenly.

10.7 Behaviour in animals

In lower animals, behaviour patterns are relatively simple and regular; in the higher animals, culminating in man, the patterns are much more complex. There are three simple responses involved in behaviour, reflex, taxis and kinesis.

REFLEX This is defined as a definite and immediate response to a stimulus by a part of an organism and depends upon the innate nervous pathways or reflex arcs (p 153). Examples of reflex action are the contraction of eye pupils, blinking and knee jerks.

TAXIS This is a *locomotive* response of an organism to a directional stimulus. Phototaxis is behaviour in response to light, positive (*i.e.* towards the stimulus) for *Euglena*, negative for an earthworm. Rheotaxis includes behaviour in response to water current, positive for *Gammarus pulex*. Other forms include chemotaxis and geotaxis.

KINESIS This is also a *locomotive* response, but in this case the stimulus effects the speed or frequency of turning of the organism, but has no effect on the *direction* of movement. The woodlouse *Oniscus* shows an increased rate of change of direction in the damp – hydrokinesis.

Practical work

An experiment to show phototaxis involves introducing into a beaker enough *Chlamydomonas* to give a green colour to the water. The beaker is then covered with brown paper except for a small slit about 5 mm long. If the beaker is placed in bright sunlight and examined after a few hours, the algae should have collected in the region through which light passed.

INSTINCTIVE BEHAVIOUR This is a complex pattern triggered by a specific stimulus, endocrines normally being involved. It is found only in higher animals. Examples are migration and courtship and mating behaviour in birds, responding to changing day-length.

LEARNING This occurs when animals modify the behaviour pattern elicited by a stimulus. It may be by *imprinting* (*e.g.* a lamb will accept a dummy as its mother if it is presented to it immediately after birth). Learning may also be *habituation*, as when birds learn to ignore scaring devices, by *association*, as in the 'reward' training of pets and domestic animals, or by *conditioning*.

The best known example of the latter is Pavlov's classical experiment on dogs. He presented food to a hungry dog; on seeing the food, the dog would salivate. If as the food was given a bell was rung, the dog would eventually learn to salivate at the sound of a bell in the absence of food. Pavlov interpreted this as showing that the dog had learnt to associate 'food' and 'bell' *i.e.* that when a bell sounded it would not be long before food appeared. Salivation was a reflex following the presentation of food. Association of the bell sound with the food resulted in a modification of the nerve pathways so as to give a new reflex arc in which the sound of the bell caused salivation. But the path is more complex than a simple reflex arc, since sense receptors are involved which are not connected by a single interneurone to the motor nerves causing salivation. There may be many intervening interneurones in the brain between a sensory input and a motor output (multisynaptic).

The capacity to learn is bound up with *intelligence*, *i.e.* the ability to reason, and solve problems by drawing on experience.

10.8 Response in plants

TROPISM Responses of plants to external stimuli are called tropisms. The various kinds of tropisms are tabulated in Table 10.3.

Table 10.3 **Tropisms**

tropism	*stimulus responded to*
chemotropism	chemicals
electrotropism	electrostatic fields
geotropism	gravity
hydrotropism	water
phototropism	light
thigmotropism	touch

GROWTH REGULATING SUBSTANCES Charles Darwin exposed the coleoptile (a sheath of tissue that protects the young leaves) of young cereal seedlings to light from one side only. He found that it bent towards the light.

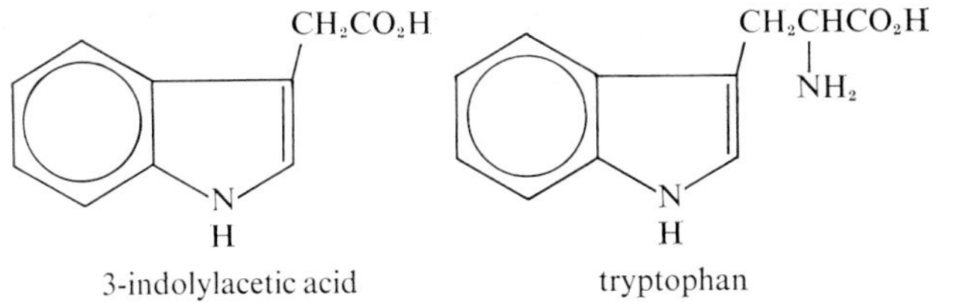

CH_2CO_2H CH_2CHCO_2H NH_2

N H N H

3-indolylacetic acid tryptophan

Fig 10.19 **3-indolylacetic acid (3-indolylethanoic acid)**

Covering the tip or cutting it off completely deprived the coleoptile of the ability to respond. The effect was not confined to the tip since regions below this bent towards the light even if it was only incident on the tip. Later workers showed that, if the excised tip was replaced, the phototropism was restored; it was thus concluded that the influence was chemical.

It was established that the tip produces hormones which control growth. These hormones are often called growth regulating substances (GRS). A principle one is 3-indolylacetic acid (IAA) whose structure is shown in Fig 10.19; it is present in fungi as well as angiosperms. It is very similar to tryptophan. It is produced by cells in the apical meristems of shoot and roots, as well as by cells in other actively growing regions, *e.g.* cambium, flowers, fruits, leaves; leaves continue to produce IAA for some time after maturing. The IAA can move from the tip of the root/shoot towards the older tissue, against gravity. If a portion of the coleoptile is removed and inverted, the IAA will be transported from the original apical region to the original basal region, but not vice-versa; this shows that as the hormone moves from tip to base it must be used, be broken down, or otherwise become unavailable.

AUXINS Auxin is the general term used for the group of growth regulating substances (such as IAA) particularly concerned with stem elongation. They are known to be involved in tropisms, secondary growth, wound healing, initiation of buds, fruit formation etc.

Auxins are present in plants in very small amounts, *e.g.* an oat coleoptile would contain about 5×10^{-11}g. An auxin can either stimulate or inhibit a process which it influences, what happens depending on the concentration. Fig 10.20 shows that growth of buds, roots and stems is promoted at low concentrations, and inhibited at higher concentrations. It is clear that an auxin concentration of between 10^{-6} and 10^{-3} mol dm^{-3} stimulates stems to grow, but inhibits buds and roots.

Auxins can control cell growth by altering the degree of elasticity of the cell wall (probably by affecting synthesis of cellulose). This enables the cell to absorb more water by osmosis, and so get larger.

Auxins are a major factor in determining plant shape (together with the angles between the branches and the main stem and the stiffness of the shoots). Auxin formed at the apical meristem travels down the shoot and accumulates at the leaf axils. If it produces an inhibitory concentration there, buds form rather than shoots. This is called apical dominance because the apex exerts control over the rest of the plant. If the apex is removed, the auxin concentration falls to the level required for shoot formation and the lateral or side shoots begin to grow out. Practical use is made of this in trimming hedges.

Plants have optimum concentration requirements of auxins for the best growth and developments.

Roots are positively geotropic and shoots are negatively geotropic. If a plant is held horizontally auxin sinks. In roots, where the normal concentration exceeds that required for maximum growth, the effect is to render the underside concentration more inhibitory and the upperside more promotory; downward curvature therefore results. In stems, where the normal concentration is sub-optimal, the effect is opposite; increased promotion below, decreased promotion above, causing the stem to curl upwards. What would happen to plants grown in zero gravity conditions aboard a spaceship?

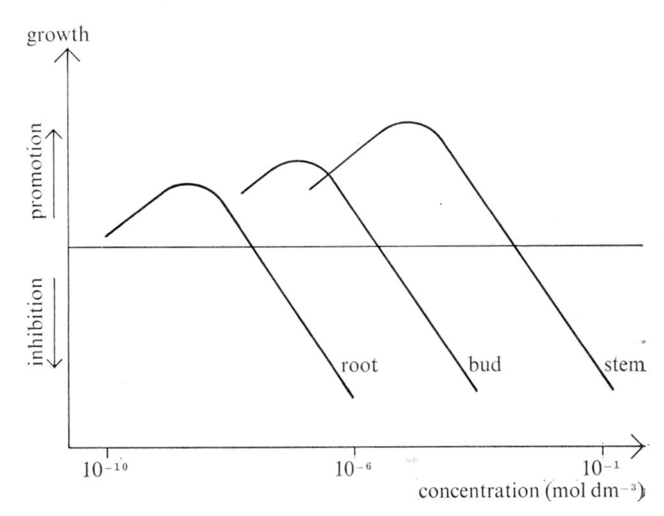

Fig 10.20 **Auxin concentration and growth**

Light deactivates IAA, which explains the behaviour of the coleoptile when exposed to light on one side only. The deactivation of IAA causes a decreased concentration of the auxin on the illuminated side, causing the coleoptile to bend over towards the light.

Auxins can promote the development of adventitious roots, aiding plants to take root in the soil; this fact is made use of in commercial rooting powders. In strawberries and other fruits development depends on the formation of auxins in the seed; removal of the achenes of the strawberry causes the receptacle to fail to swell; but development can be restored if the receptacle is treated with artificial auxins when the achenes are removed. The application of auxins to leaves can also delay leaf- or fruit-fall.

Many synthetic hormones have been made resembling IAA, including 2,4-dichlorophenoxyacetic acid (2,4-D), or MCPA as it is commercially known. This is an important selective weedkiller; its effects on grasses and cereals are slight (provided the right concentration is applied), but it is lethal to many broad leaved weeds.

KININS Another group of growth regulating substances are the *kinins* which are concerned in initiating mitosis and promoting the growth of lateral shoots. By abolishing apical dominance they oppose the effects of auxins.

GIBBERELLINS A third group of growth regulating substances are the *gibberellins* which were discovered by Kurosawa in Japan in 1926. They can be produced on a large scale by methods analagous to those used to obtain antibiotics from fungi. They have been isolated from higher plants and are known to *co-ordinate* plant processes. When artificially applied, they may effect stem and leaf elongation, promote the growth of normal plants, eliminate the growth differential between genetic dwarf strains and normal plants, or promote flowering when the usual photoperiod and vernalisation requirements are lacking. Auxins usually inhibit flowering of plants. Gibberellins can be used to produce seedless fruits; they can prevent seed dormancy but unlike auxins they do not affect bud inhibition, do not aid formation of adventitious roots, do not affect leaf/fruit removal from the plant, do not show polar transport and cannot act as herbicides.

Fig 10.21 **Masking young chrysanthemum plants to produce flowering plants all the year round**

Optimum concentrations of naturally occurring auxins and gibberellins are needed for the flowering of plants to take place.

PHOTOPERIOD Plants respond to light qualitatively and quantitatively in many ways, of which one of the most important is their response to *day-length.* Plants can be divided into three groups: those that flower naturally in

1 short days (*e.g.* chrysanthemum)
2 long days (*e.g.* iris)
3 either (*e.g.* rose). These are sometimes called day-neutral plants.

Some plants (such as the potato) can flower in total darkness because they carry a big supply of stored food.

Early research suggested that it was the length of the light period that promoted the change-over to flower buds. But later work has shown that short-day plants are really long-night plants, and that it is the length of the dark period that is most important. This was established by interrupting the dark period by a flash of light.

The day length is called the *photoperiod*, and the response to day-length the *photoperiodic response.* Often the response is only evoked if the plant is exposed to the critical day length at a critical stage of its growth. Masking of young chrysanthemum plants during the long-day periods of the year has made possible a large industry devoted to the production of flowering plants in pots all the year round (Fig 10.21).

A blue pigment called *phytochrome* has been found to be involved. It exists in two forms, one showing an absorption band in the red region and the other in the infrared. They are interconverted as shown in Fig 10.22.

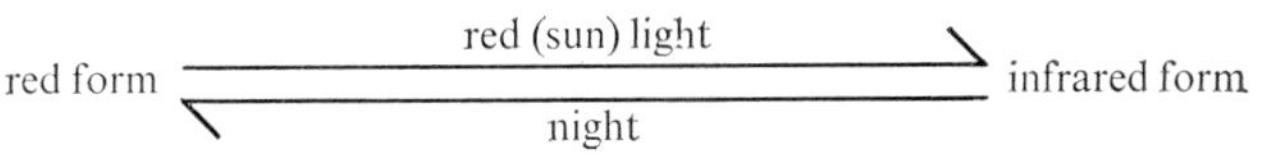

Fig 10.22 **Interconversion of phytochrome**

Light with a peak absorption in the red has been found experimentally to influence flowering, stimulating flowering in long-day plants, and inhibiting flowering in short-day plants. Red and/or infrared light also affects stem elongation, seed germination, leaf shape and size, the formation of lateral roots, and pigmentation.

By initiating ovarian activity photoperiodism affects reproduction in birds, insects and snails; it also affects fur and feather colour in some mammals and birds.

TEMPERATURE AND VERNALISATION Temperature also has an important effect on plant growth. In general growth is stimulated by higher temperatures as this affects the rate of metabolism. But in many seeds a necessary precursor to germination is a cold period; this phenomenon is known as *vernalisation*. In nature the cold period is winter, with germination following in the spring. Dormancy in the shoots of some trees and shrubs is also broken by cold. Bulbs and shoots can be refrigerated before sale so that their dormancy is already broken, so that they will flower sooner after purchase. A similar phenomenon occurs in the winter eggs of insects or mites; eggs that do not experience sufficient cold may hatch later, or not at all. Here the dormant condition is called *diapause*.

10.9 Types of movement

We have seen how organisms respond to stimuli, and how actions are initiated. In this section we shall deal with the effector systems that produce movement.

Animal movement can be broadly classified as *amoeboid flow*, *ciliary motion* or *contractile motion*.

AMOEBOID FLOW How *Amoeba* moves has been shown in Fig 3.1 on p 42. Remember that *Amoeba* consists of an outer elastic membrane surrounding stiff plasmagel and fluid plasmasol. It moves simply by changing shape: the plasmasol flows towards certain points in the elastic wall causing foot-like projections called *pseudopodia* to appear. Although the mechanism is not thoroughly understood it appears that when the plasmasol reaches the leading edge of the pseudopodium it changes state and becomes plasmagel. At the other end the reverse happens, the plasmagel becoming fluid plasmasol. ATP is known to be involved in the process.

CILIARY MOTION Cilia and flagella (Fig 10.23) are structurally similar, varying only in length (cilia are shorter and are usually more numerous). They are fine hair-like processes projecting from the surface of a cell. A cilium comprises of a membrane which is an extension of that of the cell containing two central and nine peripheral fibrils with a basal granule. The integrated activity of cilia produce a wave of motion like that in the legs of a millipede. The contractions are powered by the breakdown of ATP, but the exact function of the fibrils in producing the forceful active phase and the relaxed recovery phase (like swimming breast-stroke) is not known.

Cilia and flagella occur in many orders, sometimes to move the organism within a liquid, and sometimes to move a liquid within the organism. In the Protozoa, the Mastigophora are flagellate and the Ciliophora are ciliate. *Hydra* uses flagella in digestion; *Planaria* and free living flatworms use cilia for movement. Nematodes lack cilia, and so do arthropods, but annelids use them in excretion (nephridia p 174). Chordates have flagellate

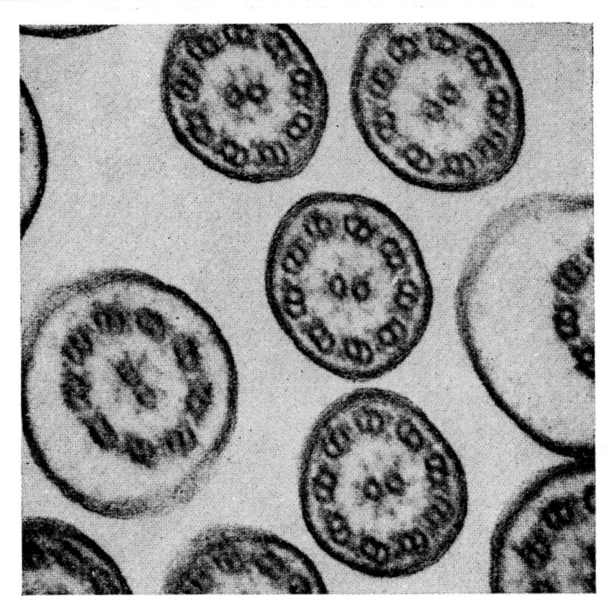

Fig 10.23 **An electron micrograph of a cross-section of flagella showing the arrangement of the fibrils in a flagellate protozoan**

spermatozoa, and ciliate tissues for feeding and digestion, in nasal, tracheal and bronchial passages, and near oviducts. Cilia and flagella occur in the algae, fungi and in the antherozoids of the spermatophyte *Gingko biloba*.

CONTRACTILE MOTION Perhaps the most important method of locomotion is brought about by the contraction (and relaxation) of tissue within a jointed frame or skeleton. This is known as *contractile motion*. The main type of contractile effector is the *muscle* (§10.10), but *myonemes* are met in protozoa. These are contractile threads, and are thus functionally if not morphologically allied to muscle.

Examples of contractile motion are given in §§10.12 – 10.15.

10.10 Muscle

Muscle is tissue consisting of cells that are elongated and highly contractile. There are three types of tissue, smooth, striated and cardiac (Fig 10.24).

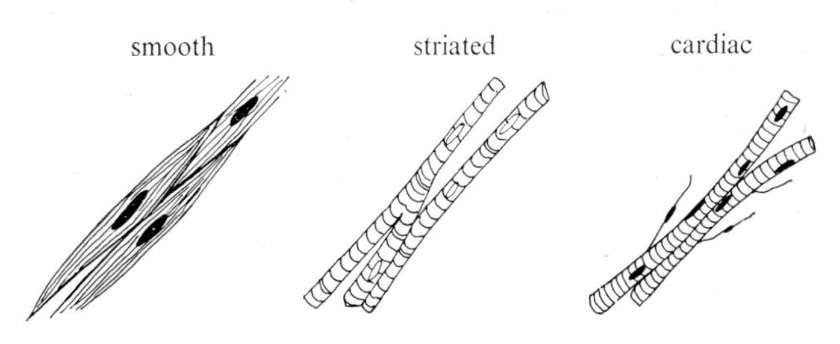

Fig 10.24 **The three types of muscle**

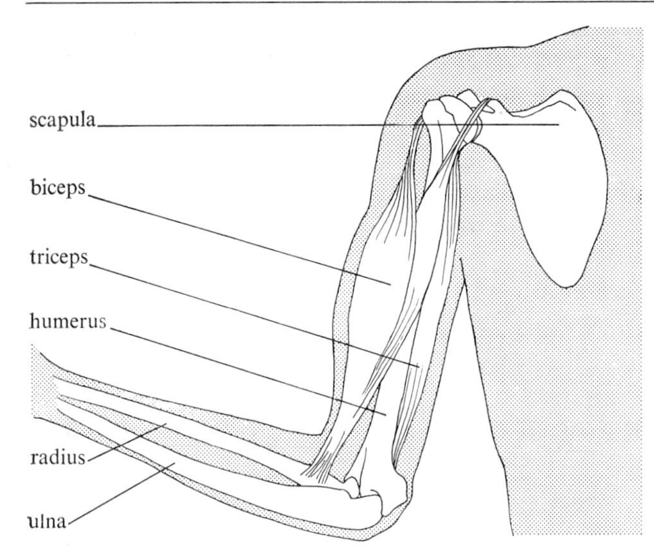

Fig 10.25 **Flexors (biceps) and extensors (triceps)**

STRIATED MUSCLE This form of muscle is controlled by the voluntary part of the nervous system. The *origin* of the muscle is fixed to the skeleton (hence the alternative name, *skeletal muscle*) and the opposite end, the *insertion*, is connected to a moveable bone. Normally striated muscles are paired so that contractions of one can bring about an extension of the other. We call the two types *flexors* and *extensors*. Thus biceps are flexors and triceps are extensors (Fig 10.25). It is striated muscles that are involved in movement.

If striated muscles are examined under a microscope they are seen to have a banded structure, a repeating pattern of light and dark bands as shown in Fig 10.26, hence the name striated. Now muscle fibres are cells whose contractile parts are longitudinally disposed *myofibrils* with a diameter of about 1-2 μm. Between the myofibrils of very active muscle there are many mitochondria and an elaborate endoplasmic reticulum system. Under the electron microscope myofibrils are found to be composed of small filaments. It is longitudinal differentiation which gives rise to the pattern of light and dark bands, which are characteristic of individual myofibrils. The dark bands are called the *anisotropic* or *A-bands.* Their particular molecular structure causes a refraction of plane polarised light known as *birefringence*; highly birefringent structures have their molecules aligned in a highly symmetrical pattern. The light bands of the muscle are much less birefringent and are called *isotropic*, or I-*bands*. The I-bands are divided down the centre by a Z-membrane; the inter-Z-membrane region is referred to as the *sarcomere* (about 2 μm long). The A-band has thick filaments (about 1 μm diameter) separated by a space of about 3 μm. Thin filaments (about 0·5 μm diameter)

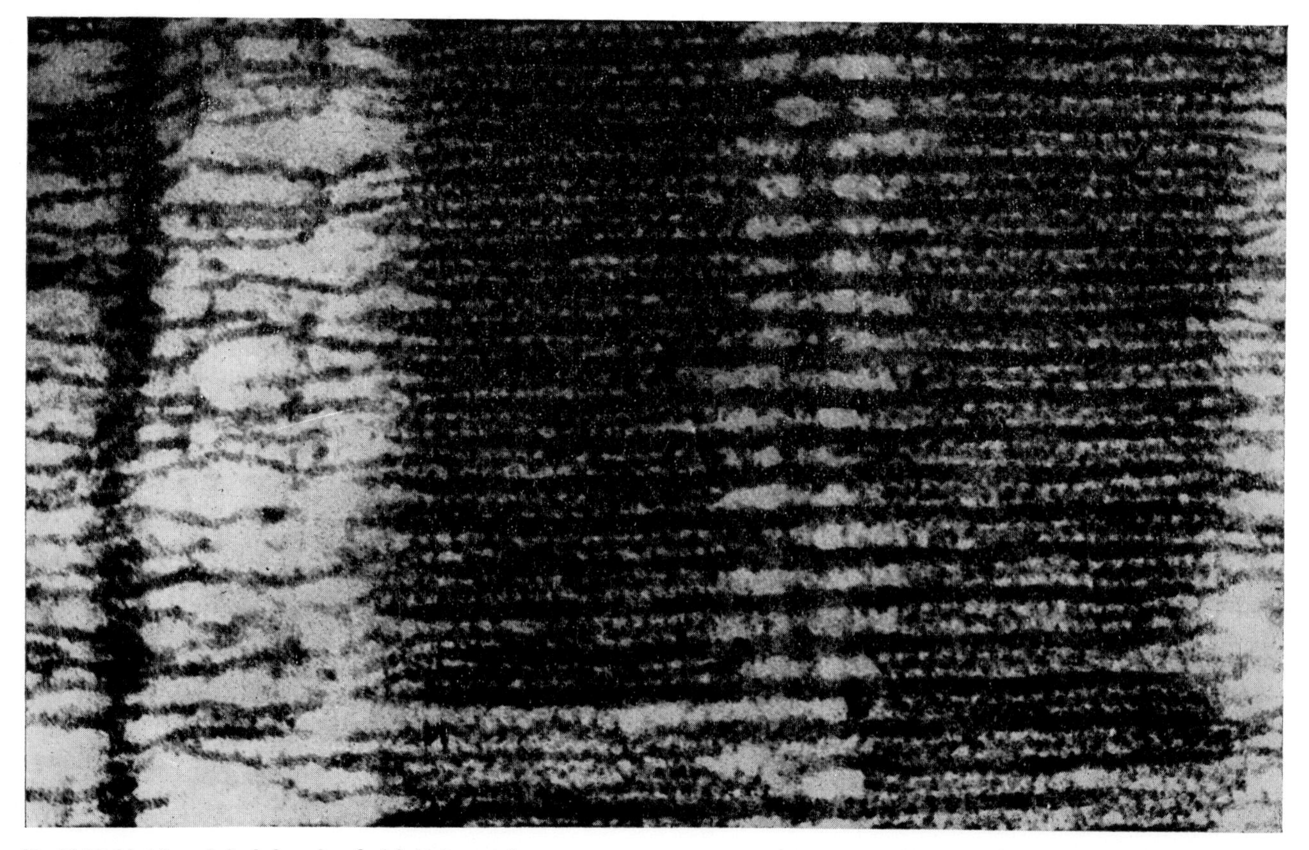

Fig 10.26 **Light and dark bands of striated muscle**

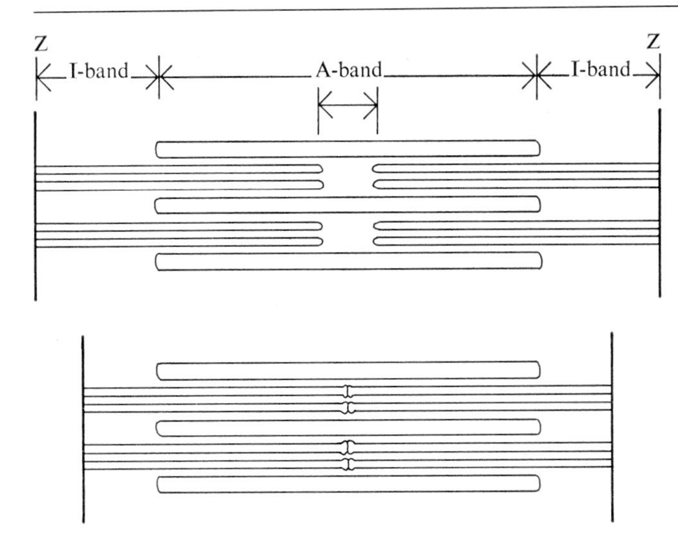

Fig 10.27 **Operation of muscle filaments**

come between the thick filaments, and they are attached to one end of the Z-membrane. The central region of space in the A-band where the thin filaments are not attached is called the H-*zone*. X-ray diffraction has revealed the fine structure.

The densest region of the A-band is caused by overlap of thick and thin filaments; the less dense regions are when thin filaments are not crossed by thick ones. The thick filaments are made of the protein *myosin* and the thin ones of the protein *actin*.

The myosin and actin are joined together by branches which project out at regularly spaced intervals of about 5 to 7 nm, each bridge being orientated at about 60° with its neighbour. A repeating helix (repeating unit six) is obtained. When the relaxed muscle is stretched, the A-band does not increase in length and the total increase is the sum for the I-bands, equal to the increase in length of the H-zones. On contraction, the A-band is found not to shorten and the inter Z/H distance stays constant; on contraction the H zone eventually disappears altogether and is then replaced by a dense region. It appears therefore that the two sets of filaments slide together as shown in Fig. 10.27. The mechanism of contraction is probably the combination of actin and myosin in the presence of ATP and Ca^{2+} ions to form another protein *actinomyosin*.

Experiments on muscle contraction using the leg muscle (gastrocnemius) of *Rana*, in which a stimulus is applied as an electrical impulse, recording being done by a kymograph (Fig 10.28), have revealed that although each fibre responds with a maximum contraction, a threshold stimulus causes excitation of only a limited number of fibres, insufficient to cause contraction. When the stimulus is increased, the number of fibres that contract increases until all the fibres in the muscle have responded. Thus, the strength of contraction of the muscle is determined by the number of fibres stimulated. The power of contraction can be increased by increasing the number of motor units in action, or by increasing the frequency of stimulus to the muscle. If the frequency is increased sufficiently, this can lead to sustained contraction, or *tetanus*.

Muscle must convert chemical potential energy into mechanical energy, and about 30% of the glucose available in our bodies is converted into mechanical energy (the remainder being given off as thermal energy). Energy for contraction comes from glucose in the blood. A muscle has energy stored as glycogen, whose molecule is formed by condensation polymerisation of glucose molecules. Energy for contraction comes from ATP (broken down to ADP) and from the conversion of phosphocreatine (PC) to creatine(C). The energy released in the latter reaction is used in the synthesis of ATP from phosphate (V) ions. Glycogen (G) in the muscle can be hydrolysed to 2-hydroxypropanoic (lactic) acid (L) with the release of energy, and this energy can be used to

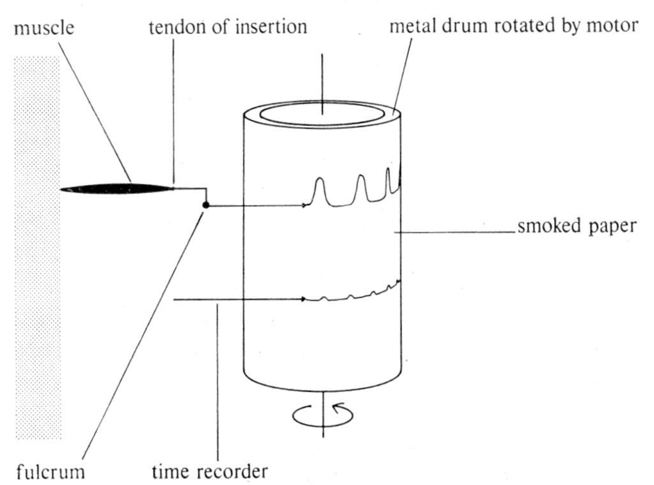

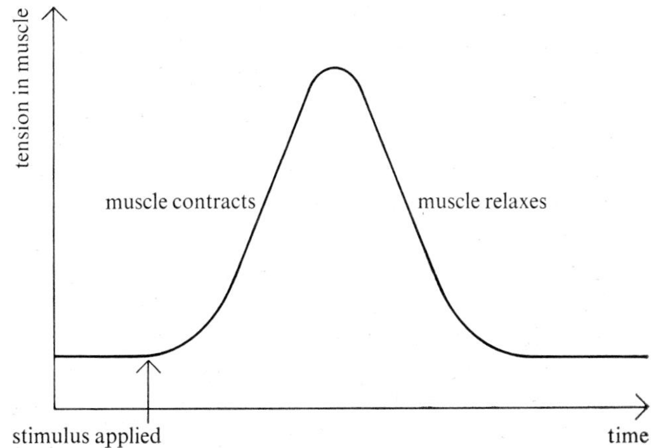

Fig 10.28 **Kymograph recording muscle contractions**

synthesise phosphocreatine (PC) from creatine (C) and phosphate(V) ions. 2-hydroxypropanoic acid, in its turn, is oxidised in respiration, forming carbon(IV) oxide, water and releasing energy. This latter energy is used in the synthesis of glycogen from more 2-hydroxypropanoic acid. All this is discussed in detail in Chapter 8.

The reactions mentioned can be summarised as follows:

$$\text{ATP} \underset{2}{\overset{1}{\rightleftharpoons}} \text{ADP} + \text{P} \quad \triangle G\text{-ve (free energy liberated)}$$

$$\text{PC} \underset{4}{\overset{3}{\rightleftharpoons}} \text{C} + \text{P} \quad \triangle G\text{-ve}$$

$$\text{G} \underset{6}{\overset{5}{\rightleftharpoons}} \text{L} \quad \triangle G\text{-ve}$$

$$\text{L} + O_2 \xrightarrow{7} CO_2 + H_2O \quad \triangle G\text{-ve}$$

Reactions 1, 3, 5 are anaerobic.
In violent exercise reactions 1, 3, 5, 4, and 2 proceed rapidly, more so than the aerobic oxidation of 2-hydroxypropanoic acid (7). The muscles have an *oxygen debt.*

The ATP actually aids combination of actin and myosin; Ca^{2+} ions are needed to neutralise charges which arise on the ends of the bridge of myosin and on the point where the bridge is made with the actin filament.

Should calcium ions not be present in the required amount, *tetany* (muscle lock) results.

It should be remembered that when an impulse arrives at the surface of a muscle an action potential (dependent on Na^+/K^+ concentrations, as for the neurone), is triggered off along the muscle: as the inside of the fibre becomes less negative with passage of the action potential it is thought that the actin and myosin contract.

SMOOTH MUSCLE Smooth muscles, or *visceral muscles* as they are often called, are much simpler than striated muscles, consisting of elongated cells bound together by connective tissue. They do not have cross-striations. They are controlled by the autonomic nervous system, hence a third name, involuntary muscles. They are found lining hollow organs such as the gut, and also around blood vessels.

CARDIAC MUSCLE Cardiac or *heart muscles*, as the name implies, are situated in the walls of the heart. They have the banded structure of striated muscles, but are not under voluntary control. They contract rhythmically, and are completely immune to fatigue. The connections with the autonomic nervous system only serve to control the strength and speed of contraction, not the contraction itself. The contractions are therefore said to be intrinsic or *myogenic*.

10.11 The skeleton

The frame in which the muscles work to produce movement is the skeleton. There are three types of skeleton: the *endoskeleton* of the vertebrates, so called because the framework is inside the muscles to which it is attached: the *exoskeleton* of the arthropods, where a hard cuticle is external to the muscles: the *hydrostatic skeleton* of animals such as annelids where there is no hard framework at all, the skeletal functions being performed by coelomic fluid under pressure.

ENDOSKELETON The main function of the endoskeleton is to provide the rigid framework to which the muscles are attached, thus acting as support for the whole body. But it also serves to protect the visceral organs.

There are two types of skeletal tissue, bone and cartilage. In some of the more primitive aquatic forms, such as the dogfish, the skeleton consists almost entirely of cartilage; but in the mammals the structure is made up almost entirely of bone. The skeleton of all vertebrate embryos is mostly composed of cartilage, which is gradually replaced by bone.

CARTILAGE The simplest form is *hyaline cartilage*, commonly known as gristle. It consists of a matrix of chondrin, secreted by special cells called chondroblasts. It is bounded by a fibrous membrane known as the perichondrium.

BONE Bone also consists of an organic matrix, but containing salts. The composition of human bone is about 48% $(Ca^{2+})_3\,(PO_4{}^{3-})_2$ and $Ca^{2+}CO_3{}^{2-}$, 22% water and 30% organic material, mainly ossein-protein. Bone forming cells called *osteoblasts* are in the periosteum, the connective tissue on the surface of the bone. The osteoblasts are situated in cavities called *lacunae*, which are connected to the nervous system and circulatory systems. Long bones are hollow, the cavity containing the *marrow* which produces erythrocytes and leucocytes. The marrow is red where the erythrocytes are formed. The rest of the marrow is yellow and contains fat, blood vessels and connective tissue.

HUMAN SKELETON There are about 200 bones in the human skeleton. The *axial skeleton* consists of the skull, the vertebrae and ribs; the *appendicular skeleton* consists of the bones and cartilages of the limbs and girdles. The *ligaments* join bone to bone, and the *tendons* join bone to muscle.

Most of the bones of the adult skull are joined by immoveable *suture* joints. The *frontal bone* is situated at the upper front part of the skull, the bone at the lower rear part is the *occipital bone*, and *parietal bones* are on

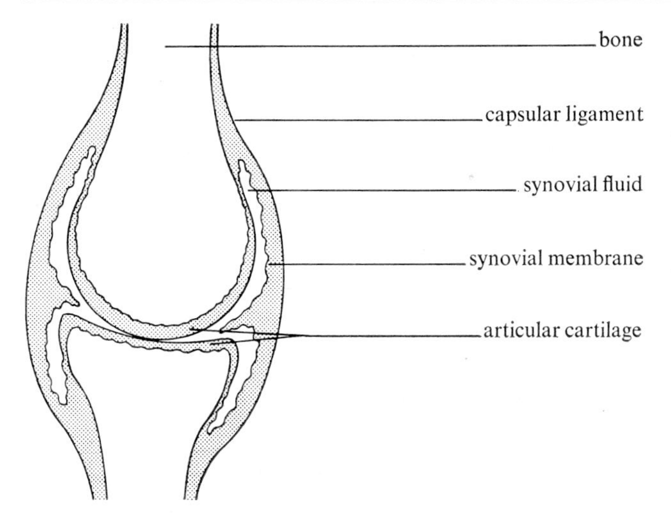

Fig 10.29 **A typical joint**

either side of the top part of the skull. Near the ear region there is the *temporal bone*. Two *maxillary bones* (upper jaw), the *mandible bone* (lower jaw) and two *malar bones* (cheeks) complete the skull. The only moveable joint is between the occipital bone and the first vertebra, the atlas.

There are 33 vertebrae in the backbone 7 *cervical* (neck), 12 *thoracic*, 5 *lumbar* (lower back), 5 fused *sacral*, and 4 fused *caudal* in the coccyx. A pair of ribs is attached dorsally* (at the back) to each of the 12 thoracic vertebrae the upper 10 pairs are attached ventrally* (at the front) to the breastbone or *sternum* and the 2 lower pairs are floating, *i.e.* not attached to the sternum. The ribs and the sternum together make up the rib cage.

The shoulder girdle or *pectoral girdle* is made up of 2 *scapulae* (shoulder blades) and 2 *clavicles* (collar bones). The scapulae lie dorsally, one on each side of the vertebral column. The clavicles lie ventrally, each one connecting the sternum to the shoulder joint.

The *pelvic girdle* is made up of 6 bones in all: 2 bones called the *ilia*, one on each side of the vertebral column and attached dorsally to the sacrum (the fused sacral vertebrae): 2 *pubic* bones attached to each other by cartilage, the *symphysis pubis*, and 2 *ischia*.

The *pectoral girdle* bears the arms. The upper arm has a large bone called the *humerus*. The forearm has 2 bones, the *radius* (small) and the *ulna* (large). The hand has 8 *carpals* (wrist), 5 *metacarpals* (hand, long bones) and 14 *phalanges* (fingers, short bones).

The *pelvic girdle* bears the legs. The thigh has a large bone called the *femur*. The lower leg has the long *tibia* and the shorter *fibula*. The *patella* (kneecap) protects the knee joint. The ankle has 7 *tarsals*, 5 *metatarsals* and 14 *phalanges* (bones of the toe).

*Dorsal usually means above, and ventral below; but man is an upright animal.

Joints occur where two or more bones meet together. The ends of bones are covered with *articular cartilages* (mainly *collagen*), which have a smooth surface lubricated by *synovial fluid*, secreted by the synovial membrane. The joint (Fig 10.29) is enclosed with a *capsular ligament* (limits movement; tough and inelastic).

The types of joint are:

1 Ball and socket joints which allow universal movement, *e.g.* hip, shoulder.
2 Hinge joints which allow movement in one plane only *e.g.* knee, fingers, elbow.
3 Compound joints which consist of two flat surfaces which glide over one another, *e.g.* most joints in the wrists and ankles.
4 Suture joints which allow no movement, *e.g.* joints in the skull.

Breakdown of the synovial membrane causes *rheumatoid arthritis*. The fluid has mainly low relative molecular mass compounds, but some larger molecules. When a load is applied to a joint, the small molecules are thought to move away and the larger ones are then able to take up the effect of the force. On releasing the load, the smaller molecules return to the proximity of the joint. If a load is applied to a joint a few times it does no harm, but continued application damages the joint ends, causing in later life *osteoarthritis*. Wearing away of the bone is caused by abrasive wear, crumbling and fractures in the substance which separates fragments of bone. Mis-shapen bones can be replaced and repaired using plastic or metal parts.

10.12 **Movement in man**

Muscles between the first limb bone and the limb girdle produces the movement of the proximal section of a limb. The muscles have six types of movement acting in antagonistic pairs:

1 forwards (protractor), backwards (retractor),
2 clockwise and anticlockwise,
3 upwards (elevator) and downwards (depressor).

Thus in walking the first bone of the limb moves backwards followed by a recovery in the forward direction. Locomotion can only be achieved satisfactorily by a co-ordinated set of movements to follow up the movement in the first part of the limb,.

When a man walks he uses all the muscles of the legs as well as some in the trunk. If he takes a long stride with his left foot, followed by an ordinary walking pace with the right foot and another with the left foot, the sequence would be as follows:

The right heel is raised (contraction of right gastrocnemius), body pivots on left leg. Right leg is now protracted (contraction of right quadriceps femoris), the right toe being lifted (contraction of right biceps femoris). The right leg moves past the left leg (relaxation of right biceps femoris), the right toes are raised (right anterior tibia contracts) and heel of right foot comes to floor; there is contraction of extensor muscles to stop the limb from collapsing (*i.e.* left quadriceps femoris and gastrocnemius contract). The left gluteus maximus also contracts to aid pivoting. In the second pace, the order of events is similar but the left leg is dominant.

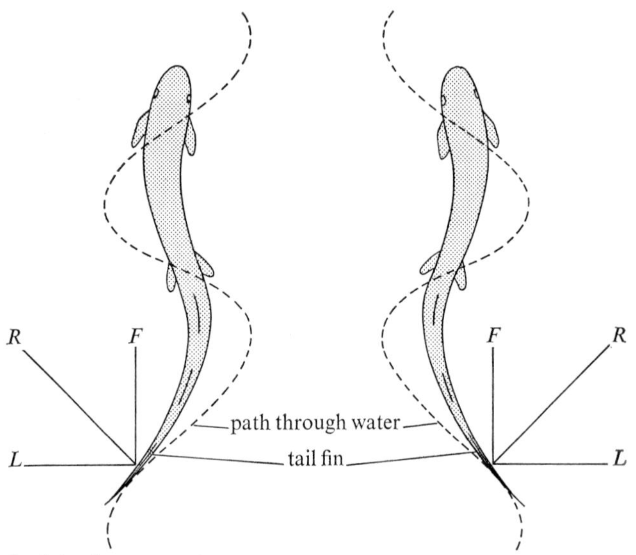

L=lateral component
F=forward component
R=reaction of water on tail

Fig 10.30 **Forces operating as fish swims**

10.13 **How fish swim**

The fish has a flexible spine and has well-developed muscles (myotomes) along the axial skeleton, these producing a wave-like motion down the body. Movements are lateral (Fig 10.30). The fish has a streamline shape and stabilising structures in the form of the fins. When the dogfish swims, there is a lateral wave of contraction which passes along the body from front to rear. These contractions push against the water and drive the fish forward, the expanded tail helping greatly here. The tail and body segments skim through the water at an angle to the surface, with the result that there is a net force at right angles to the surface of the body; the lateral component of this force drives the body sideways (counteracted by equal and opposite forces elsewhere along the body); the forward component moves the fish forward. In general, the distance travelled per second by a medium sized fish is approximately ten times the length of its body.

A fish might tend to roll about the longitudinal axis and yaw and pitch about its centre of gravity rather like a glider. All the fins of the dogfish tend to oppose rolling. The tail fin, plus the fins lying behind the centre of gravity maintain the fish in the direction in which it is travelling. Vertical fins stabilise against yawing and rolling. The tail fin gives lift. The dogfish has a *heterocercal tail*; it has a lower lobe which is larger than its upper one and, being flexible it becomes bent as it moves sideways in a sweeping arc. In contrast, the tail of a bony fish is *homocercal*; it produces a forward thrust only. Since bony fish have a swim bladder (containing a bubble of air) they can adjust their density until it is equal to that of the environment.

During lateral movement, the tail fin produces lift since the caudal fin (flexible) lags behind the spinal column. This lifting force has a moment about the centre of gravity which tends to make the head dip, but this tendancy is counteracted by an equal moment resulting from lift caused by pectoral fins. The *lateral line*, a jelly-filled tube lying just below the skin, opening via pores to the environment, is used to detect movements in the water.

10.14 **How insects walk and fly**

Arthropods have a protective *exoskeleton* whose joints are of flexible *chitin*, with the muscles inside. The way in which the muscles operate in antagonistic pairs is similar to that in the mammal, the pairs of muscles being attached to chitin projections of the skeleton.

When the locust walks or runs, the 1st and 3rd right legs push on the ground at the same time as the 2nd left leg, then vice versa. The insect therefore moves with a rotational gait.

As regards flight, the wing tips are made to move fast by a small rubbing action of the tergum or the sternum of the thorax, operating via circular and longitudinal thoracic muscles. When in flight, more powerful muscles at the wing-bases maintain powerful downward motion, presenting the greatest possible area of surface to the air; on the upstroke, the muscles turn the wings so that they present a minimum area. The high speed of rotation of the wings during flight contributes to a high rate of metabolism.

10.15 **Why animals move**

Much movement such as peristalsis and heartbeat is of course involuntary. It is worth recalling, in an ecological context, that responses and voluntary movements are related to:

FEEDING The getting of food, either in grazing or predacious activity, and avoiding being the food of predacious enemies.

BREEDING Searching for and finding a mate, fighting off rivals; in copulation, gestation, parturition and rearing young.

DISPERSAL OR MIGRATION Movements in direct response to unfavourable conditions (overcrowding in aphids causes wing formation and dispersal flight), or as an evolved response to prevent overcrowding (dispersal of caterpillars newly hatched from an egg-mass), or to reach a more favourable environment, *e.g.* for breeding. Examples of this kind of movement are the fur seal (a mammal) which migrates 5000km from California to Alaska and back every year, the swallow (Europe to South Africa), the eel (rivers to sea), the edible crab (deep water in winter to tidal zone in summer).

10.16 **Plant movement**

In general plants are sessile while animals are motile. There are exceptions, however, and some algae and the reproductive parts of many lower plants have mechanisms by which they can move. Also, parts of sessile plants move. The tropic movements have already been described in §10.8. The *nutation* of the apex of a climbing stem, which causes it to spiral around a support is an example of *autonomic* movement. The opening and closing of stomata are *turgor* movements caused by differing guard cell turgor pressures. *Nastic* movements take place in response to some external stimulus, but, in contrast to tropic movements, the response is non-directional. The opening and closing of many flowers is a nastic movement.

If crocus flower buds experience a sharp temperature increase they quickly open because of a more rapid rate of growth on the inner side of the petals. This is a *thermonastic* movement. Sleep or *nyctinastic* movements occur in leaves which fold up at night (*e.g. Mimosa*) protecting the stomata from excessive condensation but allowing transpiration to start as soon as the leaves unfold.

(a) Model of haemoglobin

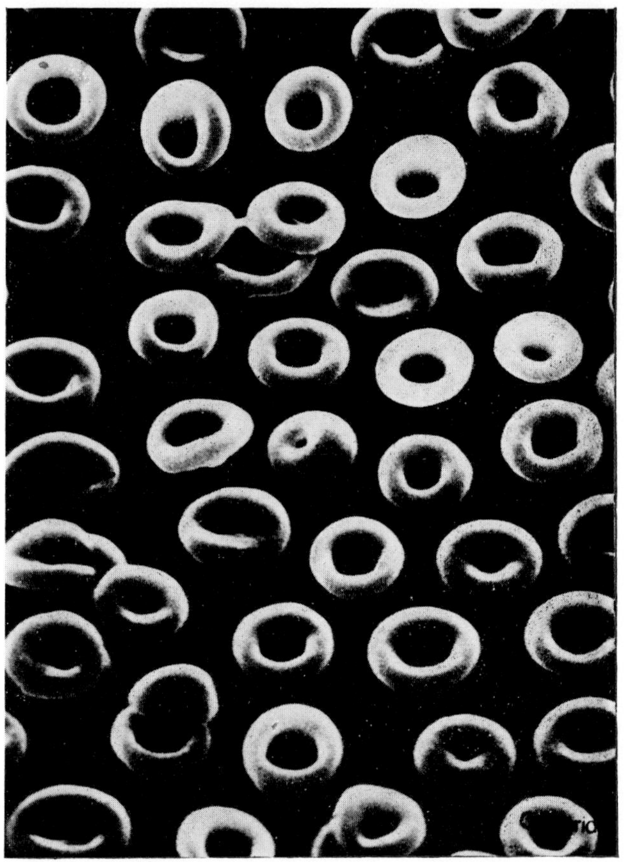

(b) Healthy red blood cells

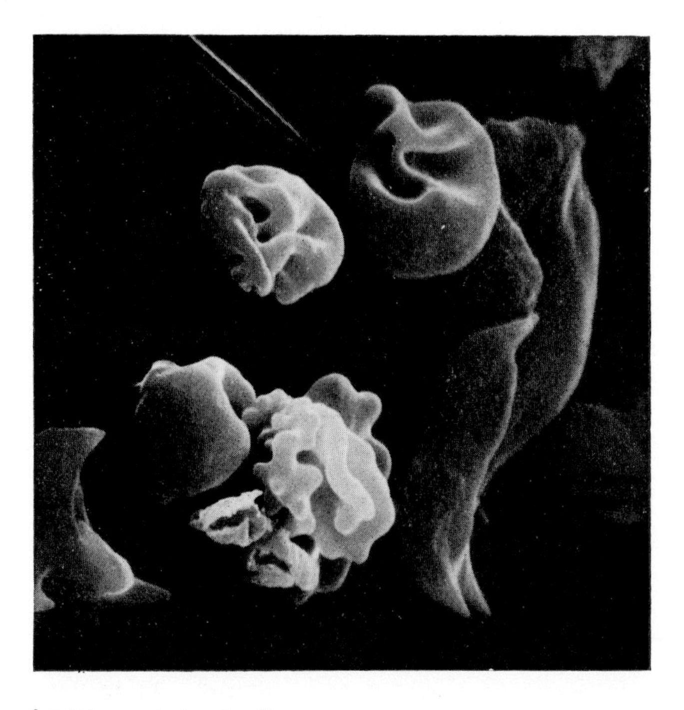

(c) Diseased blood cells

Fig. 11.1 **Haemoglobin and its effect in producing healthy and diseased cells**

(a) Haemoglobin contains 574 amino acids; these are arranged in four polypeptide chains round four haem groups. Each polypeptide chain is an α-helix and is folded and held in shape by electrostatic attraction, hydrogen bonding and covalent bonding

(b) This shows healthy red blood cells. These cells contain no nucleus; the interior of the cell is filled with haemoglobin. Red cells carry oxygen from the respiratory organ to the tissues

(c) Diseased blood cells due to defects in haemoglobin are shown. Healthy cells can be changed into sickle cells by single defects in the polypeptide chains. These defects are often genetically controlled and lead to incorrect positioning of an amino acid

11 Body Hydraulics

11.1 The physiology of body fluids

Every organism, even one as complicated as man, reacts as a unit to the external environment. It is clear therefore that all the cells which make up the organism must be co-ordinated. Most cells are specialised, and so the ability of complex organisms to carry out activities depends on the co-operation of many different types of cells. Just as a community functions by co-operation between individuals, so the body functions by the co-operation of cells. In the body this is achieved by control and feedback systems (p 161).

HOMEOSTASIS For all organisms there are optimum conditions for life, which are related to the state of the *internal environment*. By internal environment of a multi-cellular organism we mean all that is external to the cells themselves, but internal to the organism as a whole – *i.e.* the intercellular fluid. If the organism has some method of control over the internal environment such that any change is counteracted, and there is always a trend towards optimum steady state conditions, this is known as *homeostasis*. In general, changes (such as the temperature of the external environment) will tend to make conditions sub-optimal; homeostatic mechanisms will react to this change and tend to restore equilibrium.

There are two ways in which homeostasis is achieved. Firstly, when changes in conditions are regular, the counteracting mechanism may follow a similar regular pattern, anticipating the change. Examples of this are *circadian rhythms*. But the most important homeostatic mechanisms are feedback mechanisms, which respond to changes as they occur. Thus if the temperature of an animal changes the counter-control is for the blood supply to move to or from the skin, or for the animal to shiver or to sweat. Most feedback mechanisms are co-ordinated by hormones. All the systems of the body are involved at some time in homeostasis, but the excretory and circulatory systems are particularly concerned and are discussed below.

The development of efficient homeostatic mechanisms has been very important in the evolution of organisms. The better the control, the more independant is the organism of its environment, and thus the better it is protected from environmental change. Frogs for instance must live in damp places because they cannot control efficiently enough the evaporation of water from their skin. The reptiles are poikilothermic (cold blooded) which means that the body temperature rises and falls with the temperature of the external environment (p 182). Such animals can only live in warm or temperate climates; in the latter they have to hibernate in winter when the external environment becomes too cold for active life The body temperature of the polar bear, on the other hand, is constant whether it is living in arctic conditions or in the relatively warm conditions of a zoo.

11.2 Excretion

In very simple organisms the waste products of metabolism diffuse into the surrounding medium from active cells; this constitutes their *excretion*. In complex organisms, however, the waste products have to be removed from tissue fluids and blood. Notice the difference between excretion and elimination. The former is the waste produced by chemical and metabolic changes; the latter is the bulky waste from the gut, which has undergone no such change.

11.3 Excretion in Amoeba

A large part of the nitrogenous waste in *Amoeba* escapes by *diffusion* through the ectoplasm, the continuous covering of the entire body surface of the animal. The ectoplasm allows oxygen to enter the animal by diffusion and, of course, carbon(IV) oxide to escape by a similar process.

Some nitrogenous waste materials accumulate in the contractile vacuole of the *Amoeba* and is eliminated along with fluid which leaves the ectoplasm. But since the elimination of waste by this method is purely incidental we cannot say that the contractile vacuole is an excretory organ.

11.4 Excretion in platyhelminths

Flatworms have special hollow cells called *flame cells* (Fig 11.2) for excretion. Metabolic wastes diffuse from surrounding cells into these flame cells and ciliary motion then removes the waste from the body.

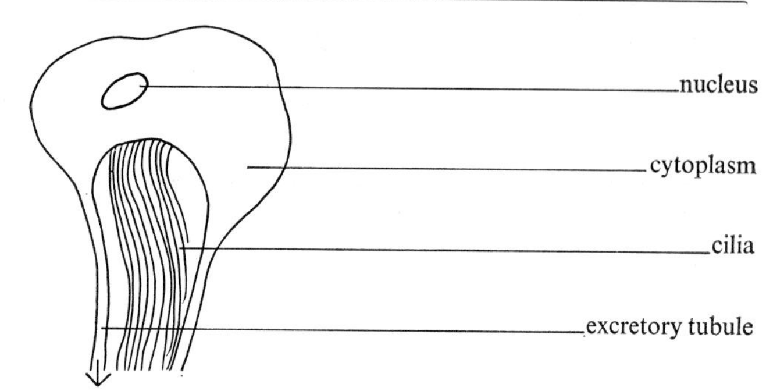

Fig 11.2 **Flame cell**

11.5 Excretion in the earthworm

NEPHRIDIA There are several ways in which *Lumbricus* can excrete waste; the most important of these is by means of *nephridia.* These consist of a series of tubes which are found in pairs in nearly all segments of the worm. They are not connected with one another internally, but each opens to the outside through what are known as *nephridiopores.* There are five distinct sections of the nephridial tube, the *nephrostome*, the *narrow tube*, the *middle tube*, the *wide tube* and the *muscular tube.* All along its length the nephridial tube is well supplied with capillaries. Each nephridium originates in the coelomic cavity (Fig 11.3) with a ciliated open ended tunnel (the *nephrostome*). Fluid containing mineral ions, water and nitrogenous materials is first moved into the nephrostome by ciliary currents, and passes into the coiled, ciliated *narrow tube*; this tube gets larger in diameter (middle tube) and leads to the uncoiled wide tube. More waste materials enter at the narrow end of the tube from capillaries. Reabsorption of proteins and ions required by the body take place in the wide part of the tube; water is also taken up here, especially if the worm is living in dry soil conditions. In moist conditions the urine which passes out of the worm is hypotonic to the body fluids, which is a way of getting rid of excess water. But in dry conditions the urine becomes hypertonic and small in amount.

What do the terms hyper- and hypotonic mean? The term *isotonic* is used to mean that no water passes across a membrane separating two solutions. A solution is *hypertonic* to another when water passes into it across a membrane (*i.e.* it has a higher osmotic pressure); a solution from which water is lost is *hypotonic* (*i.e.* it has a lower osmotic pressure).

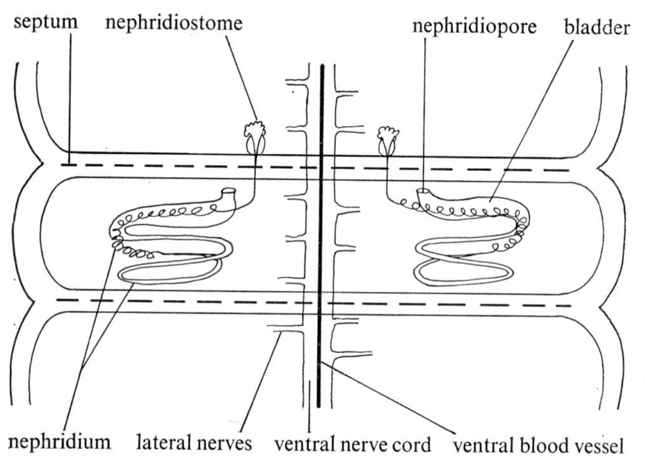

Fig 11.3 **Nephridia**

Waste passes out of the worm through the nephridiopores under the control of a *sphincter muscle.*

OTHER METHODS OF EXCRETION Worms have a skin which is permeable to the entry of water, (which is why it can easily become dehydrated in a dry environment and swollen in a wet one). The skin is constantly secreting mucus which contains nitrogenous waste.

The *chloragogenous cells* perform the same kind of function as the liver does in mammals. Deamination and the formation of ammonia and urea as excretory substances takes place there.

The main function of the *calciferous glands* is the excretion of excess calcium. The calcium as $Ca^{+}CO_3^{2-}$ passes into the lumen of the gut where it helps to neutralise acids in the food.

11.6 Excretion in insects

The main organ of the insect excretory system consists of a series of tubules (called the *Malpighian tubules*), which project into the body cavity from the hind gut. The far ends are unattached and float in blood. Nitrogenous waste in the form of potassium urate is liberated into the blood, from where it is absorbed by the cells at the distal end of the Malpighian tubules; this process is helped by the fact that the tubules make writhing movements which 'stir' the blood. The potassium urate reacts with water and carbon(IV) oxide to form uric acid and potassium hydrogen carbonate. The latter is reabsorbed by the blood. As the uric acid flows towards the gut, water also may be reabsorbed into the blood so that by the time it reaches the proximal end of the tubule it may be in solid crystalline form. In the locust the control of the reabsorption process is via diuretic hormone. Control of water loss is very important in hibernating insects and in soil insects where the protective waxy cuticle is constantly being abraded by soil particles.

Body waste is also used to form pigments – notably in the scales of butterflies and moths; reds and browns come from the degradation of tryptophan, yellows and whites from urates, and greens from the breakdown of cytochromes and haemoglobin. Waste materials deposited in the cuticle during its formation and hardening are shed at moulting.

11.7 Excretion in molluscs

In a mollusc, such as the *freshwater clam*, there is a pair of kidneys lying ventral to the pericardial cavity. Excretory substances are extracted by these from the fluid in the cavity and from the blood, and are then sent out into the outgoing water current (the excurrent siphon).

11.8 Excretion in man

In man the main organs of excretion are the kidneys, but the salivary glands, liver, skin and lungs also play a part.

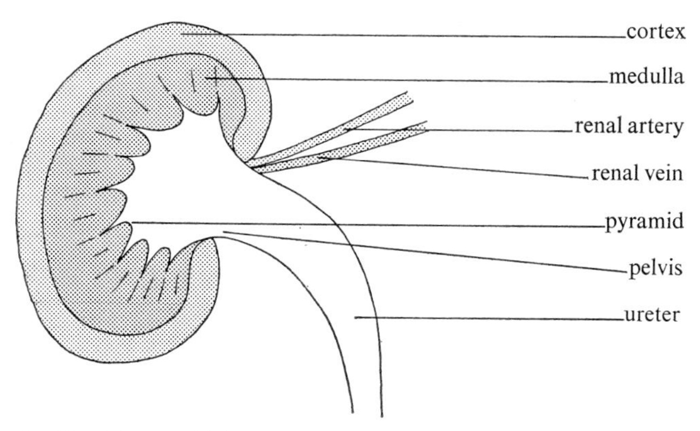

Fig 11.4 **Human kidney**

KIDNEYS The human kidney (Fig 11.4) consists of a pair of bean-shaped organs about 120 mm long and 75 mm wide surrounded by the connective tissue (the *fibrous capsule*). The outer cortex surrounds the medulla, from which the ureter passes to the bladder. Blood supply to and from the kidney is via the renal artery and vein which enter in the pelvis area of the kidney. The relationship between these pathways and the *aorta* and *vena cava* (p 180) is shown in Fig 11.5.

The basic unit of excretion in the kidney is the *nephron* consisting of a *Malpighian body* composed of a capillary network (the *glomerulus*) surrounded by a double-walled sac, *Bowman's capsule* (Fig 11.6). Cells in the inner wall of Bowman's capsule stick tightly to the glomerulus, and consequently substances can pass easily from the

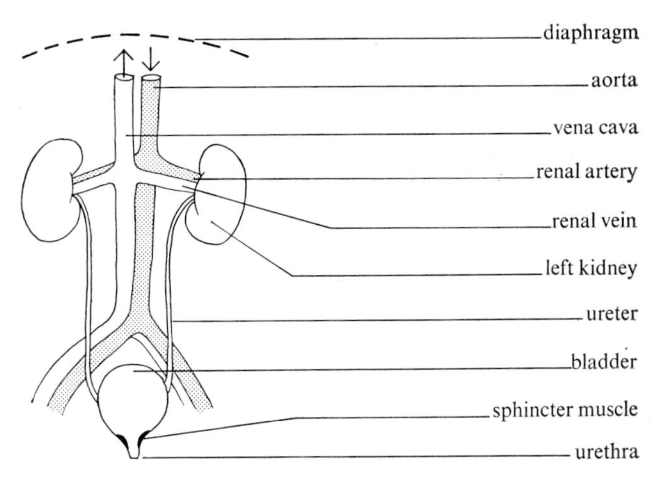

Fig 11.5 **Blood supply pathways to and from the kidney**

capillaries into the capsule. The capsule opens into three regions, the *proximal convolution*, *Henlé's loop* and the *distal convolution*. The latter region opens into a collecting tubule leading to the kidney pelvis.

The renal artery divides up into a series of arteries (arcuate arteries) with further subdivisions from which the glomeruli derive. The capillary leaving the glomerulus is of narrower bore than that entering it; this causes pressure to develop which enables water and dissolved substances to overcome osmotic pressure. Thus low relative molecular mass substances like glucose, water, nitrogen compounds and ions pass across the walls of the glomeruli into the capsule, but high molecular mass substances such as fibrinogen and other proteins remain

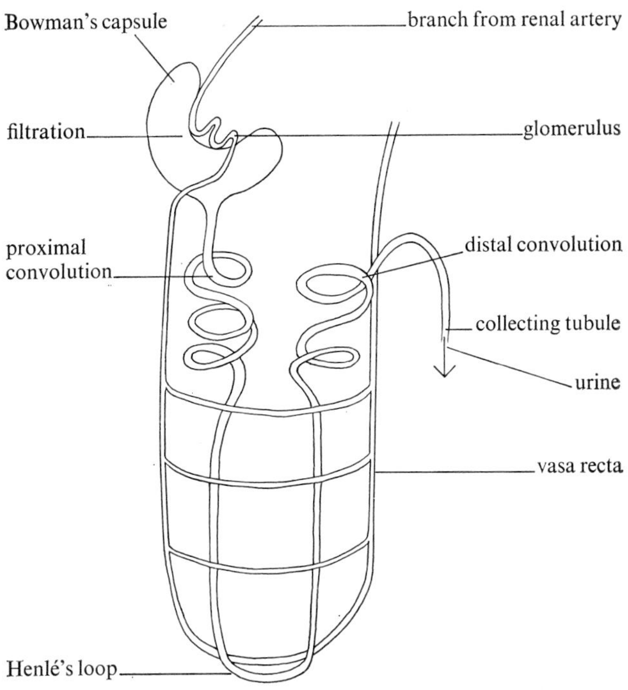

Fig 11.6 **The kidney tubule and blood supply**

in the blood. The walls of the glomeruli thus act as a filter. It is estimated that 180 dm^3 per day of filtrate pass into the Bowman's capsules.

As the filtered serum passes down the tubule all the sugar, some of the ions and most of the water are reabsorbed into the network of capillaries surrounding the tubule. The walls of the tubules have an epithelial cell lining with a greatly increased surface area (rather like villi in the ileum). They also have many mitochondria. The tubules are thus very actively exchanging materials. Glucose is returned to the blood – if under normal circumstances any remains in the urine there is something wrong (probably *sugar diabetes*). The reabsorption of ions is under the control of *mineralocorticoid hormones*. Reabsorption makes sure that no useful substances are lost from the blood serum.

LOOP OF HENLÉ Both birds and mammals have a special loop in the tubule called the *loop of Henlé* (Fig 11.7). Its purpose is to conserve water, and enables organisms that have it to produce hypertonic urine. It is thought that the loop of Henlé acts as a counter current multiplier system which works as follows. The filtrate from the Bowman's capsule and the proximal convolution flows down the descending arm of the loop, and passes into the ascending arm. As the fluid passes up the ascending arm Na^+ ions are transported out of the tubule by the lining cells and pass via the vasa recta back into the descending arm. This cycle, taking place all along the loop of Henlé, results in a high concentration of Na^+ and Cl^- in the region of the loop. The collecting tubules pass through this region of high Na^+ and Cl^- concentration. Therefore, as the renal liquid passes down the collecting tubules water molecules move outwards by osmosis, causing a hypertonic urine.

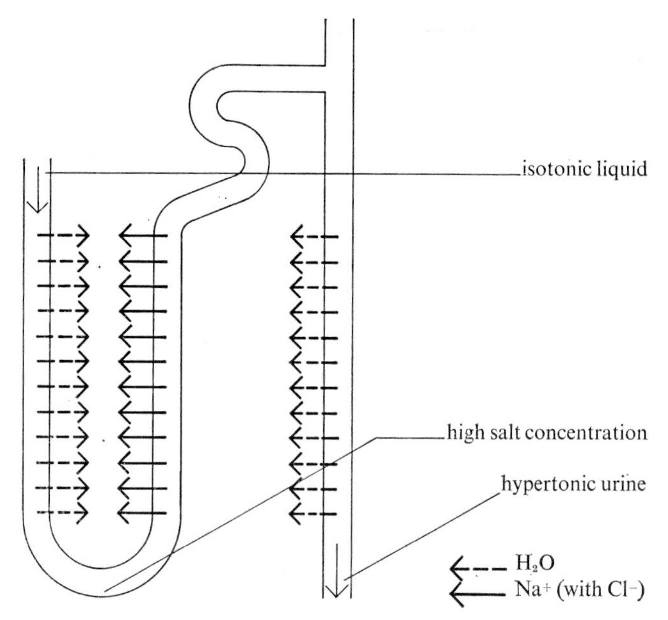

Fig 11.7 **Action of the loop of Henlé**

The tubules of the kidney are not influenced directly by osmotic pressure. Instead there are cells in the brain (*osmoreceptor cells*) which are sensitive to changes in osmotic pressure, so that as the osmotic pressure rises a mineralocorticoid hormone called antidiuretic hormone (ADH) is secreted into the blood by the pituitary gland. It is the presence or absence of ADH in the blood that affects the permeability of the cells and thus controls

Table 11.1

type	*example*	*blood concentration relative to environment*	*urine concentration relative to blood*	*notes*
teleost fish	herring (marine)	hypotonic	isotonic	loses H_2O through any permeable surface by osmosis. Drinks lots of H_2O and gets rid of ions via special cells in gills. Small capsule/glomerulus
elasmobranch fish	dogfish (marine)	isotonic (retains carbamide (urea), making higher concentration in blood than for teleost)	isotonic	does not drink sea H_2O. Large capsule and developed tubule
amphibian	frog (fresh water)	hypertonic	hypotonic	takes in H_2O via permeable surfaces. Has to eliminate H_2O; lots of dilute urine. Long tubule, large capsule and glomerulus
reptile	lizard (land)	hypotonic	isotonic	problem of desiccation; loses H_2O from permeable surfaces. Concentrated urine; urates excreted, semi-solid
bird	pigeon (land)	hypotonic	isotonic	like reptile but has large glomerulus and capsule and loop of Henlé to return H_2O to bloodstream
	gull (marine)	hypotonic	weakly hypertonic	drinks sea H_2O and has large salt intake. Excess salt lost via nasal glands which exude hypertonic solution
mammal	man (land)	hypotonic	isotonic	large capsule and glomerulus and loop of Henlé
	whale (marine)		strongly hypertonic	does not drink H_2O

reabsorption. This is therefore an excellent example of a homeostatic feedback mechanism.

The countercurrent multiplier system described above results in an *active transport* of salt. The net effect is relatively large owing to the length of the loop.

It can be seen that osmo-regulation in animals is extremely important. It is interesting that the salmon and the eel are able to adapt their osmo-regulation to both fresh-water and marine environments – they have regulatory mechanisms of both marine and fresh-water fishes. The desert rat has a kidney with a deep medulla and a long pyramid with especially long loops of Henlé reaching the tips of the pyramid. The urine produced is about 10 times as concentrated as that of a dog. Why can man not use sea water as his main source of liquid?

ORIGIN OF WASTE PRODUCTS Where do the waste products that are excreted by the kidney come from? Metabolism results in the conversion of carbohydrates, fats, proteins, etc, to small molecules such as carbon(IV) oxide (from the Krebs' cycle – p 125) and water (from the combination of hydrogen and oxygen using cytochromes – p 38). Deamination of proteins resulted in the formation of ammonia(NH_3), which is only excreted as such by animals such as crustaceans. In man and fish it is converted via the ornithine cycle (p 143). Reptiles, birds and insects excrete mainly uric acid or its salts, the urates. Ions not needed by the body are also part of the waste.

11.9 Excretion and secretion in plants

EXCRETION Organic wastes in plants are classified according to chemical nature.

1 Resins These are aromatic and acidic with a pungent taste that is distasteful to herbivores. They have antiseptic properties.

2 Tannins These occur in bark and have an astringent taste. They reduce Fehling's solution and are turned red by ammonia. They are used with iron(III) salts to make ink.

3 Alkaloids These were discussed in §1.11.

4 Glucosides These were discussed in §1.2 on p 9. The appearance of such pigments in the leaves of deciduous trees in autumn helps the removal of certain waste products at leaf fall. But autumn colouration is not completely associated with excretion. The sequence of events in the ripening of fruits follows much the same pattern as autumn colouration.

5 Essential oils (*e.g.* terpenes).

6 Calcium ethandioate (*oxalate*) This is one of the most common wastes. It can occur as single crystals or as bundles of smaller ones lying parallel to one another or radiating from a common hub; such arrangements are called *raphides*. This is formed in the leaves of deciduous trees prior to leaf fall.

7 Calcium carbonate This is deposited in large cells called *lithocysts* which occur in epidermal and parenchymatous leaf cells. The lithocyst contains an amorphous mass of $CaCO_3$ called a *cystolith* which is attached to the cell wall by a stalk.

SECRETION Like excretion, secretion involves separation of substances from the rest of the organism. Unlike excretion the products of secretion *can* be useful.

In plants secretory structures can be external or internal. *External structures* include *hydathodes* in leaves through which water is sometimes exuded, *nectaries* in insect pollinated plants, which secrete nectar, and *digestive glands* in insect eating plants, which secrete sticky substances which both trap the insect and help in decomposing it. *Internal secretory structures* include *lactiferous tissues*, which are cells containing an emulsion of waste products of metabolism called latex, and *multicellular glands*. There are three types of glands, *schizogenous glands* which are enlarged intercellular spores enclosed by whole cells (*e.g.* resin canals in conifers), *lysigenous glands* which come from the breakdown of cells, and *schizolysigenous glands* where whole cells separate with further enlargement by cell rupture.

11.10 Composition of blood

Blood consists of two types of cells, *red cells* and *white cells*, and *platelets* or cell fragments, suspended in plasma. Human blood is 60% plasma and 40% cells. The function of the red cells is to transport respiratory gases, of white cells is to fight disease, and of platelets is to help the blood to clot.

PLASMA The plasma consists mainly (90%) of water and a complex mixture of inorganic salts and proteins, as well as the substances such as glucose, fats, amino acids, enzymes, hormones and excretory products like urea, which the blood transports from one part of the body to another. There are three important types of blood protein present, albumen, globulin and fibrinogen.

RED CELLS Red cells (*erythrocytes*) contain the red pigment *haemoglobin*; this has an extremely important function as most of the oxygen from the lungs is carried in combination with it. There are about 5×10^{15} red cells per cubic metre of blood. In the mammals these

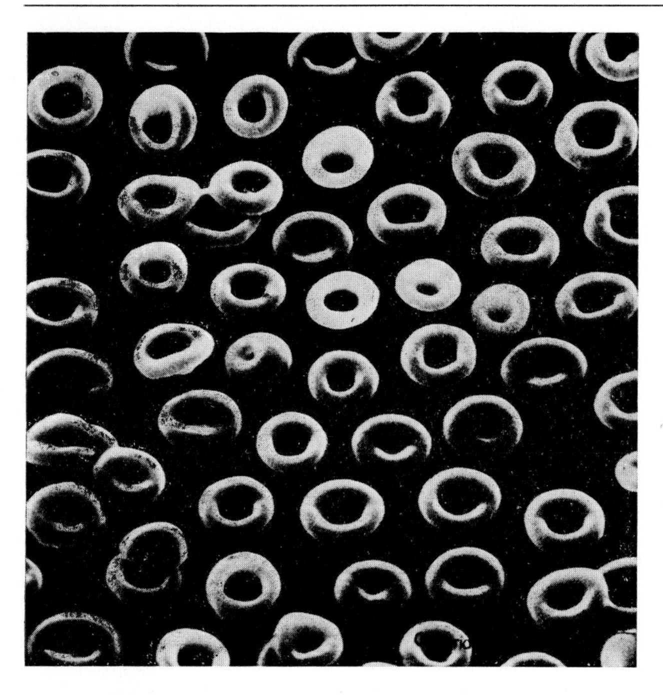

Fig 11.8 **Red blood cells**

have no nuclei, and are flattened biconcave discs, as can be seen from Fig 11.8, with a diameter of about 7·5 μm. The outside sheath of the red cell is made up of a complex combination of fats and proteins. The structure of haemoglobin which is rather like chlorophyll in plants was discussed on p 24; see also p 172.

Red cells are formed in the red marrow of the bones. They have however a short life (about 100 days in man), after which time they are engulfed by other cells and are destroyed in the liver and the spleen. The white cells that ingest them are *phagocytes*, which return the iron from haemoglobin to the bone marrow.

WHITE CELLS The white cells, *leucocytes*, have a nucleus and are true cells. They are motile, moving by pseudopodia (cf *Amoeba*). There are about 7×10^{12} per cubic metre of blood; they are of three types, lymphocytes, monocytes and granulocytes.

Lymphocytes (23% of the white cells) have large spherical nuclei surrounded by a narrow ring of cytoplasm. Their size is usually the same as for red cells. They are formed by cell division in lymph nodes and lymph tissue of tonsils, spleen and thymus gland. They develop into cells for connective tissue (important in healing) and they are helpful in the formation of antibodies.

Monocytes (7% white cells) are produced in the same place as lymphocytes but are about twice the size of red cells. The nucleus is larger too. They are involved in ridding the blood of infection by ingesting bacteria.

Granulocytes (about 70% of white cells) are formed in the red bone marrow and have diameters of about 10 μm. Their nuclei can vary in shape from one cell to the next, and according to this shape (and the extent of granulation of the cytoplasm) they can be divided into neutrophils, eosinophils and basophils. They engulf various particles in the blood.

Thus the main function of the white cells is the destruction of bacterial and other foreign cells. By these means the body may become immune to diseases. *Natural immunity* is where the body has become immune on its own to some disease by exposure to the pathogen and the building up of antibodies against it. Thus a second attack of measles in humans is a rare event. *Artificial* immunity is when immunity is built up by a doctor's intervention. In 'active' artificial immunity a weak pathogen is introduced into the body so as to cause the formation of antibodies but without causing the symptoms of the disease. When an active pathogen of the same type then invades the body, the antibodies in the blood destroy it. Once the body has had occasion to make a particular antibody, it can do so much more easily than otherwise would be possible. In 'passive' immunity, an antibody produced in another animal is injected into the body.

PLATELETS These are very small bodies, being about 25% the size of erythrocytes. They are not cells but cytoplasmic cell fragments, and are formed in the red marrow. They are involved in clotting. The plasma contains prothrombin and anti-prothrombin combined in such a way that clots are prevented from forming in the blood. If platelets come in contact with certain types of surface or if they are in contact with injured cells, they secrete the enzyme thromboplastin, which deactivates anti-prothrombin, causing prothrombin to be released. This with Ca^{2+} in the blood forms an active enzyme thrombin (note that Ca^{2+} is vital), which causes the soluble fibrinogen to change to a precipitate of fibrin. The fibres spread forming a clot.

LYMPH At the arterial end of a capillary the pressure (about $4{\cdot}0 \times 10^4\,N\,m^{-2}$) is large enough to cause plasma molecules to be squeezed out into the surrounding *lymph*. The lymph is the fluid, similar to plasma but with a lower protein content, which bathes the tissues. White cells can, likewise, leave the blood and travel via the lymph to a source of infection. At the venous end of the capillary however the osmotic pressure of the blood is greater than the capillary pressure, and so molecules are sucked back into the blood. Every organ and limb has its complex network of lymphatic vessels. These vessels open into two main collecting ducts, the left-hand one being known as the *thoracic duct* and the right-hand one as the *right lymphatic duct*. Lymph always flows towards the collecting ducts, its direction of flow being controlled by valves. The lymphatic system in man is concerned only with return of fluids to the heart. The capillaries of the system are called *lacteals* and are closed at one end. Fluids diffuse into these lacteals and are then conveyed to the collecting ducts. The latter empty into the large veins before they reach the right auricle. The flow of the lymph is dependent on breathing movements and on muscle (of skeleton) action. At injunctions between some of the lymph vessels are *lymph nodes* which *1* filter and trap bacteria, *2* produce lymphocytes, *3* produce antibodies. Remember that although the sugars and amino-acids from carbohydrate and protein digestion are absorbed directly into blood capillaries, the products from fats go into *lacteals*.

11.11 Functions of the blood

The many varied functions of the blood are summarised in Table 11.2.

Table 11.2 **Functions of the blood**

function
transport of O_2 and CO_2
transport of food and ions
transport of waste materials
transport of hormones
transfer of thermal energy
erection of penis in reproduction
clotting
prevention of infection

11.12 Circulatory system of blood

In simple organisms transport of substances between cells in different parts of the body is achieved by diffusion. But in complex organisms this would be too slow to be effective over distances; instead transport is carried out by the blood vascular system. The transporting fluid is *blood*, the pump is the heart, the tubes for circulation are the *arteries*, *veins* and *capillaries* (blood vessels of different bores with different sorts of walls). In many ways it is like a central heating system.

The *arteries* with muscular walls (which help to pump the blood) and large diameters, take the blood from the heart to the capillaries. The veins have walls with far less muscle, but with valves at intervals so that the blood flows in one direction only. Veins collect blood from the capillaries and take it to the heart. *Capillaries* are very fine tubes through whose walls exchange can take place between blood and body cells. Some vessels form portal systems which collect blood from one set of capillaries and take it to another without going via the heart.

In comparing the circulatory system with a centra heating system we can make an analogy between

capillaries and radiators;
arteries and pipes leading from the pump to the radiators;
veins and pipes leading from the radiators to the pump;
heart and the pump.

A simple blood circulation system as in the fish is shown in Fig 11.9; the arrows show the direction of blood flow.

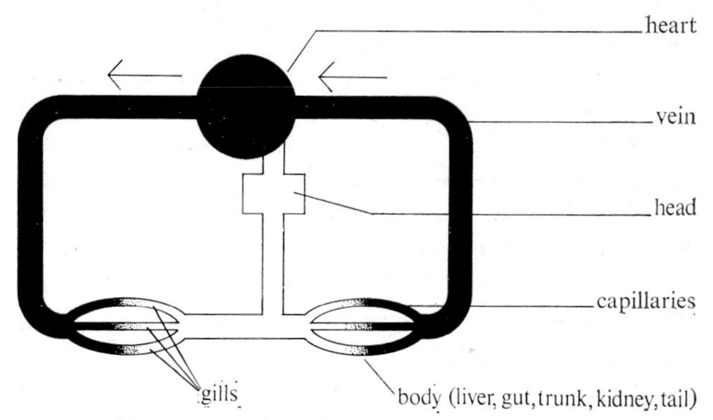

Fig 11.9 **Circulation of the fish**

The heart of man Fig 11.10 consists of four chambers and has two streams of blood going through it simultaneously (double circulation); the two left chambers are quite separate from the two right ones. The left-hand chambers have received blood from the lungs (rich in oxygen) whereas those on the right receive blood from the rest of the body (little oxygen but lots of carbon(IV) oxide); this blood is sent on to the lungs whereas the blood from the left-hand side passes round the body to do its work.

The right auricle receives blood from the region of the head through the superior vena cava and from the remainder of the body through the inferior vena cava. Oxygenated blood from the lungs returns to the left auricle through the pulmonary vein. The two auricles are pumps which contract almost at the same time. Contraction of the right auricle forces its blood through the tricuspid valve into the right ventricle. The tricuspid valve is prevented from being turned inside out by the chordae tendinae, attached to the ventricle walls.

Contraction of the left auricle forces blood through the bicuspid valve into the left ventricle. Expulsion of blood from auricles to ventricles makes the ventricle extend and this makes the heart more efficient as a pump. The walls of the ventricles are thicker than those of the auricles and both are of cardiac muscle (longitudinal and transverse striations). Heart muscle can conduct impulses like nervous tissue. The blood then goes from the right auricle to the right ventricle via the tricuspid valve.

From the right ventricle the blood is pumped through the semi-lunar valve to the lungs, via the pulmonary artery. Blood returns from the lungs to the left auricle

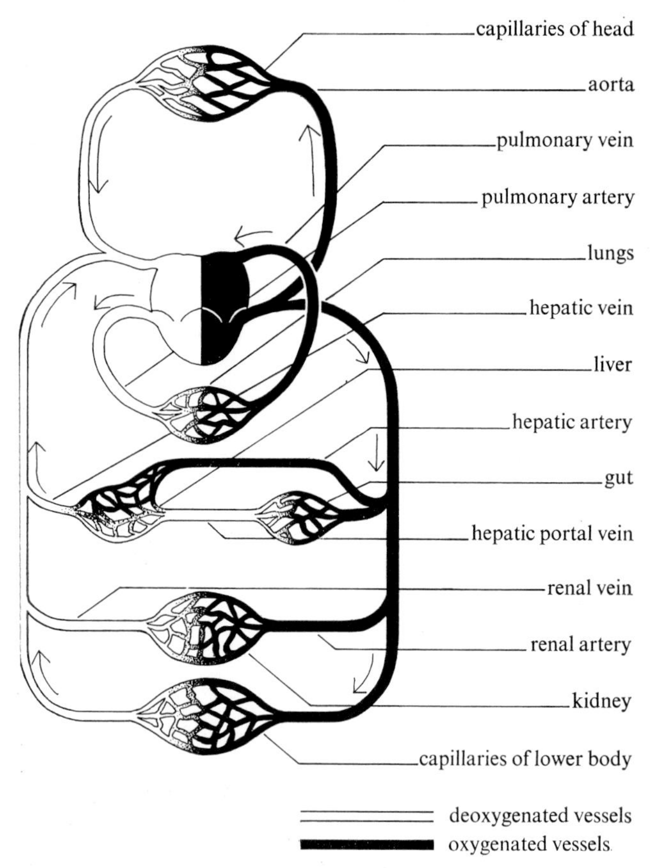

Fig 11.10 **Blood circulation of man**

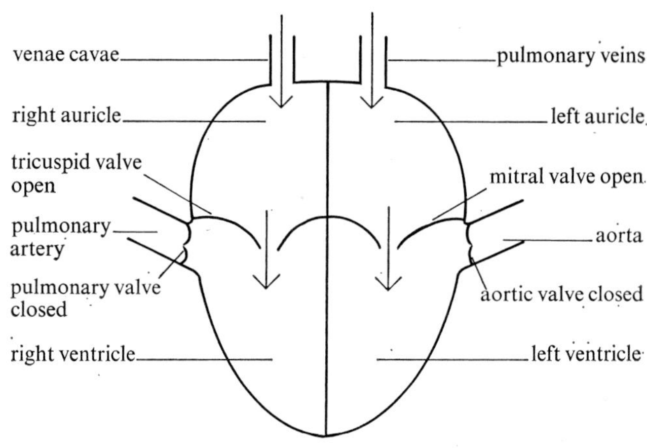

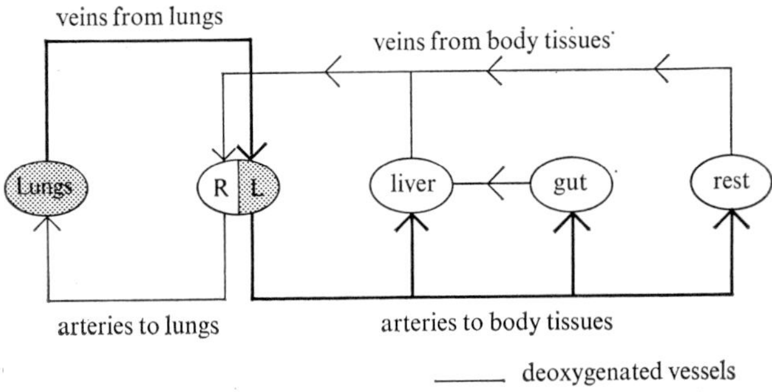

through the pulmonary vein. All this part of the circulatory system is known as the *pulmonary circulation.* The remainder, the *systemic circulation*, starts like this. From the left auricle the blood goes through the bicuspid valve to the left ventricle. From here it is pumped into the aorta and from there to all parts of the body. The aorta is separated from the left ventricle by a semi-lunar valve. The aorta branches into pathways which all eventually lead to the right auricle.

Each complete beat of the heart (a cardiac cycle) can be described as follows. In the right auricle wall there is a centre which starts each cycle, called the sinoauricular node (SA node); it consists of muscle tissue and sends out intermittent impulses, each impulse making a wave of contraction spread over both auricles. The period during which the blood goes from auricles to ventricles is called the period of contraction. When the wave of contraction and its impulse reaches the auriculoventricular (AV) node, the latter responds by sending out impulses down the AV bundle to the apex of the heart. These impulses stimulate the ventricles which contract. This is the period of relaxation of the auricles. The onset of contraction (ventricular systole) involves the closing of the bicuspid and tricuspid valves and the opening of the semilunar valves. During this systole blood is forced from the ventricles through the pulmonary artery into the aorta. Then in ventricular diastole, the semilunar valves close and the bicuspid and tricuspid valves open. Blood comes into the ventricles (from the auricles) as they relax. Another beat then starts.

The cardiac cycle can be observed in life on an *electrocardiogram* (Fig 11.11), which records electrical changes taking place. It can also be traced mechanically by means

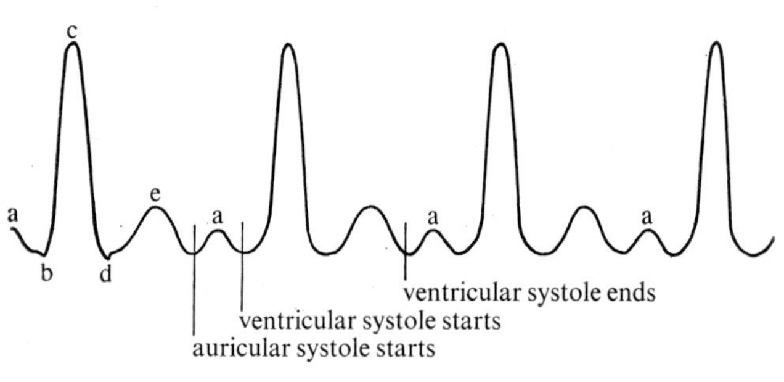

a=auricular contraction

b, c, d, e=ventricular contraction

Fig 11.11 **Electrocardiogram**

of a *kymograph* (Fig 10.28); a frog's heart is attached to the lever so that a trace is made on a smoked piece of paper set on a revolving drum.

11.13 Control of body temperature

Man is able to maintain an almost constant temperature inside his body, which is important for the functioning of his enzymes; most enzymes have an optimum temperature of about body temperature (310 K). Once again, as with circulation, we can draw parallels with a central heating system. We might expect to find in the body equivalents of thermostats, heaters, coolers and insulation.

INSULATION The hair on the surface of the skin which is the main site of heat loss traps a layer of still air. Because air has low thermal conductivity this layer acts as a layer of insulation. Fat layers (Fig 11.12) are also good insulators.

COOLING SYSTEM Heat is lost from the skin (Fig 11.12) by radiation. The number of joules of energy lost per square metre depends on the state of the blood vessels at the skin. When these are dilated, a lot of blood flows through them and much energy is lost; when you have just had a hot bath or have just run a race you look pink. If the small arteries are constricted, the skin assumes a bluish appearance and the energy radiated is small. Flow of blood in this way is controlled by the sympathetic nervous system. Heat energy can be lost from the lungs and this is increased by panting (the dog pants a lot). Evaporation of sweat is the main part of the cooling

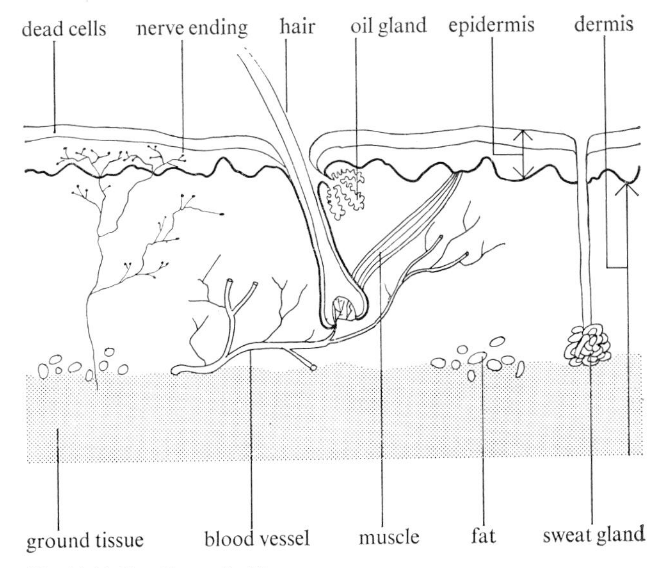

Fig 11.12 **Section of skin**

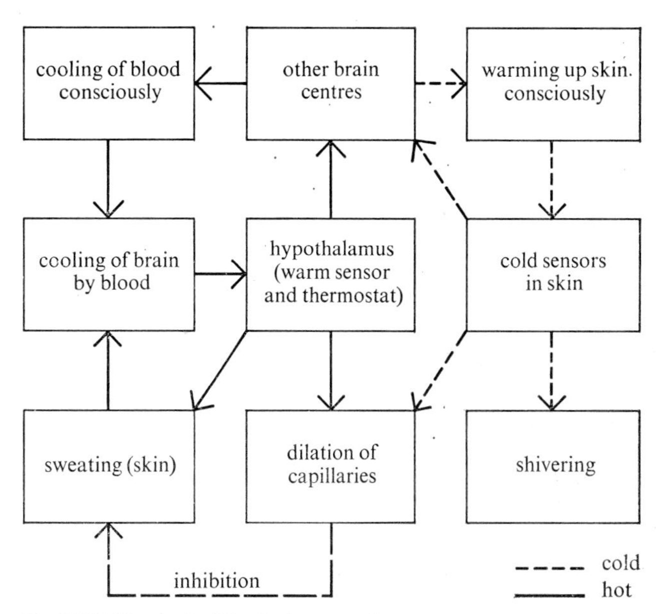

Fig 11.13 **Control of body temperature**

system. With dilated blood vessels, sweat is exuded on to the surface of the skin and evaporates, removing from the skin the specific latent heat of vaporisation of the watery sweat. The capillaries become cooled and the blood from them cools the remainder of the blood. The sweat glands are also controlled by the sympathetic nervous system.

The body has surface area of the dimensions $[L]^2$ but volume is $[L]^3$. Small animals have large surface areas compared with their volumes and therefore find it difficult to keep the heat in their bodies, whereas large animals, with a relatively small surface area, find it difficult to lose heat. A humming bird flies rapidly, and has an extremely high metabolic rate; an elephant flaps its ears and a hippopotamus basks in cool water.

HEATING SYSTEM Apart from exothermicity ($\triangle H$-ve) in many bodily processes, muscular activity also produces heat.

THERMOSTATIC CONTROLS Fig 11.13 shows how the system works. The sensors which detect and respond to an increase in body temperature are in the *hypothalamus* of the brain.

Homoiothermic (warm blooded) animals such as man, rabbit, horse, can control their body temperature, whereas *poikilothermic* (cold blooded) animals such as lizard and frog cannot. The way in which their internal temperatures are related to external temperatures is shown in Fig 11.14. Homoiothermic animals have a much higher basic rate of metabolism and this means a bigger

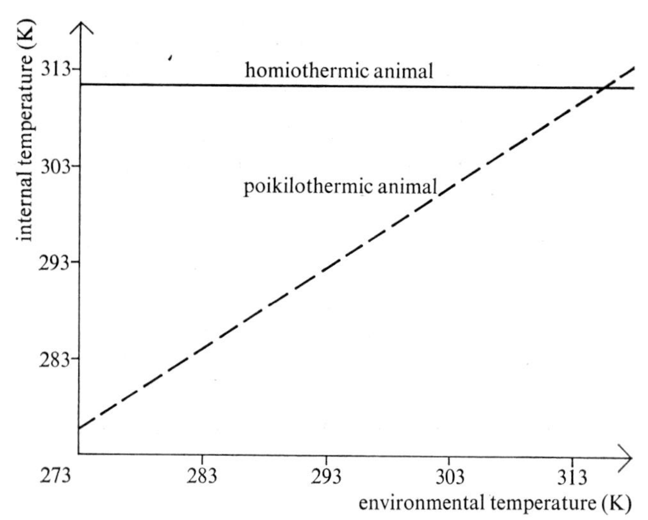

Fig 11.14 **Relation of internal to external environment in homoiothermic and poikilothermic animal**

food intake and other disadvantages. Despite this the ability to control temperature was a great evolutionary step forward. Since metabolic rates are a function of temperature it can be seen clearly from Fig 11.14 that poikilotherms are very limited as regards the type of environment they can colonise. This is not true of homoiotherms which can become adapted to live in environments as different as the hottest parts of the desert and the coldest parts of the arctic.

Questions

Questions on chapter 1

1·1 What are the important differences between carbohydrates and proteins? Briefly outline the part played by each in the structure and metabolism of living organisms. (O and C)

1·2 What are the characteristics of enzymic action? State the biochemical changes brought about by two animal and two plant enzymes and comment on the physiological importance of each. (C)

1·3 **a** Discuss some of the contributions electron microscopy has made in recent years to our knowledge in biology.
b Review briefly the role of either chromatography or spectrophotometry in biological investigations. Give examples. (NI)

1·4 Categorise the following into carbohydrates, fibrous proteins, lipids, globular proteins, nucleic acids, and 'others': starch, fructose, RNA, cellulose, glycogen, keratin, collagen, vitamin A, chlorophyll, haemoglobin, olive oil, DNA, amylase, dextranase, ribose, ethanoic (acetic) acid, glycine, glyceryl tristearate, beeswax.

Questions on chapter 2

2·1 By means of a large labelled diagram illustrate the structure of a generalised living cell. Relate the structure of any two named organelles to their function. (O and C)

2·2 In what ways do the cells in the phloem and nervous tissue differ from a 'typical' plant cell or a 'typical' animal cell? How are the modifications related to the functions of the tissue? (C)

2·3 Describe the interaction between the nucleus and the endoplasmic reticulum during the synthesis of proteins within a cell. (AEB)

2·4 Describe cell division (including nuclear division) as it occurs in a root tip. How does this process differ in plants and animals? (L)

2·5 What is the evidence for:
(a) The control of cellular activities by DNA,
(b) replication of DNA,
(c) nucleotides occurring in matched pairs in DNA molecules? (JMB)

2·6 The majority of cells have diameters ranging from 0·5 to 10 μm.
a Discuss the factors which control the upper and lower limits of cell size.
b Give three examples of cells which are exceptionally large and describe how each is adapted to this condition. (JMB)

2·7 Using labelled diagrams only, show how chromosomes behave during mitosis. List the differences between mitosis and meiosis. (O and C)

Questions on chapter 3

3·1 Why do we classify living organisms? By reference to a single animal or plant phylum, discuss the categories used to distinguish classes within the phylum, orders within the classes and species within the genera. (AEB)

3·2 Choose any *one* of the free-living unicells you have studied and describe its **a** reproduction, **b** nutrition, and **c** adaptations to its environment. (O and C)

3·3 In the life cycles of a named liverwort and a named fern say how the plants which develop from **a** the spore **b** the zygote differ.
Give drawings to show the microscopic appearance of the sex organs in the liverwort, and describe how fertilisation takes place. (C)

3·4 Give labelled drawings to show the microscopic structure of **a** *Euglena*, **b** *Paramecium*.
For each organism, describe how a rapid increase in numbers is brought about. Under what conditions might this occur? (C)

3·5 Compare, by means of annotated diagrams only, the main features of the life-history of a named fern with that of a named angiosperm (monocotyledon or dicotyledon).
What features in the life history of the fern would you regard as likely to make it more difficult for it to increase in numbers than it is for the angiosperm? (C)

3·6 By reference to a fern and an amphibian show how the life cycles of plants and animals may be related to their environment. (L)

3·7 Describe the floral structure of either a named cereal or a named grass.
Compare and contrast the characteristics of wind pollinated with insect pollinated flowers. (C)

Questions on chapter 4

4·1 Give an account of the production of spermatozoa or ova in a mammal.

4·2 Discuss the advantages and disadvantages of the different systems of reproduction found in organisms. (Exclude flowering plants and mammals from your answer.) (O and C)

4.3 Describe one example of fertilisation in a named vertebrate and show how the method used is related to the animals' environment.

State briefly the uses of the embryonic membranes and placenta in mammalian reproduction. (C)

4·4 By means of labelled drawings only, show the structure of a named endospermic seed. Describe and illustrate the development of the embryo during seed formation. (C)

4·5 What do you understand by the term germination? What factors determine whether a seed will or will not germinate? Explain in detail how you would investigate the influence of temperature on the rate of germination of seeds. (O and C)

4·6 Write an illustrated account of the dispersal of seeds and fruits. Is wide dispersal of the propagules an advantage to a species? Give reasons for your answer. (O and C)

4·7 **a** Distinguish between the terms 'gamete' and 'spore'.
b Give named examples of organisms in which
1. the gametes are morphologically alike,
2. both male and female gametes are liberated into external water,
3. only the male gamete is liberated into external water.
c What is the male gamete in a flowering plant?
d Give a labelled diagram of a longitudinal section through the male gamete of a mammal.
e Give **two** early (pre-cleavage) effects which result from the penetration of a sperm into the ovum of a vertebrate.
f For an amphibian give labelled diagrams of vertical sections through
1. the mature unfertilised egg,
2. the blastula,
3. the gastrula.
g Explain each of the following observations.
1. When amphibian tadpoles are maintained in an iodine-deficient medium they fail to undergo metamorphosis.
2. Removal of the corpus allatum from a young insect larva triggers pupation. (JMB)

4·8 Describe, using examples, the essential differences between budding, sporulation and parthenogenesis. What are the relative merits of these means of reproduction? (L)

4·9 What is meant by growth? Compare this process in a protozoan, an insect, a mammal and a flowering plant. (L)

Questions on chapter 5

5·1 A mouse homozygous for black coat and long ears is crossed with a mouse homozygous for white coat and short ears, black coat and long ears being dominant characters. What ratios of offspring would you expect in the F2 generation,
a if coat colour and ear size are controlled by genes on different chromosomes?
b if coat colour and ear size are controlled by genes carried on the same chromosome?
c If **b** is the case, would the distance apart of the two genes on the chromosome affect the ratio? You should explain your reasoning. (AEB)

5·2 Explain the term 'cross-over frequency'. In a certain organism, genes A, B, C and D were studied and it was found that the cross-over frequency between A and B was 20%, A and C 5%, B and D 5% and C and D 30%. What is the probable sequence of these genes on the chromosome and what would you expect the cross-over frequency between genes A and D to be? Discuss the significance of crossing-over. (O and C)

5·3 In a particular breed of cattle, the hornless condition is dominant to the horned condition. A farmer, who has a herd of such cattle in which occasional horned animals appear, wishes to eliminate the occurrence of horned animals in subsequent generations.
Describe clearly the possible breeding programme he should adopt. Explain fully the genetical principles underlying the steps in the programme. (L)

5·4 In man, aniridia (a type of blindness) is due to a dominant allele. One form of migraine (a type of sick headache) is the result of a **different** dominant allele.
Answer the following questions, indicating as fully as possible the genetical basis of your reasoning.
a A man with aniridia, whose mother was not blind, married a woman who suffers from inheritable migraine but whose father did not. What is the chance of one of their offspring having both aniridia and this type of migraine?
b A woman with no family history of aniridia or migraine married a man who is not blind but both of whose parents were blind as a result of aniridia. He does suffer from migraine which his father also had. What are the chances of her children having aniridia or migraine? (JMB)

Questions on chapter 6

6·1 What does the theory of organic evolution mean to the biologist? What, in your opinion, are the two most convincing kinds of evidence for evolution? Describe each as fully and as critically as you can. (C)

6·2 The word 'species' is commonly used in biology. Explain with examples from both plants and animals, what this means. When two or more populations of a species become isolated over a long period of time they may become separate species. Explain how this may occur and give examples to support your statements. (C)

6·3 Variation is a very obvious characteristic of living things. What are the causes of variation? Discuss its role in Evolution. (O and C)

6·4 Give some account of the plants and animals to be found as fossils from the main geological periods. Of what value to biologists is such knowledge of the past? (L)

6·5 A new island is pushed up from the sea floor. Describe what you think would be the probable course of events in its colonisation by living organisms. (C)

6·6 Evolution depends upon a number of factors. Amongst these are **a** mutation rate, **b** the size of the population, **c** the size and availability of ecological niches, **d** the nature and extent of environment change, **e** eventual reproductive isolation. Give an account of the ways in which each of these factors may affect the rate of evolution and the part it plays in the production of new species. (JMB)

Questions on chapter 7

7·1 How would you prepare to study the ecology of a small habitat? If you were asked to give quantitative results, what work would you try to carry out? (C)

7·2 It has been said of fieldwork that it gives a most stimulating opportunity to develop an approach of intrinsically biological character, for the study is not merely one of plant and animal life, but of the integrated activity of the organisms in the natural community. Justify this statement from your own knowledge. (O)

7·3 Review the relative contributions of the biological, chemical and physical properties of soil to the maintenance of healthy growth in a named flowering plant. (NI)

7·4 How far can ecological considerations usefully be applied to food production? (O and C)

7·5 Explain the meaning of the following, giving examples from ecological situations *which you yourself have studied:*
a community,
b population,
c niche,
d pyramid of numbers,
e trophic levels,
e food web. (L)

7·6 Explain the physiological importance of water and humus to plants. Describe how you would test a soil for **a** water content, **b** humus content, and **c** pH. (L)

7·7 Describe **under the following headings** a well-defined ecological problem or study you have tackled (personally or with a group), in the field or in the laboratory.
a The problem investigated or the aims of the study.
b The techniques used. Give full and precise details of each technique you describe.

c A summary of the results obtained. Give some indication of how these were recorded.
d The conclusions you were able to draw from the results.
e Assuming that you were able to return to the problem or study, which aspect would you choose to follow up? Give your reasons. (JMB)

7·8 Describe precisely some of the sampling techniques you have used in a field study and discuss the value of the results you obtained. (L)

7·9 The pH of a shallow rock pool in the intertidal zone was recorded from early dawn until the tide came over it at 1100 hours GMT. The results were as follows:

time	0500	0600	0700	0800	0900	1000	hrs GMT
pH	6·2	6·6	7·1	7·6	8·1	8·4	

Assuming that these changes were due to the metabolic activities of the plants in the pool, explain how they might have come about. With reasons, state whether you consider that the Algae on the surrounding rocks would be the same as those in the pool. (AEB)

Questions on chapter 8

8·1 Compare and contrast the ways in which oxygen reaches the tissues in a plant and in a terrestrial vertebrate. (C)

8·2 In the breakdown of starch and glycogen in cells during respiration the following kinds of change occur: hydrolysis; phosphorylation; oxidation; reduction. For each kind of change, describe one example to indicate its importance in the respiratory cycle as a whole. How do you account for the different quantities of useful energy released in aerobic and anaerobic respiration? (C)

8·3 Describe, with labelled diagrams, the respiratory surfaces of a dogfish and a named mammal. Why do gills not function in air and lungs not function in water? How do gases pass to and from the blood to the cytoplasm of the cells? What part do mitochondria play in cellular respiration? (NI)

8·4 Discuss the basic principles involved in the release of energy in a cell. A detailed account of the Krebs' cycle is not required (O and C)

8·5 What is meant by the term respiration? Outline the ways in which animals provide a supply of oxygen for their respiring tissues, and point out the features that contribute most to the efficiency of the process. (C)

Questions on chapter 9

9·1 What do you understand by nutrition? Give an account of the chief differences in the nutritional requirements of a named mammal and a typical flowering plant.

9·2 Describe fully one experiment in each case that you could carry out to investigate the effect of varying **a** the intensity of light, **b** the concentration of carbon(IV) oxide on the rate of photosynthesis of a named plant. What results would you expect? (C)

9·3 Discuss the importance in photosynthesis of chlorophyll, photolysis, and light and dark reactions. (L)

9·4 Tabulate the main differences between xylem and phloem. How does material move in the phloem? Give your reasons.

9·5 What are the functions of the root system of a flowering plant? How would you set out to discover what factors influence the direction in which roots grow? (O and C)

9·6 What fluctuations would you expect to take place in the transpiration rate of a well-watered plant during the 24 hours of a calm summer's day? To what extent would the loss of water from a permeable plaster surface of the same area and water surface as the plant be similar? (AEB)

9·7 Describe in detail a potometer of the type with which you are familiar. For what sorts of experiment would you use this instrument and what are its limitations? (O and C)

9·8 In what respects do parasites **a** differ from saprophytes and **b** resemble saprophytes? Briefly describe the importance of saprophytes to man. (L)

Questions on chapter 10

10·1 To what extent do you think a simple spinal reflex action, such as withdrawal of a limb from pain, either resembles, or differs from co-ordination involved in riding a bicycle? (AEB)

10·2 The endocrine system has been referred to as the endocrinal 'orchestra' with the pituitary as its 'conductor'. Discuss as far as possible the reasons for this analogy. (O and C)

10·3 Describe the experiments you would perform to investigate the response of a shoot to **a** the stimulus of gravity, **b** the stimulus of light. Design and then describe experiments by which you would attempt to ascertain which of these stimuli is the stronger. (O and C)

10·4 Give a comparative illustrative account of the brains of two of the following:
a a dogfish, **b** a frog, **c** a named mammal.
Stress the features that they have in common and show how far differences may be related to mode of life or evolutionary level. (NI)

10·5 State briefly the chief differences distinguishing the central and autonomic nervous systems in mammals. Contrast a named reflex action with the action of a person answering a door bell. (C)

10·6 What is meant by phototropism and geotropism? By reference to experiments explain how the same auxins might bring about both types of response. (AEB)

10·7 Describe the fundamental organisation of a mammalian vertebra. How do these vertebrae vary in different parts of the vertebral column, and what is the functional significance of these variations? (AEB)

10·8 **a** Show the position of, and label, eight different endocrine glands on a large outline drawing of a human male.
b Give an account of the role of two hormones which play a part in carbohydrate metabolism in the mammal. (JBM)

Questions on chapter 11

11·1 One of the characteristics of the living organism is the maintenance of the steady state (homeostasis). Explain briefly how this is achieved in three of the following cases: **a** temperature control in mammals, **b** growth form in trees, **c** the response of *Euglena* to light, **d** human posture. (AEB)

11·2 Discuss the transport of oxygen and carbon (IV) oxide in the blood of a mammal. If the haemoglobin of an earthworm is inactivated with carbon monoxide, the animal does not die, whereas in the same situation a mammal would. Account for this. (AEB)

11·3 Describe the internal structure of the heart of a named mammal. How is the heartbeat controlled? What important changes take place in blood circulation at birth, and what are their significance? How are arteries, veins and capillaries structurally adapted to their respective functions? (NI)

11·4 Explain how both birds and mammals are able to maintain a constant temperature despite wide changes in the temperature of the environment. (C)

11·5 Give a labelled diagram of a complete kidney tubule, together with the blood supply. In cases of severe bleeding, no urine is formed. Why not? Sea water contains about 3% salt, the salt content of blood is about 1% and, as a maximum, the kidneys can excrete only a 2% salt solution. Give a brief explanation for the various consequences of drinking a great mass of sea water. (C)

11·6 In the mammal, blood leaves the heart under high pressure and with uneven flow, but reaches the capillaries under low pressure and with even flow. How is this brought about? Discuss the necessity for such a transformation. (O and C)

11·7 Briefly describe the components of mammalian blood. How does blood carry out its protective functions? Explain the significance of the high proportion of water present in plasma. (L)

BIBLIOGRAPHY

Reading Lists for A and S Level Biology G. W. Shaw *School Science Review Vol. 52 p 97, p 358. Vol 53 p 119. Vol 54 p 90, p 742. Vol 55 p 735. Vol 56 p 746.*
Animal Biology A. J. Grove and G. E. Newell *UTP, 9th ed. 1974*
Biology, A Functional Approach M. B. V. Roberts *Nelson, 1971*
A Dictionary of Biology M. Abercrombie, C. J. Hickman and M. L. Johnson *Penguin, 4th ed. 1969*
Enquiries in Biology S. W. Hurry and D. G. Mackean *John Murray, 1968*
The Experimental Basis of Modern Biology J. A. Ramsay *CUP, 2nd ed. 1969*
Integrated Biology L. Hill, D. Bellamy, I. Chester Jones *Chapman Hall, 1971*
Intermediate Biology W. F. Wheeler *Heinemann Educational, 6th ed. 1962*
Introduction to Biology D. G. Mackean *John Murray, 5th ed. 1973*
Plant and Animal Biology A. E. Vines and N. Rees *Pitman, 2 volumes, 4th ed. 1972*
Principles of Biology *Ed.* W. G. Whaley and O. P. Breland *Harper & Row 1965*
Borradaile's Manual of Elementary Zoology W. B. Knapp *OUP, 14th ed.*
Lowson's Textbook of Botany *Ed.* E. W. Simon, K. J. Dormer and J. N. Hartshorne *UTP, 15th ed. 1973*
Organization in Plants W. M. M. Baron *Edward Arnold, 2nd ed. 1967*
Biology of the Mammal A. G. and P. C. Clegg *Heinemann, 3rd ed, 1970*
The Life of Vertebrates J. Z. Young *OUP, 2 volumes 2nd ed. 1962*
Animals without Backbones R. Buchsbaum *Penguin, 2nd ed, 1971*
Soil Animals E. D. K. Kevan *H. F. & G. Witherby, 1968*
A General Textbook of Entomology A. D. Imms *Methuen, 9th ed. 1965*
Studies in Biology (48 titles) Sponsored by the Institute of Biology *Edward Arnold*
The Chemistry of Life S. Rose *Pelican, 1970*
A Guidebook to Biochemistry K. Harrison *CUP, 3rd ed. 1971*
Understanding the Chemistry of the Cell G. R. Barker *Edward Arnold, 1968*
An Introduction to the Chemistry of Carbohydrates R. D. Guthrie and J. Honeyman *OUP, 3rd ed. 1968*
Carbohydrates of Living Tissues M. Stacey and S. A. Barker *Van Nostrand Reinhold, 1962*
An Introduction to Chromatography D. Abbott and R. S. Andrews *Longman, 2nd ed. 1970*
Enzymes D. W. Moss *Oliver & Boyd, 1968*
The Biochemistry of the Nucleic Acids J. N. Davidson *Chapman & Hall, 7th ed. 1972*
The Living Cell C. A. Stace *Scientific American, Freeman, 1965*
Cell Physiology A. C. Giese *W. R. Saunders, 4th ed. 1973*
The Thread of Life J. Kendrew *G. Bell, 1966*
The Double Helix J. D. Watson *Penguin, 1970*
Looking at Chromosomes J. McLeish and B. Snoad *Macmillan, 2nd ed. 1972*
The Electron Microscope in Molecular Biology G. H. Haggis *Longman, 1967*
Viruses K. M. Smith *CUP, 1962*
Ultrastructure of Fertilisation C. R. Austin *R. & W. Holt, 1969*
Outline of Human Embryology H. Wang *Heinemann, 1968*
The Chromosomes M. J. D. White *Chapman & Hall, 6th ed. 1973*
Elememtary Genetics W. George *Macmillan, 2nd ed. 1965*
General Genetics A. M. Srb and R. D. Owen *W. H. Freeman, 2nd ed. 1965*
Genetics for 'O' Level J. J. Head and N. R. Dennis *Oliver & Boyd, 1968*
The Mechanism of Evolution W. D. Dowdeswell *Heinemann Educational, 3rd ed. 1963*
The Origin of Species C. Darwin *Everymans Universal Library, Dent, 1972*
The Vertebrate Story A. S. Romer *University Chicago Press, 1959*
Micro-Ecology J. L. Cloudsley-Thompson *Edward Arnold, 1967*
Introduction to Animal Ecology W. H. Dowdeswell *Methuen Educational, 2nd ed. 1966*
Woodland Ecology E. G. Neal *Heinemann Educational, 2nd ed. 1958*
An Introduction to Parasitology R. A. Wilson *Edward Arnold, 1967*
Parasitology: The Biology of Animal Parasites E. R. Noble &. G. A. Noble *Lea & Febiger, 3rd ed. 1971*
First Course in Statistics R. Loveday *CUP, 2nd ed. 1966*
Statistics for Biology O. N. Bishop *Longman, 1971*
Animal Physiology K. Schmidt-Nielsen *Prentice-Hall, 3rd ed 1970*
Physiology of Mammals and other Vertebrates P. T. Marshall & G. M. Hughes *CUP, 1967*
A Physiological Approach to the Lower Animals J. A. Ramsay *CUP, 2nd ed. 1968*
A Textbook of Histology A. W. Ham *Lippincott, 6th ed. 1969*
An ABC of Modern Immunology E. J. Holborow *Lancet, 2nd ed. 1973*
Photomicrographs (Non-Flowering and Flowering Plants) A. C. Shaw, S. K. Lazell and G. N. Foster *Longman*
The Concise British Flora in Colour W. K. Martin *Michael Joseph & Ebury Press, 2nd ed. 1969*
Flora of the British Isles A. R. Clapham and T. G. Tutin and E. F. Warburg *CUP, 2nd ed. 1962*
Grasses C. E. Hubbard *Penguin, 1968*
Name This Insect E. F. Daglish *Dent, 3rd ed. 1972*
Land Invertebrates J. L. Cloudsley-Thompson and J. Sankey *Methuen, 1961*
Microbial Life W. R. Sistrom *Holt, Rinehart & Winston, 2nd ed. 1969*
Classification of Plants Cambridge University Botany Department
British Amphibians and Reptiles M. Smith *Collins, 1969*
Pocket Guide to the Sea Shore J. Barrett and C. M. Yonge *Collins*
Comparative Physiology of Vertebrate Respiration G. M. Hughes *Heinemann, 1963*
The Study of Instinct N. Tinbergen *OPU, 1969*
Animal Locomotion J. Gray *Weidenfeld & Nicolson, 1968*
Muscle D. R. Wilkie *Edward Arnold, 1968*
Animal Body Fluids and their Regulation A. P. M. Lockwood *Heinemann Educational, 1963*

INDEX

A-band 167
α-glucose 8
achene 75
acid peat 111
actin 167
adaptation 96
adaptive radiation 102 et seq
adenosine triphosphate (ATP) 11, 123
adrenaline 161
adsorbent 19
aerobic respiration 125
Agnatha 53
aglycone 9
algae 56
alimentary canals (in vertebrates) 144-145
 anus 145
 bile duct 145
 caecum 145
 ileum 142, 145
 large intestine 143, 145
 liver 145
 oesophagus 145
 pancreas 145
alkaloids 16
allantois 79
allopolyploidy 93
amino acids 11-13
amnion 79
Amoeba 42, 155, 173
Amphineura 52
Amphibia 54
amylopectin 10
amylose 10
anabolic change 7
anaerobic respiration 124
anaphase 33
androsterone 77
Angiospermae 64
Annelida 48
annual cycle 88
annual rings 87
annuals 68
antherozoids 72
anthocyanins 15
anthoxanthins 15
Arachnida 51
arithmetic mean 116
Arthropoda 49
Artiodactylia 55
Ascomycetes 60
ascorbic acid (vitamin C) 11
asexual reproduction 71
astacin 15
Asteroidea 52
Australopithecus 97
autecology 107
autolysis 28
autonomic nervous system 152, 153
autoradiograph 21
auxins 163
Aves 54

β-glucose 8
bacteria 56
bases (in nucleic acids) 16-17
Basidiomycetes 61
behaviour 162
berry 75
biennials 68
bile acids 14
binary fissions 71
biotic factors — study of 107, 113
birth in — mammals 80
blind alley evolutions 97
blood 89, 177 et seq
 groups 89
Bowman's capsule 175
Brachiopoda 48
brain of mammal 154-155
breathing 127
bronchial tubes 127
bronchioles 127
bronchus 127
brown earth 111
Brownian motion 24
Bryophyta 61
buffering 128

calyptra 62
Calvin cycle 136
Campanularia — see *Obelia*
canines 141
capsanthin 15
capsule 62
carbohydrates 7-11
carbon cycle 138
Carnivora 55
carcinogens 25
cardiac poisons 14
carotenoids 14
catabolic changes 7
cellobiose-structure of 9
cells 27
 comparison of animal and plant 30
 orders of magnitude of 40
cellulose-structure of 10
central nervous system 151
centrifugation 23

centriole 33
centromere 33
Cephalochordata 53
Cephalopoda 52
cercaria 46
cerebellum 154
cerebral cortex 154
Cetacea 55
chemical symbols, nomenclature, terminology 7
chemosynthesis 138
Chiroptera 55
chitin 10, 170
Chlamydomonas 57
Chlorella 135
Chlorophyceae 56
chlorophylls 134
chloroplasts 27, 29, 133
cholesterol 14
Chondrichthyes 54
Chordata 52
chorion 79
chromatid 33
chromatography 18-21
chromosome 30
chrysin 15
chyme 142
chylomicrons 142
circadian rhythms 173
circulatory system 179
  aorta 180
  arteries 179
  blood 177 et seq
  capillaries 179
  heart 179-180
  pulmonary circulation 180
  systemic circulation 180
cisternae 29
classes 41
clone 71
codon 31
Coelenterata 44
co-enzyme 37
colloids 24
colon 143
commensalism 148
community 107
compensation point 134-135
Compositae 10
co-ordination 151 et seq
copulation 77
corpus luteum 78
cortex 86
covariance 118
Crick 17
Crinoidea 52
cristae 27
crossing over 93
Crustacea 50
cybernetics 7
cyclopentanoperhydrophenanthrene 14
cypsela 75
cytochromes 38
cytology 7, 27

Darwin 95 et seq
defaecation 143
defects of the eye:
  accommodation (loss of) 157
  astigmatism 157
  far sightedness 157
  near sightedness 157
dehiscent fruit 75
dehydroandrosterone 77
demes 95
dendrites 152
dendrons 152
deoxyribonucleic acid (DNA) 16-17, 89 et seq
deoxyribose 8
diakinesis 35
diastema 141
diatoms 58
Dicotyledons 65
diet 140
digestion 123, 140 et seq
  in animals other than mammals 144-146
diplosome 29
diplotene 35
dominant gene 91
dragnet 114
Drosophila 89, 92
Dryopteris filix-mas 63
drupe 75
duodenum 142
Dutch elm disease 72

ear (of mammal) 158
ecdysis 162
Echinodermata 52
Echinoidea 52
ecological niche 107 et seq
ecology 107 et seq
ecosystem 107
edaphic factors 111
ejaculation 77
electrocardiogram 180
electron microscope 18
electrophoresis 11, 25
embryonic development in animals 81-84
  animal pole 81
  archenteron 82
  blastophore 82
  blastula 81
  gastrula 82
  neural groove 82
  neural tube 82
  notochord 82
  somite 82
  vegetal pole 81
emulsin 9
endoplasmic reticulum 27
endoskeleton 168
endosperms 73
energy of activation 7
Enteromorpha 57
environment 107
enzymes 7, 37-39, 126, 142
epididymis 77
epiglottis 141
Equisetum 63
era 99
  mesozoic 99
  caenozoic 99
  paleozoic 99
esters 13
ethanoyl (acetylcholine) 153
ethanoyl (acetyl) co-enzyme A 125
Eudorina 57
Euglena 42
Euglenophyceae 59

Eutheria 55
evolution 95 et seq
excretion 173 et seq
 in mammmals 175
 in plants 177
eye — compound 157
eye — of mammal 156, 157

Fallopian tube 77
families 41
Fasciola 46
fats 13, 126
feedback 161
fen peat 111
fertilisation in mammals 79
fimbriate funnel 77
flame cells 174
flavone 15
floor (of brain) 154
floral diagrams 66
floral formulae 66
flower — parts of 66
 androecium 66, 67
 anther 66
 axillary flower 66
 bracteole 66
 bracts 66
 calyx 66
 carpels 66
 epigynous arrangement 66
 filament 66
 floral axis 66
 gynoecium 66
 hypogynous arrangement 65
 inflorescence 66, 67
 pedicel 66
 peduncle 66
 perigynous arrangement 66
 sessile flower 66
 stigma 66
 style 66
fluke 46
foetus 79
forebrain 154
form (and growth) 85
fossil 98
frequency distributions 117
fucoxanthin 15
Fucus 58
Funaria 62
fundus 141
fungi 59
furanose ring 8
fructose — structure of 8
fruits 75

gall bladder 142
gaseous exchange:
 in man 127
 in animals other than man 128-129
 in plants 129-130
gaseous exchange in plants:
 guard cells 129
 lenticels 129, 130
 stomata 129, 130
Gastropoda 52
gene pool 95
genes 89 et seq
genera 41
genetic code 31
genetics (heredity) 89 et seq
genotype 89
germination of seeds 72
 epigeal 73
 hypogeal 73
gestation in mammals 79
gibberellins 164
gills 128
Ginkgo 64
glands 160 et seq
 adrenal 160
 pituitary 160
 sex 161
 thyroid 160
gley 111
glucose — structure of 8, 11
glucose phosphate(V) 11, 137
glucoside 9
glycerides 13
glycerol (propantriol) 13
glycogen — structure of 10, 124
glycolysis 123, 124
glycoside 8
Golgi bodies 29
gonads 161
Gonium 57
grana 29
growth and form 85
growth curve 85
 accumulated 85
 cell and organism, level of 85
 grand period of 85
 rate of 85
guttation 133
Gymnospermae 64

H-zone 167
haemoglobin 13, 24
haemophilia 93
haustoria 60
hemicelluloses 10
heparin 11
Hepaticae 62
heteropolysaccharides 11
heterospory 63
heterozygote 91
hexokinase 137
hexose 8
hindbrain 154
histology 27
Holothuroidea 52
holozoic 41
homeostasis 173
homoiotherms 126
homopolysaccharides 11
homozygote 91
horse-types of in evolution
 Eohippus 105
 Equus 105
 Merychippus 105
 Mesohippus 105
Huxley 85
hyaluronic acid 11
hybridisation 90
hydathodes 177
Hydra 44

hydrolases 37
hydrolysis 37
hydrophytes 133
hygrometer 110
hyphae 59
hypothalamus 154

I-band 167
ileocolic valve 143
ileum 142
impulse — transmission of 152
incisors 140
indehiscent fruits 75
ingestion 140
Insecta 50, 174
Insectivora 55
instinct 162
insulin 161
interneurone 151
inulin 10
ion-exchange 24
islets of Langerhans 161
isoelectric point 25
isomerism 8

jejunum 142
jelly-fish 44
juvenile hormone 162

kidney of mammal 175
kinesis 162
kinins 164
kingdoms 41
Knoll 27
Krebs' cycle 29, 125-126

lactose — structure of 9
Lagomorpha 55
Lamellibranchiata 52
larynx 127
learning 162
legume 75
lenticel 129
leptotene 35
lignin 10
light traps 114
Lymnaea 109
linkage 92
lipids 7, 13
liver 142
liverworts (Hepaticae) 62
living organisms — features of 7
locust—structure of 50
loop of Henlé 176
Lumbricus 49, 155, 174,
luteinising hormone 78
lutein 78
Lycopodiales 63
lyosomes 29

Malpighian body 175
maltose 9
maltose — structure of 9
Mammalia 54-55
mammary glands 78

medulla 154
meiosis 35-36, 91, 95
melanism 97
Mendel 90 et seq
menstrual cycle 77
 hormones in 78
meristems 85
 primary 85
 secondary 86
mesentery 142
mesophyll 133
mesophytes 133
metabolic pathways 7
metaphase 33
Metatheria 54
micelles 29
micro-climate 107
microfibrils 29

microscopes 18
 electron 18
 light 18
 phase contrast 18
 polarising 18
 stereoscan 18
midbrain 154
mimics 96
mineralocorticoid hormones 176
miracidium larva 46
mitochondria 27, 126
mitosis 32-33, 71, 89
molars 141
Mollusca 52, 175
Monocotyledons 65
Monocystis 43
monosaccharides 8
mosses 62
motor neurone 151
movement 165
 amoeboid 165
 ciliary 165
 contractile 165
 fish (swimming) 170
 insects 170
 man 169
 plants 171
mucopolysaccharides 11
Mucor 60
muscle 165 et seq
mycelia 59
myofibrils 166
myosin 167
Myriapoda 51

nematoblasts 44
nematocysts 44
nervous systems 151 et seq
neurones 151
ninhydrin 20
nose — of man 159
nuclear magnetic resonance (NMR) 22
nucleic acids 16, 30 et seq
nucleus 27
nutrient-deficiency symptoms 140
nutrition 131 et seq
 autotrophic 131
 heterotrophic 131
 holozoic 131
 parasitic 131, 146-148

Obelia (*Campanularia*) 44
oils 13
oligosaccharides 9
oomycetes 60
oospheres 72
open-chain structures 8
Ophiuroidea 52
orders 41
organelles 27
organisers 84
osmo-regulation 176
  hypertonic 176
  hypotonic 176
  isotonic 176
osmosis 131-132
Osteichthyes 54
ova 77
ovaries 77,
ovulation 77
oxidoreductases 38
oxytocin 78

pH 110
pachytene 35
palisade layer 133
Pandorina 57
Paramecium 43, 155
parasites 59, 146-148
parasympathetic nervous system 154
parenchyma 130
parietal peritoneum 141
pectic acids 11
Pellia 63
pentadactyl limb 104 et seq
  carpals 105
  forelimb 105
  humerus 105
  metacarpals 105
  phalanges 105
  ulna 105
pentose 7
peptide link 13
perennials 68
periods of geological time
  Cambrian 99
  Carboniferous 99
  Cretaceous 99
  Devonian 99
  Jurassic 99
  Permian 99
  Quaternary 99
  Silurian 99
  Tertiary 99
  Triassic 99
Perissodactyla 55
peristalsis 141
phagocytes 178
pharynx 127
phellen 130
phellogen 130
phenotype 89
phloem cells 39, 138
phospholipids 14
phosphorylase 137
phosphorylation 124
photosynthesis 131 et seq
Phycomycetes 60
photoperiod 164
phyla 41
Phytophthora infestans 60
pinocytosis 27, 142
pitfall traps 114
placenta 79
plasmalemma 29
plasmodesmata 30
plasmolysis 132
Plasmodium 43
Platyhelminths 45
pleural cavity 127
Pleurococcus 56
podsols 111
poikilotherms 126
polarimeter 8
pollination 64
  by insects 64
  by wind 64
polymer 11
polymerase 11
polyploidy 93
polysaccharides 9-11
Polytrichium 62
Polyzoa 48
pons 154
population 93
  fluctuations 107
population size 114
  marking samples 114
  recapturing samples 114
  releasing samples 114
Porifera 43
potometer 133
predators 71
pregnancy hormones 77
Primates 55
propantriol (glycerol) 13
prophase 33
proteins 13, 30, 126
Prototheria 54
Protozoa 42
Pteridophyta 63
pyloric sphincter 141
pyramid of numbers 107
pyranose ring 8
pyrrole ring 134

quadrat 114
quercetin 15

rabbit — dissection of 55
radioactive dating 99
recessive gene 91
redia 46
reflex 162
  arc 153
rendzina 111
reproduction
  amphibians 80-81
  birds 81
  fishes 80
  invertebrates 81
  mammals 77
  reptiles 81
Reptilia 54
resistance mechanisms 113
respiration 123 et seq
  cell 123
  tissue 123
respiratory quotient 126
response systems (primitive) 155-156
Rhodophyceae 59
rhodopsin 157

ribonucleic acid (RNA) 16-17
information 30
messenger 30
ribose 8
ribosomes 13, 27
Rodentia 55
roof (of brain) 154
roots 85
structure of 86
cortex 86
endodermis 86
pericycle 86
phloem 86
xylem 86
rotary evaporation 25
Rotifera 48
Ruska 27

sacculus rotundus 143
saliva 141
sampling
destructive 114
sequential 114
saponins 14
saprobiosis 148
saprophytes 59
Scaphopoda 52
scurvy 11
scrotum 77
secretin 142
secretion (plants) 177
seta 62
seed dispersal 75
animals 75
plants 75
water 75
wind 75
seed formation 74
seeds 75
Selaginellales 63
selection 94, 96, 97
pressure 94
selective advantage 96
semen 77
seminiferous tubules 77
sense organs 156-157
sensory neurone 151
Sertoli cells 77
sex hormones 14, 77
sex linkage 93
sexual reproduction 72 et seq
sickle-cell anaemia 13
Sirenia 55
skeleton (mammal) 168
soils 111
chalks 112
clay loams 112
loams 112
minerals in 112
normal 112
peat 112
sandy loam 112
surveys 113
temperature of 113
somatic nervous system 152
Spallanzani 7
species 41, 95
spectrophotometry 22
spectroscopic analysis 22
spermatocytes 77
spermatogenesis 77
spermatogonia 77
Spermatophyta 64
spermatozoa 77
Sphagnum 62
spindles 33
Spirogyra 56
spirometer 127
sporocyst 46
standard deviation 117
standard error 117
starch — structure of 10
statistics 114 et seq
statistical diagrams 114-115
charts 115
correlation graphs 115
graphs 115
histograms 115
pie charts 115
stems 86 et seq
cortex 86
endodermis 86
epidermis 86
pericycle 86
pith 86
phloem 86
xylem 86
stereoisomerism 8
steroids 14, 17
bile acids 14, 17
cardiac poisons 14, 17
cholic acids 14
favoured positions for groups 14
general nomenclature of 14
saponins 14, 17
sex hormones 14, 17, 77, 161
toad poisons 14, 17
sterols 14
strobili 64
stroma 29
succulent fruits 75
sucrose — structure of 9
suction trap 114
sugar diabetes 176
sugars 8
sweep net 114
symbiosis 148
sympathetic nervous system 154
synapse 153

Taenia solium 46
taxis 162
teeth (of mammals) 141
telophase 33
temperature — body-control of 181
terpenes 15
testes 77
thalamus 154
Thallophyta 56
toad venoms 14
tongue (of man) 159
tonoplast 139
transcription 30
transect 114
transport — of minerals and ions 139
active, in cells 139
Trichonympha 42
Trilobita 50
triplet code 30, 31
Trypanosoma 42
Tswett 19
turgidity 132

ultramicroscope 24
Ulva 57
umbilical cord 79
urethra 76
Urochordata 53
uterus 77
uvula 141

vagina 77
van Leeuwenhoek 18
variance 116
vas deferens 77
vegetative reproduction 71 et seq
vernalisation 164
Vertebrata 53
vertebrate-nutrition of 140
vesicles — secretion 29
virus 68-69
virus — structure of 68
  base plate 68
  bacteriophages 68
  head 68
  parasitic 69
  prophages 69
  tail fibre 68
  tail sheath 68
  transduction 69
visceral peritoneum 141
vision 156-157
vitamin C (ascorbic acid) 11, 141
vitamins 141
viviparity 71
Volvox 57
Vorticella 43
vulva 77

Wallace 95, 100
water diffusion potential 132
Watson 17
waxes 13
Woodland Ecology (by Neal) 108

X-ray diffraction 23
Xanthophyceae 59
xanthophyll 15
xerophytes 129
xylem cells 39, 85-87

yellow carotenoid pigments 14
yolk 42
yolk sac 79

Z-membrane 167
zone electrophoresis 21
Zygomycetes 60
zygote 78
zygotene 35